Studies in Systems, Decision and Control

Volume 332

Series Editor

Janusz Kacprzyk, Systems Research Institute, Polish Academy of Sciences, Warsaw, Poland

The series "Studies in Systems, Decision and Control" (SSDC) covers both new developments and advances, as well as the state of the art, in the various areas of broadly perceived systems, decision making and control–quickly, up to date and with a high quality. The intent is to cover the theory, applications, and perspectives on the state of the art and future developments relevant to systems, decision making, control, complex processes and related areas, as embedded in the fields of engineering, computer science, physics, economics, social and life sciences, as well as the paradigms and methodologies behind them. The series contains monographs, textbooks, lecture notes and edited volumes in systems, decision making and control spanning the areas of Cyber-Physical Systems, Autonomous Systems, Sensor Networks, Control Systems, Energy Systems, Automotive Systems, Biological Systems, Vehicular Networking and Connected Vehicles, Aerospace Systems, Automation, Manufacturing, Smart Grids, Nonlinear Systems, Power Systems, Robotics, Social Systems, Economic Systems and other. Of particular value to both the contributors and the readership are the short publication timeframe and the world-wide distribution and exposure which enable both a wide and rapid dissemination of research output.

Indexed by SCOPUS, DBLP, WTI Frankfurt eG, zbMATH, SCImago.

All books published in the series are submitted for consideration in Web of Science.

More information about this series at http://www.springer.com/series/13304

Rajalakshmi Krishnamurthi · Anand Nayyar · Aboul Ella Hassanien
Editors

Development and Future of Internet of Drones (IoD): Insights, Trends and Road Ahead

Springer

Editors
Rajalakshmi Krishnamurthi
Department of Computer Science
and Engineering
Jaypee Institute of Information Technology
Noida, India

Anand Nayyar
Graduate School, Faculty of
Information Technology
Duy Tan University
Da Nang, Viet Nam

Aboul Ella Hassanien
Scientific Research Group, Faculty of
Computers and Artificial Intelligence
Cairo University
Giza, Egypt

ISSN 2198-4182 ISSN 2198-4190 (electronic)
Studies in Systems, Decision and Control
ISBN 978-3-030-63338-7 ISBN 978-3-030-63339-4 (eBook)
https://doi.org/10.1007/978-3-030-63339-4

This Springer imprint is published by the registered company Springer Nature Switzerland AG
The registered company address is: Gewerbestrasse 11, 6330 Cham, Switzerland

Preface

In recent years, the widespread drone usage for handling real-life problems has gained popularity and has been studied extensively by research community. The unmanned aerial vehicles (UAVs), also known as drones, are widely used in applications such as military, government and non-government system. In the past, the applications of drones were restricted to the military purposes such as monitoring battlefield, rescue and relief operations, and cross-border surveillance. However, with recent intervention of emerging technology such as Internet of Things (IoT), the usage of drones has unwrapped to extensive application solutions of real-life scenarios. The Internet of Drones provides widespread opportunities to enhance the quality of life in day-to-day activities of common people.

This book targets to explore wide range of IoD applications towards improving quality of life (QoL) in civilians. To mention few applications of IoD are package delivery system, traffic monitoring, agriculture, environment monitoring, etc. There are several factors need to be addressed towards IoD, such as compliance with existing mechanisms. Hence, this book explores various architectures and frameworks to meet the specific requirements of IoD. In order to meet on-demand applications, value-added services and user-specific requirements, these IoD are integrated with IoT devices like sensors, actuators, cameras, etc., according to application requirements. This book explores mechanism towards data collection, data processing and data analysis based on IoT devices deployed over IoD systems. Further, to provide performance enhancement through cognitive and intelligent solutions, IoD requires machine learning, deep learning, artificial intelligence and nature-inspired algorithm-based solution approaches. Also, the IoD has several open challenges like deployment strategy, collision-free manoeuvring, power consumption, technology integration and security and privacy.

The objective of this book is to provide technical insight about IoD, such as the current state-of-the-art requirements, performance, evaluation and challenging aspects, to the readers at one place.

The book is organized as follows: the first chapter titled "The Swarm Is More Than the Sum of Its Drones" identifies several gaps in existing nature-inspired algorithm-based solutions towards the group behaviour of swarm of drones.

This chapter presents evaluation mechanisms such as tools and metric towards swarm of drones. The second chapter titled "Underwater Drones for Acoustic Sensor Network" presents the underwater drones. The study and survey on onshore and offshore functionalities associated with underwater drones are presented. This chapter presents architecture, applications, characteristics and challenges associated with underwater drones. The third chapter titled "Smart Agriculture Using IoD: Insights, Trends and Road Ahead" focuses on the applications of Internet of Drones in agriculture sector by incorporating smart farming. The capturing crop data through IoT sensors, crop image-based data analysis and crop management mechanism are discussed in this chapter. The details on remote-sensing platforms, such as spaceborne, airborne and ground based, are elaborated. Further, the drone technology and developments towards smart and precision farming through Internet of Drones are presented. The fourth chapter titled "Towards a Smarter Surveillance Solution: The Convergence of Smart City and Energy Efficient Unmanned Aerial Vehicle Technologies" proposes an energy-efficient framework for autonomous intelligent drones for smart city surveillance. Further, this chapter addresses various strategies that allow the development of new strategies to the Internet of Drones frameworks and the effective utilization of drones in surveillance for smart cities. The chapter highlights the different areas of research towards IoD in smart city concepts. The fifth chapter titled "Efficient Design of Airfoil Employing XFLR for Smooth Aerodynamics of Drone" presents the various types of drones, manufacturing practices and flight equipment incorporated. The chapter presents XFLR-software-based simulation of designing air foil and aircraft, along with study and analysis of its flying operations. The sixth chapter titled "Drone-Based Monitoring and Redirecting System" presents the parallel processing capability in drone systems to avoid overloading of drone computation. The integration of cloud computing platforms for handling various drone tasks is elaborated in this chapter. The performance metrics such as accuracy and reliability are analysed through simulation environment. In addition, the lightweight primitives and protocol for efficient security in drone computation are also addressed. The seventh chapter titled "Drone Application in Smart Cities: The General Overview of Security Vulnerabilities and Countermeasures for Data Communication" addresses the security vulnerabilities and cyber-attacks on the drone networks. This chapter provides insight awareness and approaches towards Wi-Fi security, network security and software security within the drone networks. The eighth chapter titled "Cloud-Based Drone Management System in Smart Cities" proposes cloud computing-based drone network for providing various services in smart cities. The cloud computing provides storage and computation capabilities to assist drone network to offer real-time environment monitoring in smart cities. This chapter presents study on monitoring and controlling of Internet of Drones, and experimental results are interpreted with performance metrics for IoD in delivering various services under smart city concept in real-time environment. The ninth chapter titled "Smart Agriculture: IoD Based Disease Diagnosis Using WPCA and ANN" proposed IoD systems for enhancing the agriculture sector through IoD sensor-based image processing mechanism. This chapter presents study on recognition and classification of crop leaf diseases incorporating machine learning and image-processing mechanisms. The study is presented based on enhanced artificial neural network for estimating parameters such as probability, similarity and decision-making towards the image-based leaf disease identification through IoD systems. The tenth chapter titled "DroneMap: An IoT Network Security in Internet of Drones" presents the Map planner for drones that provides admittance to fog computing resources to drones such that heavy load computations can be performed in efficient manner. Further, this chapter summarizes the overview of the Internet of Drones (IoD), UAV applications, 5G communication challenges towards IoD, and security & privacy aspects of the drone's communication.

This book will provide clear insight about Internet of Drones for graduates, researchers and engineers in the fields of computer science, information & communication technology, mechatronics and business management. In addition, this book will engage readers including advanced students, instructors and academicians in Data Scientists, Data Analysts, and Applied Science with a professional need for materials on Internet of drones. This book also provides basic platform for developers and designers of UAV platforms, government departments under aviation and hobbyist for drones.

Editors

Noida, India — Rajalakshmi Krishnamurthi
Da Nang, Viet Nam — Anand Nayyar
Giza, Egypt — Aboul Ella Hassanien
September 2020

Contents

About the Editors

Dr. Rajalakshmi Krishnamurthi is a Senior Member of IEEE, Professional Member of ACM, SIAM, IET and CSI. She is serving as Treasurer, Delhi ACM-W chapter. She is currently working as Assistant Professor, Senior Grade, Department of Computer Science and Engineering, Jaypee Institute of Information Technology, Noida, India. She has more than 17 years of teaching experience. She has more than 50 research publications in various reputed peer-reviewed international journals, book chapters and international conferences. She is serving as Guest Editor in Springer Nature. Her research interest includes Internet of Things, cloud computing, optimization techniques in wireless mobile networks, e-learning applications using mobile platform and advanced fuzzy approaches. She has introduced and developed several courses at B.Tech. and M.Tech. levels. She has refereed in reputed international journals and book chapters. She has been technical program member in several international conferences.

Dr. Anand Nayyar received Ph.D. (computer science) from Desh Bhagat University in 2017 in the area of wireless sensor networks. He is currently working in Graduate School, Duy Tan University, Da Nang, Viet Nam. He is Certified Professional with 75+ professional certificates from CISCO, Microsoft, Oracle, Google, BeingCert, EXIN, GAQM, Cyberoam and many more. He published more than 400 research papers in various national and international conferences, international journals (Scopus/SCI/SCIE/SSCI Indexed). He is member of more than 50+ associations as senior and life member and also acting as ACM distinguished speaker. He has authored/co-authored cum edited 30 books of computer science and associated with more than 400 international conferences as Programme Committee/ Advisory Board/Review Board member. He has two patents to his name in the areas of Internet of Things and speech processing. He is currently working in the areas of wireless sensor networks, MANETS, swarm intelligence, cloud computing, Internet of Things, blockchain, machine learning, deep learning, cyber-security, network simulation and wireless communications. He was awarded 30+ awards for Teaching and Research—Young Scientist, Best Scientist, Young Researcher Award, Outstanding Researcher Award, Publons-Top 1% Reviewer Award (computer science, engineering and cross-fields). He is acting as Editor-in-Chief of IGI Global, USA Journal titled "*International Journal of Smart Vehicles and Smart Transportation (IJSVST)*".

Dr. Aboul Ella Hassanien is the Founder and Head of the Egyptian Scientific Research Group (SRGE) and a Professor of Information Technology at the Faculty of Computer and Artificial Intelligence, Cairo University. Professor Hassanien has more than 1500 scientific research papers published in prestigious international journals and over 50 books covering such diverse topics as data mining, medical images, intelligent systems, social networks and smart environment. His other research areas include machine learning, optimization, medical image analysis, space sciences and telemetry mining. Prof. Hassanien won several awards including the Best Researcher of the Youth Award of Astronomy and Geophysics of the National Research Institute,

Academy of Scientific Research (Egypt, 1990). He was also granted a scientific excellence award in humanities from the University of Kuwait for the 2004 Award and received the superiority of scientific in technology—University Award (Cairo University, 2013). Also, he was honored in Egypt as the best researcher at Cairo University in 2013. He was also received the Islamic Educational, Scientific, and Cultural Organization (ISESCO) prize on Technology (2014) and received the State Award for Excellence in Engineering Sciences (2015). Professor Hassanien received Abdul Hameed Shoman Arab Researchers Award (2015). Prof Hassanien holds the Medal of Sciences and Arts from the first-class from President Abdelftah Al-Sisi, and also he wins the Scopus International Award (2019) for meritorious research contribution in the field of computer science.

The Swarm Is More Than the Sum of Its Drones

A Swarming Behaviour Analysis for the Deployment of Drone-Based Wireless Access Networks in GPS-Denied Environments and Under Model Communication Noise

Hanno Hildmann, Khouloud Eledlebi, Fabrice Saffre, and A. F. Isakovic

Abstract This chapter reports on a family of nature-inspired algorithms that were successfully used to deploy a drone-based wireless sensor network into (simulated) indoor environments. The algorithms are designed to work in a decentralised manner, with each drone operating autonomously, using only local (subjective) information about a drone's own position as well as the inferred positions of its neighbours. The aim of the chapter is not to propose these algorithms but to highlight a currently under-represented research area in the field of *Internet of Drones*, namely the study of group behaviour for collectives (swarms) of cyber-physical devices (here: drones). The availability of swarms of devices has only recently grown to a level where the deployment of such collectives is actually becoming a reality. Even today, for most researchers operating swarms of drones is still impossible simply due to the lack of legal guidelines and frameworks. In the near future we will see increasing reports in the literature reporting results collected not on the basis of simulations but generated from the actual deployment of a swarm in the real world. The case this chapter makes is that the time has come to start looking at such swarms of cyber-physical

---

Hanno Hildmann and Khouloud Eledlebi contributed in equal parts to this manuscript.

H. Hildmann (✉)
TNO Dutch Organization for Applied Scientific Research, The Hague, The Netherlands
e-mail: hanno.hildmann@tno.nl

K. Eledlebi
Khalifa University, Abu Dhabi, United Arab Emirates
e-mail: green_leav13@hotmail.com

F. Saffre
VTT Technical Research Centre of Finland, Espoo, Finland
e-mail: fabrice.saffre@vtt.fi

A. F. Isakovic
Colgate University, Hamilton, USA
e-mail: aisakovic@colgate.edu; iregx137@gmail.com

R. Krishnamurthi et al. (eds.), *Development and Future of Internet of Drones (IoD): Insights, Trends and Road Ahead*, Studies in Systems, Decision and Control 332,
https://doi.org/10.1007/978-3-030-63339-4_1

systems with the tools and metrics used in theoretical biology, such as finer-grained simulation that acknowledges inter-individual variability.

Keywords Internet of Drones · Swarming behaviour · Adaptivity · Self-organisation · Emergence · Biology · Noise · Wireless access network · GPS-denied environments · Exploration · Network coverage · Node failure

1 Introduction

To our knowledge, the first mention of drone related research at the Netherlands Organization for Applied Scientific Research (TNO) was at the end of 1937, when the Royal Dutch Navy requested the lab (TNO) to develop a method to "*control a plane [...] from the ground*" (see Fig. 1). As such, this might mark the first official study into drones in Europe, it almost certainly constitutes the first such effort in The Netherlands. As the Rijksverdedigingsorganisatie TNO (National Defence Organisation, RVO-TNO) has grown since its early years, so has its involvement with unmanned vehicles. Alongside its Finish counterpart, the Technical Research Centre of Finland (VTT), TNO is leading or participating in a wide variety of projects related to a plethora of aspects and challenges for the operation of unmanned vehicles.

As device autonomy for commercially available drones increases, so does our ability to deploy intelligent self-organizing swarms. Off-line deployment planning is restricted to known environments while centralized approaches can pose certain requirements (such as connectedness) on the ad-hoc network [90, 91, 118]. The ability to deploy a swarm [116] in which the members can make decisions on the fly and based on locally available information massively increases the practical applications for the swarm in the context of an internet of drones. However, when the swarm size increases or when the members of a swarm are operating in a non-trivial environment, these individual decisions [136] can lead to complex emergent behaviour [66] of the entire swarm [29].

This motivates the need to study the collective behaviour of swarms. This chapter does so with regard to the dimensions of spatial-temporal deployment, but the approach can be transferred to investigate other dimensions such as, for example, the respective computational load of a drone in comparison to its peers or its relevance as a node in the network.

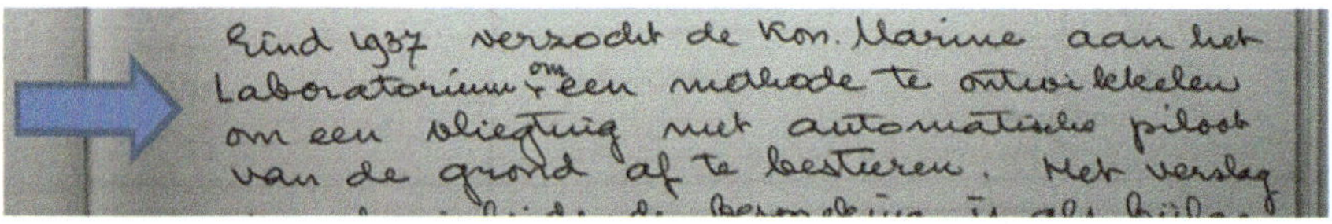
Eind 1937 verzocht de Kon. Marine aan het
Laboratorium om een methode te ontwikkelen
om een vliegtuig met automatische piloot
van de grond af te besturen. Het verslag

Fig. 1 Entry (in Dutch) in the reconstructed lab records by Van Soest in 1947 and archived at TNO Museum Waalsdorp. Translation (by author HH): "*At the end of 1937, the Royal Dutch Navy requested the lab* [TNO] *to develop a method to control a sea plane* [...] *from the ground*"

1.1 From Drones to Swarms

Unmanned vehicles and autonomously operating cyber-physical systems [22] continue to gain popularity for a wide range of applications [4]. Especially unmanned *aerial* vehicles (UAVs), often referred to as *drones*, are a trending technology [82], for which the number of applications is rapidly rising [110]. Generally speaking, autonomous devices can be used as (a) mobile sensing platforms [61], (b) actors [44] or (c) service providers [62]. For all three, the deployment of multiple UAVs together as a single collaborative multi-robot systems (MRS) [73], known as a *swarm* [129], has become possible due to increasing device performance combined with more favourable unit costs. For military applications, the geo-spatial intelligence [7] can be crucial [25]; in the civilian domain, using swarms in Search and Rescue (SAR) operations [44] or to augment a Smart City's sensing capability [61] is receiving increasing attention in the literature. Especially for high risk scenarios and operations, in the context of homeland security [26] the use of autonomously operating devices is becoming a viable alternative to relying on human operators.

1.2 The Challenge of Deploying a Swarm Indoors

For drone swarms, unknown [24] or indoor [67] environments are considered potentially challenging due to a number of restrictions and constraints and because of additional complexities they can create: besides the need for a deployment strategy, indoor manoeuvring may be more complicated in the absence of precise GPS-signals [104] or due to the presence of signal distorting obstacles. In ([47]), we study the behaviour of a decentralized and self-organizing swarm of e.g., drones tasked with deploying into, and providing coverage for, a large, GPS-denied, 2D indoor environment.

The example application used is that of a drone-based (mobile) wireless sensor network (WSN) [128] or wireless local area network (WLAN) in which communication is realized node-to-node (N2N) using ad-hoc [17] routing [31] for messages between nodes and where the topology is dynamic and adaptive [4]. The article provides details on the swarming/deployment algorithms, how collisions are avoided and the way in which message routing is implemented.

1.3 Contributions of This Chapter

We discuss the modelling of the problem and the implementation of our approach but the focus of this chapter is on the evaluation of *the behaviour of the collective*, in relation to *the performance* of the swarm. We define a variety of performance measures, e.g., actual movement of the nodes is used to assess the cost of a solution while, e.g. area coverage and the evolution thereof are used as an indicator for performance. With regard to the group behaviour, we use measures discussed in the literature (theoretical biology) to capture the diffusion of the swarm into an environment as well as the drift of the individual swarm members.

This allows the drawing of conclusions regarding the delivered quality of service (of the WSN/WLAN), the efficiency and speed of the deployment thereof as well as the cost incurred by the individual device in the swarm in order to do so. Drone movement is recorded as *drift* and *diffusion*; these measures are used to evaluate and compare the performance of the swarm. Specifically, this study focuses on the swarming behaviour observed (a) for the proposed approach as well as its augmentation with genetic algorithms or PSO [8], (b) when using a variety of environment types / when enriching the environment with different types (and numbers) of obstacles, and (c) when the drone-to-drone communication/local sensing is subjected to model noise.

1.4 Situating the Chapter in the Context of the Literature

This chapter is discussing the behaviour analysis for swarms of drones under specific conditions and constraints. This investigation is based on work applied to drone-based WSN deployment that was already presented to the community through a number of peer-reviewed publications ([44–47]), presentations as well as demonstrations (simulations). Furthermore, this chapter builds on the insights gained during a (recently defended and awarded) PhD dissertation [48]. This chapter adds to these by looking at the results with a done-behaviour specific view. The contribution made is the combined analysis of the swarming behaviour, which in this form has not been published or discussed previously.

Therefore, this chapter will only discuss the approaches in as much as detail as needed for the analysis and we refer to the specifics of the individual implementations to the respective journal papers.

1.5 The Intended Audience for this Chapter

The results presented constitute an analysis of the collective behaviour of the swarm. As the use of increasingly large swarms is becoming a realistic option, we argue that *drift* and *diffusion* are a measure of increasing importance and relevance. As devices are increasingly acting (semi-)autonomously, group behaviour is an aspect growing in relevance, with the study of emergence and emergent behaviours a field of growing significance for the community.

Therefore, this chapter is of interest to practitioners interested in understanding the (potentially) complex group behaviour of a swarm, as a *passive* observer. For example, when evaluating the performance of a swarm to report findings to the community, some comparable metrics are required. Furthermore, in addition to proposing a measure to the community we also expect that this chapter is useful for those of our colleagues *actively* trying to realise, deploy and fine-tune implementations of swarms. For those we hope to be able to offer some way to assess properties of the swarm behaviour over time and thereby a means to gain a better understanding of the impact individual parameter settings may have for global performance.

We believe this chapter to be of interest to the theoretical scientist in the field, as well as the engineers working on actually getting the drones out to fly in a world full of uncertainties, friction, resistance and noise.

2 Background

2.1 An Internet of Drones

The idea of an *Internet of Drones* (IoD) is relatively new [54, 77] but it is slowly coming of age. As it is usual for novel technologies or new applications for existing technologies (e.g., the operation of UAVs for non-recreational purposes [23]), legal issues [110] are among the plethora of obstacles encountered. Like the internet (of computers), an internet of drones needs to address concerns about jurisdiction, ownership and accountability [54]. The fact that drones can move across borders or simply between land plots means that the legal frameworks put in place for the internet alone are unlikely to cover all relevant aspects of drone swarms forming flying communication networks. In its current state, legal factors restrict what can be achieved and allowed [59].

Due to recent geopolitical events, the potential impact, and dangers, of social networks and the information they can aggregate (directly or as meta-data) has become a mainstream media topic. An IoD can be seen as a social network connecting not beings but machines, sensors and data storage devices [151], i.e., a mobile Internet of Things (IoT) [117], by some referred to as the Internet of Drone Things (IoDT) [95]. And just like the static internet, a drone based network will incur significant operational cost through connectivity requirements [31] (the fact that the end nodes are moving will not make the process simpler) as well as pose challenges with regard to computation heavy tasks that one prefers to offload to more suited devices [152]. This is especially so because the nodes in a drone-based wireless sensor network (WSN) are composed of energy-constrained devices [128].

The benefits are, however, also evident: the mere ability to move computation, sensing as well as acting devices physically around means that the entire application can become reactive and potentially adaptive. The literature lists application areas as diverse as e.g., precision agriculture [112, 129], manufacturing and industry [77], government [57], municipalities and smart cities [129] as well as private organizations [95, 140, 142].

2.2 Application Areas for Drones, and Swarms of Drones

We have previously argued that UAVs are ideally suited to serve as relatively low-cost mobile sensing platforms [61] e.g., in the aftermath of disasters [6] (when existing infrastructure is wiped out) or when the environment is dangerous for humans (e.g., fire fighting [5] or detecting and clearing land mines [38, 87, 125, 126]). Especially indoors [67], where GPS positioning is not reliable [80] as well as in military scenarios, were the GPS infrastructure may be compromised or destroyed [44], but also in large outdoor areas such as cities as well as rural areas. There are many applications of drones in the area of civil security or public safety [61] where drones, equipped with a variety of sensors [25, 26, 80, 141] can offer fast [50] and reliable monitoring [32] and surveillance [50, 157] capabilities and offer a means to explore and map unknown environments [24]. The ability to perform some data analysis and evaluation allows for drones to be reactive, that is to use sensed information collected in real-time to adapt to changes in the environment [62].

Drones are also increasingly used in established fields such as (precision) agriculture [61] where they are deployed for tasks such as crop-monitoring [140, 142]. In addition, drones can do more than just move technology (sensors, actuators) and create mobile infrastructure, they can also be used to physically carry products or information to e.g., distant destinations [4].

Due to advances in the field of autonomous cyber-physical devices (in our case, drones) there is now increasing interest to deploy collectives of such devices in a manner that enables them to operate as a whole (to some extent) [50]. A common term applied to this is Swarm Intelligence (SI) [53], which has been described as the "*study of combined behavior* [sic] *of systems designed by varied components [...]*" [97]. The next section will briefly discuss this field and its inspiration from nature.

For applications of reasonable complexity involving multiple objectives, tasks [73] or large/unknown/unstructured [9] environments [35], collectives of drones in some formation [132] or network [39, 98] have been shown to provide a significant advantage over using single devices [9, 39]. In most (or all) of the areas where individual drones have been shown to be useful, especially search and rescue [50], surveillance [157] and monitoring [32], a strong case can be made for using swarms.

While still mainly a future vision for potential applications, there is already work being done targeting swarming applications in locations so far away that a human control is prohibited simply due to the long turnaround time of signals (i.e., deployment in space or on other planets in our solar system [12, 122, 133]). While there are very few examples of this yet (NASA has used collectives of deep space probes as opposed to sending a single, larger probe [33]) there are clear and immediate parallels with the use cases (discussed above) for an Internet of Drones on earth.

2.3 Swarming Behaviour in Natural Sciences

Self-organization and the decentralised control are important properties of swarm intelligence [96]. In this section, we discuss some background from natural sciences where it is standard practice to look at the individual as well as the whole flock, school, pack, etc. Natural sciences can be divided into life sciences and physical sciences and in both there is significant literature with regard to collective motion and self-organization.

Swarm Intelligence and self-organization have been studied for decades [94] by practitioners in different fields in life sciences [36, 56, 100, 124], but also physics [143] and chemistry [20] and even social sciences [102, 105, 106] as well as more recently computer science [42, 65], artificial intelligence [18] and cyber-physical systems (aka robotics) [120, 150] where collective decision making mechanisms found in group-living animals from insects to vertebrates are nowadays used to coordinate robot swarms [60]. Recently, studies reporting on mixed groups of robots and animals to study self-organized collective behaviour [21] appearing in top journals.

The term self-organization can be traced back to physics and chemistry [100]. In these fields of physical sciences, researchers have attempted to explain interactions at microscopic level can give rise to the emergence of complex patterns or structures at macroscopic level [20].

In life sciences the concept of self-organization has been applied (a) to model the creation of global order through local interactions [98] between, for example, molecules or cells [56], (b) for viral particle assembly [124], (c) to help evolve specialised organ patterns [36], as well as maybe the most popularly known example, namely (d) to understand complex behaviours of groups of animals [30] such as fish, birds or social insects. For example, the concept of order by fluctuation [100], which can arise as a result of amplifications of non-linear feedback interactions between independent entities and can be considered almost synonymous to self-organisation, has provided a workable explanation for various stages of nest construction in termites [27, 40] and is another, maybe counter-intuitive, example of the positive effect of noise amplification.

Swarming behaviour in biology has long since been the subject of study, and more recently the models proposed by theoretical biologists [27, 40] have inspired solutions in the field of computer science [42, 65, 94, 115, 122, 123] capitalizing on the benefits of decentralised approaches. For example, the framing conditions of an agent's limited range and view on the environment combined with positive feedback from the surroundings in which the agent exists have been successfully applied to among other things, dynamic client-server allocation in wireless access networks [65], operational control for swarms of autonomously operating UAVs [63, 64], self-assembly into 2-dimensional shapes of large (1000+ members) collectives of small robots [120] and the automated construction of user-specified [150] or self-determined [131] 3-dimensional structures through ground based vehicles.

The concept of self-organization has recently been used for large swarms of autonomous cyber-physical devices [120]. With regard to the use case application used in this chapter, researchers in the field of mobile networks have long since taken inspiration from nature [1, 3, 130, 149] using, e.g., animal strategies of self-organization to reach a dynamic steady state distribution and successfully recreated such behaviours in mobile networks [1, 3, 97, 130, 149]. This chapter aims to, on one hand, propose the investigation of swarms of cyber-physical systems using the view of a theoretical biologist but on the other hand cautions that merely copying biological swarms for its own sake has no intrinsic value. This is discussed in the conclusion, cf. Sect. 7.

2.4 Summary

Drones are rapidly becoming a pervasive technology, and the deployment of multiple devices (studied theoretically for decades now) is starting to happen for a variety of applications. In the context of drone swarms forming a network for the collection and distribution of information, the term *Internet of Drones* is gaining attention from the community. Individual decision making in groups (potentially on the basis of local, incomplete or subjective information) is well studied, with numerous models proposed in the literature [155]. In science, we study the world around us, often using strong assumptions or simplifications. Up to a certain point, this serves us well but eventually lacks precision or detail to provide insight into some complex details. We know that in complex systems, processes occur simultaneously on different scales and levels. The resulting behaviour of a swarm depends on its individual members, and does so in a non-trivial way [144]. Once the interactions between members become sufficiently complex, a new theory is required as the laws that describe the behaviour of the whole are qualitatively different from those that govern the individual. In other words, and although mean-field approximation remains a powerful tool to model certain collective phenomena: **as complexity increases, the swarm becomes more than the sum of its drones**.

3 Modelling Indoor WSN Deployment

The application to which our investigation was applied was the use of multiple drones to deploy a drone-based sensor network into indoor environments, where localization through GPS or Galileo was not possible and where the exact layout of the environment is unknown or at least uncertain (e.g., due to some disaster having occurred). If this were not the case (if position and environment were known), then the problem would essentially be reduced to moving the assets into their positions as fast as physically possible without collision and interference [52, 111]. In our application, neither of these is known and therefore even the number of required devices has to be determined during deployment (as the drones explore the environment).

Table 1 Notation and the relevant aspects in our model for our drones

n_i	Drone i, constituting a node in the network	
R_{S_i}	The sensing range R_S of drone i	Set for all i to $R_S = x\ units$
R_{C_i}	The communication range R_C of drone i	Set for all i to $R_C = \sqrt{3}R_S$

The nature of our disaster relief scenario implies that the deployment time is a critical factor. In addition, upon completion of the deployment and once a WSN is established, routing and communication in general are a challenge.

3.1 *The Model for the Drones*

We focus our investigations on swarms of drones, that is, collectives of individual devices. For simplicity, we assume that all drones are homogeneous with identical sensing-, communication-, computation-, and mobility-abilities. The reason for this assumption is that we want to measure and compare the impact of external influences on the behaviour of the swarm and by modelling the swarm as a collective of identical devices we can make stronger claims about the observed behaviour. In line with this simplification, we assume that all devices share the same communication and sensing ranges (which may differ) and and we simplify the coverage of both to be circular.

The actual flight performance of the devices is subject to ongoing innovation in the industry and is therefore not of interest to us. We abstract from device specific properties and simply model a drone as a mobile device that carries a battery (without quantifying the energy storage) and a wireless access point/sensing equipment. This suffices to investigate their performance as an arbitrary nodes n_i in a WSN.

The choice for setting the communication range to a specific ratio of the sensing range (cf. Table 1) is motivated by the literature (cf. [130, 154, 156]) because it is proven to minimize overlap and redundancy. Any consideration regarding the energy supply for the devices is omitted, the battery charge is not included in the model. Nevertheless, the energy consumed is estimated (Eqs. 8 and 9, in Sect. 3.4.5) and included in a performance measure (cf. Sect. 3.4), both for the individual drone as well as for the entire swarm.

3.2 *The Model for the Application*

Given the emergency aspect of the application, the following critical requirements for the application can be stated: with lives potentially at stake, complete coverage should be achieved as fast as possible. That being said, keeping things simpleand achieving coverage in a cost-efficient manner is also important since the drones have a limited energy budget. With regard to the deployment of the WSN into the target environment, we consider iterative deployment, that is, we will release drones iteratively (one by one) instead of in batches (in groups) at the same time. The entry vector (the *angle* θ_i) is chosen for each drone at random, between 0° and 90°. We also assume that deployment into the environment happens from a single point of entry. This means that the WSN will grow by one node at a time, and will effectively be anchored at one point (coordinates 0, 0). It cannot be guaranteed that the network remains fully connected at all time, i.e., that a valid sequence of 'hops' can be continuously maintained between the entry point and every participating unit/relay. Accordingly the drones' collective ability to re-establish communication after an interruption is a highly desirable feature and a critical performance measure. We will discuss the conditions that (a) triggers the release of an additional done into the environment or (b) terminate the application later.

3.3 The Model for the Scenario/Environments

The target environments for the application are GPS-denied locations with unknown layout. Despite the fact that 3 dimensions of movement are a great advantage of drones, our simulations are simplified to 2D environments (basically, we do not consider very high environments that require multiple drones above each other to provide coverage).

In this chapter, 4 types of environments are considered:

1. **Empty** environments, framed only by the 4 bounding walls (cf. Fig. 2).
2. Environments with obstacles such as **pillars** or debris (cf. Fig. 2).
3. Environments with **walls** in various configurations (cf. Fig. 4).
4. Environments with **narrow hallways/large cracks** (cf. Fig. 4).

The 4 scenario types are meant to capture a diverse range of challenges arising from various amounts of signal blocking obstacles, as well as degrees of connectivity between these obstacles. Pillars, or debris, physically block the path of a drone but can easily be side-stepped. Their impact on signal propagation is probably more dramatic as the shadows created extend outwards (i.e. in the shape of cones). Walls are fundamentally different from pillars, especially when they almost enclose parts of the target environment (i.e., rooms), in which case walls are obstacles that primarily restrict physical access. Note that our simulations use abstract spatial dimensions (we often speak of *distance units* instead of e.g., *meters* or *kilometres*). When deployed into e.g., a floor of an office building, the drones used would be mini- or nano-drones [89] or small drones such as the NEC-001 which is the size of an A4 paper (see Fig. 3, left), as opposed to e.g., an USAF Predator drone or TNO's NEO AceCore drones (see Fig. 3, right). These larger drones would be used for entire factory buildings, industrial parks or underground installations.

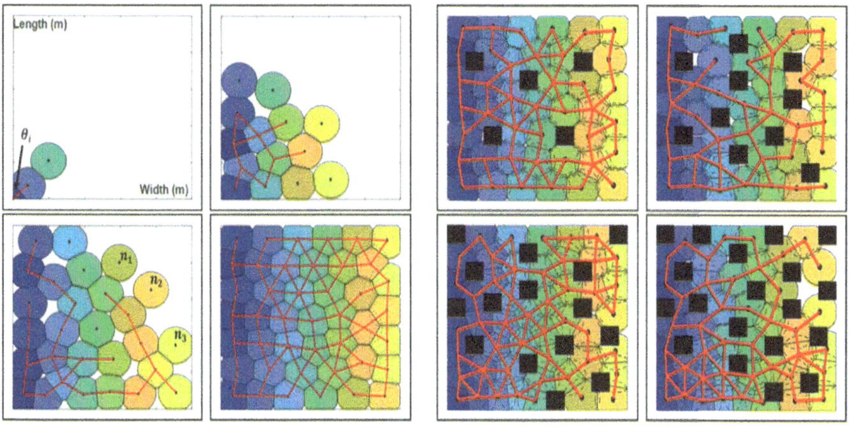

Fig. 2 A visual representation of the two simple target environments: (left) an empty hall enclosed by walls (not shown) with various drones inside; circles are sensing range and red lines indicate drones being able to communicate with each other. (right) similar to the above, but with 5, 10, 15 or 20 pillars or other blocking obstacles

Fig. 3 Different commercial civilian drones: (left) a small home-made drone capable of outdoor flight, used at NEC Research Labs Europe [61–64]; (middle) a small home-made drone of TNO for testing of indoor operations in GPS-denied environments, shoe for scale; (right) one of the professional (from a commercial provider) NEO drones of TNO capable of sustained flights in adverse weather conditions, BLOS (beyond line of sight) flight operations as well as carrying 8–10 payloads, again: show for scale. The presented results and considerations can be applied to all of these as well as larger and/or fixed-wing drones

The results presented and the views proposed hold true for small as well as for large scenarios and drones. For example, TNO's NEO drones (shown for comparison in Fig. 3, above) could be deployed into a large industrial complex after a disaster, with warehouses and plants extending for hundreds of meters. We argue that results scale well because larger drones can move longer distances in larger environments analogous to small devices in small spaces. This does, of course, not entirely hold for e.g., pillars as those tend to not grow linearly in size and density with the location where they are found. However, we argue that the abstraction level taken in our environment models is general enough to allow us to make general claims on the basis of our results.

With the exception of the first scenario (for which no variations on the obstacles are possible), all scenario types were represented by 4 variations.

Note that the basic layout of a square does not affect the results, we could have easily used a circular environment instead. The 4 scenario types are:

Empty: the environment, set to a simulated 10×10 distance units, is bounded only by the 4 surrounding walls (cf. left panel in Fig. 2). In a way, this is the base scenario within which all the others are situated (as all scenario types are bounded by walls).

Pillars: in environments with dimension 10×10, pillars are objects oriented in the x- and y-axis with a dimension of 1×1 (cf. right panel in Fig. 2). Their placement is random with some guiding conditions to prevent pillars from touching each other or touching the outer walls. 4 configurations were investigated, differing only in the number of pillars: we simulated 10×10 environments with 5, 10, 15 and 20 pillars.

Walls: as before for pillars, the thickness of walls is set to 1 distance unit (cf. left 4 screenshots in Fig. 4). The 4 different variations were designed and chosen using insights gained during the project. For example, we did not simulate long hallways with many rooms, only variation (d) "*Three Rooms*" captures the specific challenges of entering and exploring rooms. The other three variations were designed to pose specific challenging wall configurations: (a) (which we named "*H Shape*") offers the challenge of walls protruding into the environment, (b) ("*T Shape*") creates a small hallway open to both sides as well as one that is a dead end while (c) ("*C Shape*") encloses a small area inside an otherwise empty space.

Cracks: we simulated two variations of cracks/crevices, one where there is a single elongated open space (left side of the environments) and one where there are two

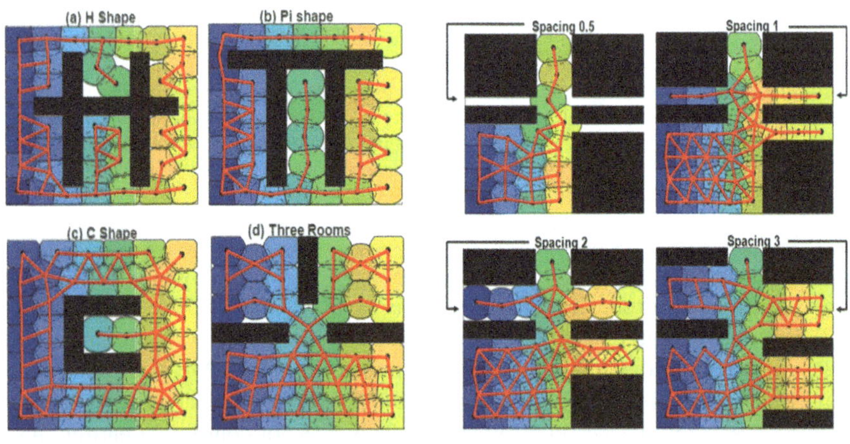

Fig. 4 Two complex types of environments (cf. Fig. 2): (left) environments with various wall configurations. These environments consist of large open areas separated by walls. (right) environments with hallways, cracks or crevices: the consist mainly of obstacles with the available space possibly being too constraint for drones to even enter into them

close to each other (right side) (cf. the 4 screenshots on the right in Fig. 4). In addition, the quarter of the environment that has the drone injection point as lower left corner is left empty entirely. This is to allow the approach to deploy into the environment and then to *discover* the cracks instead of directly started near or even inside them. The 4 variations used differ only in the width of the cracks, which is ranging from 0.5 distance units over 1, 2 to a maximum value of 3. As the screenshots in Fig. 4 show, a width of 0.5 is too small for the devices to enter.

3.4 Performance Measures

WSNs are comprised of devices which are, in all likelihood, constrained by their energy reserves [17, 128]. As such, service coverage and energy conservation are two key issues for WSNs [10]. Under this assumption, IoD systems are generally assumed to be energy aware [151]. Node movement, the operation of the sensing- or communication-equipment [4] are examples of processes that incur an energy cost. In the literature, clustering algorithms based on node energy have been shown to perform well with regard to an IoD [83].

3.4.1 Algorithm Runtime (`steps`)

The number of simulation iterations (`steps`) required to achieve optimum performance (as determined by the termination criteria). This is a straight forward metric to measure the speed with which the algorithms converge on the optimal/best found solution. Since the execution of the algorithms happens on-board each individual drone and because the actual flying from one location to another takes a different amount of time (by orders of magnitude) we evaluate the performance of candidate distributed the algorithms by comparing the number of simulation steps needed to achieve full (best) coverage. This performance measure can be regarded as independent of the actual size of the environments or the drones.

3.4.2 Percentage Area Coverage (`PAC`)

The first and most obvious performance measure is the aggregated area coverage achieved by the swarm [135]. The coverage for the individual drones is pre-defined and, unless blocked by an obstacle, automatically achieved. Therefore the *percentage* of the *area* that is *covered* (`PAC`) is a measure that only applies to the performance of the collective. To calculate the value we use information available in the simulation environment that is not available to the individual drones: we know the complete area to be covered (i.e., we know what 100% coverage would mean, area wise). In line with the literature [78, 79, 159], we compare the actual coverage (thus *not* what the drones *think* they are covering) to this value:

$$\text{PAC} = \frac{\textit{Combined coverage by the WSN}}{\textit{Maximum area that can be covered}} \tag{1}$$

3.4.3 Average Distance Travelled (ADT)

Drones incur an energy cost for their movement. To keep track of this we measure the *average distance travelled* (ADT) by the nodes in the network. Note that, due to the iterative insertion of nodes into the environment, some nodes may contribute massively to this value while others (i.e., the ones deployed last) might just have moved a few meters before full coverage is achieved. We used equations found in the literature [78, 79, 159] to analyse the ADT of the sensor nodes:

$$\text{ADT}(t) = \frac{1}{n(t)} \sum_{i=1}^{n(t)} d\big(origin, (x_i, y_i)\big) \tag{2}$$

where $n(t)$ is total number of nodes at time t, *origin* is the initial location/the injection point of node i (which in our simulations is fixed to location (0, 0)), (x_i, y_i) are the current x-/y-coordinates of node i and $d\big((x_1, y_1), (x_2, y_2)\big)$ the absolute Euclidean distance between two locations:

$$d((x_1, y_1), (x_2, y_2)) = \sqrt{(|x_1 - x_2|)^2 + (|y_1 - y_2|)^2} \tag{3}$$

Note that, unlike in [78, 79, 158], we do not require the number of nodes to be constant over time as our application allows for the iterative injection of nodes. In our simulation all nodes enter at the same injection point (0, 0)). Therefore, the ADT (at any time t) can be easily calculated as:

$$\text{ADT}(t) = \sqrt{\left(\sum_{i=1}^{n} x_i(t)\right)^2 + \left(\sum_{i=1}^{n} y_i(t)\right)^2} \tag{4}$$

3.4.4 Cumulative Distance Travelled (CDT)

The *cumulative distance travelled* (CDT) is measured recursively: at time t_0 the combined movement of all nodes is zero, at any other time t_k the CDT is the sum of all CDTs in preceding time-steps plus the movement since t_{k-1}:

$$\text{CDT}(t_0) = 0 \tag{5}$$

$$\text{CDT}(t_k) = \frac{1}{n(t)} \sum_{i=1}^{n(t)} d(pos_{t-1}, pos_t) + \text{CDT}(t_{k-1}) \tag{6}$$

With $d(\_, \_)$ again the absolute Euclidean distance, as defined in Eq. 3.

3.4.5 Energy Cost Incurred for Deployment (E_M, E_S)

While there is increasing research into the continuous powering of aerial devices though e.g. *high energy lasers* (HEL) [88], we assume that all drones operate with a limited energy budget. Since the energy consumption of individual drones (as well as that of the entire network [39]) are of primary interest to engineers, this is also included as a performance measure. In accordance with the literature [2, 55] energy efficiency is measured over the lifetime of a network:

$$\textit{WSN efficiency} = \frac{E_{total}}{E_M + E_S + E_{other}} \tag{7}$$

with E_{total} the total energy that sensor nodes have access to, E_M the energy cost (incurred through node movement) and E_S the operating energy cost (incurred by the node's sensing equipment). The value E_{other} is added to allow for additional power requirements (which, in our simulation, are none). Regarding the value E_M: the movement cost E_{M_i} (based on kinetic energy) of an individual node i is calculated as follows [74, 154]:

$$E_{M_i} = \frac{1}{2} m_i v_i^2 = \frac{1}{2} m_i \left(\frac{\Delta d_i}{\Delta t_i} \right)^2 \tag{8}$$

with m the mass of the drone, t the time, d the distance moved and v the corresponding velocity of the node at, both during time step t (again, calculated as in Eq. 3). For simplicity we assume that the work-energy theorem applies and re-write Eq. 8 to:

$$E_{M_i} = m \frac{\Delta^2 d_i}{\Delta t_i^2} d_i \tag{9}$$

We focus on the kinetic aspects of the application, therefore we will use mass-normalised energy: $\frac{E_{M_i}}{m}$.

3.4.6 Uniformity of the Coverage (U_A)

We prefer solutions where the overlap of the coverage provided by the individual drones is minimised. The argument for this is that the surface of a circular area increases non-linearly with its radius. Furthermore, the signal strength of a broadcast diminishes with distance to the source. The cost of providing coverage therefore also increases non-linearly with the radius and one can state, generally, that if complete coverage can be achieved with all nodes in a network operating at the same coverage radius then this is guaranteed to be (as far as signal cost is concerned) the best possible solution. Since we simulate environments with obstacles, and due to the fact that our approach uses Voronoi tessellations, we do not compare the coverage radii but instead the areas inside the Voronoi regions: following the literature [78, 79] we calculate a value U_A, capturing the uniformity of the distribution of the drones individual coverage.

$$U_A = \frac{1}{mean(A_V)} \sqrt{\frac{1}{|N|} \sum_{i \in N} \left(A_{V_i} - mean(A_V)\right)^2} \tag{10}$$

where A_{V_i} is the area of Voronoi region i and N is the total number of nodes in the network. The smaller the value of U_A the better the system is towards uniform distribution.

3.5 Defining Characteristics of the Approach

Our work specifically addresses the problem of obstacles in the environment. This is mostly ignored in the literature with many approaches implementing the deployment process through repulsive forces acting between drones as well as between a drone and an obstacle. Often this requires a prior scanning of the environment, with the resulting map of the environment a required input for the drones and their algorithms. This is, or can be, quite costly in terms of the incurred energy cost as well as rather inefficient since the mapping of environments in disaster scenarios may take a long time [52, 86, 111]. In our approach, drones are only required to scan their local immediate environment (defined by their sensing range). Furthermore, many approaches in the literature assume perfect knowledge with regard to the actual location of the drone as well as of its immediate neighbours or required so-called *anchoring nodes* which remain at known locations and facilitate localization efforts by the swarm [1]. Our approach does not require this simplification of the scenario.

4 A Decentralised and Self-organising Approach

In [44, 47], we proposed a Voronoi tessellation-based algorithm called BISON (for Bio-Inspired Self-Organising Network). In this approach, nodes in a dynamic (wireless) network use their own coverage range, as well as deduced information about the position (and thus coverage range) of other drones to calculate a local partitioning of the observable area, effectively allocating areas in overlapping fields to drones.

4.1 *Inspiration from Nature*

As already pointed out in Sect. 2.3, researchers in Computer Sciences have been taking a look at nature for inspiration on how to solve complex problems for a long time [65, 115, 122]. In the field of mobile networks this has also led to positive results [1, 3, 130, 149]. For example, strategies to reach a dynamic steady state distribution have been found in social animals and subsequently been applied to/implemented for mobile networks.

Studies of *social* insect behaviour provided computer scientists [121] with rather powerful methods for the design of distributed control mechanisms [123] as well as optimisation algorithms [18] where the *collective intelligence* is not born from complex individual abilities but rather a product of the networks of interactions that exist among individuals (as well as, as in the case of our example application of drone based wireless access networks, between the individual devices and their environment [19]).

Among the biggest advantages of many of these approaches is the fact that they can perform well when operating on limited, incomplete or highly subjective/local knowledge. This has been shown for geometric phenomena [101], as well as for biological ones [43, 75, 103], with Voronoi tessellations being among the techniques that were applied successfully to achieve uniform distribution of nodes in a WSN. In the past, Voronoi algorithms have been used to enhance network-coverage as well as -connectivity. This can be done using only the Voronoi approach [81, 109, 145] or in the form of a hybrid approach, pairing Voronoi with some nature-inspired approaches [11, 13, 71, 113, 114].

With the application of indoor exploration and surveillance/monitoring in mind, coverage of the network is a fundamental aspect of performance that has been addressed in the literature: e.g., for known areas and a fixed number of homogeneous sensor nodes, a number of different coverage-optimising algorithms have been proposed [47, 127, 147, 148]. Specifically, the moving of a sensor towards its centroid (i.e., the *centre of mass* of an area that is actually covered by the sensor) has been investigated [81, 109, 134, 145]. One particular algorithm (used for comparison by us) is the Node Spreading Voronoi Algorithm (NSVA) [58, 78, 79, 158, 159] which (contrary to our approach) assumes a fixed number of nodes and obstacle-free environments.

Voronoi diagrams have also been used for the deployment of WSNs consisting of heterogeneous sensors [14, 68, 70, 84, 85], for scheduling mechanisms in a WSN [145], as well as in security algorithms where they have been found to increase robustness against malicious identity attacks during the deployment phase [15].

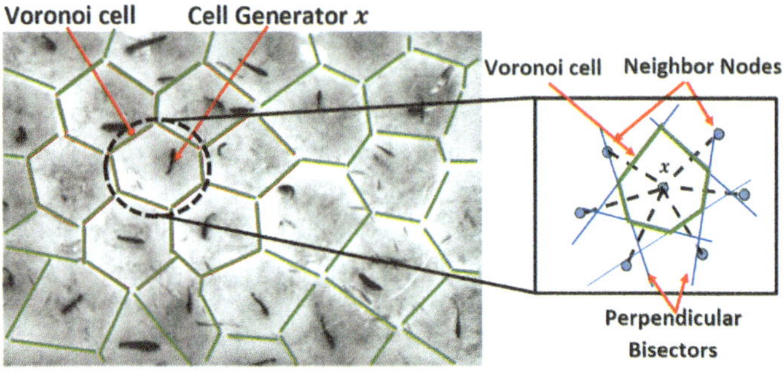

Fig. 5 Using Voronoi tessellations (applied in the left image to identify the territories of fish (*tilapia mossambica*) in a school) to allocate coverage to nodes with the aim to keep moving the centre of the nodes to the centre of their coverage. We are using inspiration from biology (left) to use neighbouring nodes and overlapping coverage to derive a partitioning of the available space. The panel on the right illustrates how this can be applied to nodes in a network. Compare the right image to the one in Fig. 6, specifically to panel (d)

4.2 A Voronoi Tessellation-Based Approach

In biology, to neighbouring cells may share a common border (i.e., they might touch each other). In the physical world, cells cannot overlap and thus the internal pressure of the cells, combined with the elasticity of the cell walls (and other factors) result in the formation of polygons (see left panel of Fig. 5). For drones, and their fields of sensing coverage, overlap is of course possible. However, when we ignore the scenario where a single drone alone cannot provide sufficient service to an area (and thus requires a second drone to cover the same area) we can say that such overlap/redundancy constitutes a waste of resources. Our aim (and the philosophy behind the BISON approach) is to allocate any location in the space to exactly one device (see Fig. 6).

This results in the virtual partitioning of an area (right panel, Fig. 5), not unlike the physical partitioning happening in cells (left panel, Fig. 5). Once such virtual areas are allocated, BISON simply calculates the centre of gravity for each and then moves the node (the device at the centre of the respective sensing range) towards this centre. This is illustrated in Fig. 6. Note that we compute the nodes' next positions iteratively and based solely on local information (i.e., to perform the required calculations a drone only needs to have information about its neighbours, that is, all nodes within its sensing range) [43, 47]. As we have shown, this alone suffices to create inherent movement in the swarm to deterministically converge towards complete (as far as possible) coverage. As we will discuss in Sect. 5, BISON's performance actually improves in the presence of various levels of Gaussian noise, which makes it attractive from the point of engineering and in the context of an actual implementation of the approach.

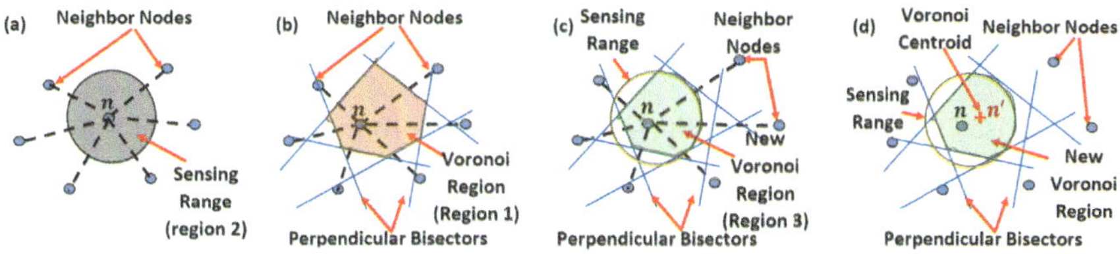

Fig. 6 When the communication range R_C of a drone is larger than the sensing range R_S, the device can *see* devices outside its sensing coverage (panel **a**). When (for practical reasons) this is limited appropriately (see Table 1 for our relative setting for R_C) then this can be restricted to those neighbouring drones whose coverage can overlap with the drone's own. Using the perpendicular bisectors (panels b, c) we can then calculate the Voronoi region (see Fig. 5 on Voronoi tessellations) which will be either a polygon (as shown in panel b) or, in case some neighbouring drones are too far away, will have some circular boundaries (see panel c). Either way, this area is likely to have a different centre of gravity (n' in panel d) than the circle (n, i.e., the drone's position) and BISON is based on the idea of moving the node towards this centre of gravity. Since all nodes do this, the system will be subject to continuous change until the WSN has achieved a stable state

4.3 Defining Characteristics

The approach has a number of defining characteristics:

1. **Decentralised**: the approach can be implemented in an entirely decentralised fashion, meaning that an actual physical realisation of BISON could be deployed with devices communicating only with their visible neighbours and without the need of any central control system. This does of course not apply to the application-inherent need to have a receiving node that connects the network to the outside world, the higher-level requirements for the application may very well be such that a fully decentralised approach is not possible but the algorithm itself can be entirely decentralised.
2. **Self-organising**: the actions and decisions of any individual device in the swarm will affect its immediate neighbours, potentially causing ripple effects that can ultimately affect the entire network/swarm. Local decisions can lead to emergent behaviour at system level.
3. Using **local Information only**: the approach is agnostic to the environment into which the swarm is deployed as well as the size of the swarm itself (though the network can derive whether there are deployed drones that are currently not connected to the swarm). Each device solves the problem locally and independently of the rest of the network. This makes the approach scalable (growing linearly with the number of devices as well as the number of neighbouring drones for each).

4.4 The BISON Algorithm Family

We proposed a number of algorithms, which together form the BISON (bio-inspired self-organising network) family. In this section we will briefly explain how and why these work, and illustrate their kinship relation to one another.

4.4.1 The Original, Voronoi Tessellation Inspired BISON Algorithm

The initial BISON algorithm only relies on Voronoi tessellations ([45–48]). Using this algorithm, a collective of drones can be deployed automatically and autonomously to self-organise into a WSN covering the deployment area.

Automatic deployment: the drones (network nodes) are deployed iteratively into the environment using clearly defined triggering conditions for the deployment of an additional device into the swarm. This happens entirely independently of the actual environment in which the swarm is operating and is agnostic to the number of already deployed drones.

Autonomous operation: the drones and the deploying mechanism operate independently. That means that decisions are taken using the agent's local information only. This information can be generated by other devices, and the trigger for an action can be the absence of some information, but the entire approach operates on the basis of each individual device acting autonomously.

Self-organisation: devices continuously evaluate their own situation and make decisions that affect their own position. In the swarm, this individual behaviour leads to a collective *effort* to construct a WSN and to ensure coverage and connectivity in the network without, guidance or assumptions. At the level of the WSN, it appears as if the network purposefully adapts to the environment as it expands while, in reality, this adaptation is an emergent property resulting from the accumulation of many local individual decisions.

The impact of obstacles: as stated above, the algorithm relies on each node calculating their effective coverage as the area allocated to them to cover. In case of overlap, only one drone will be allocated the overlapping area, the other drone is therefore free to move without considerations for that area. The approach to obstacles is similar: the drone's signal is blocked by the obstacle, so by subtracting the area that is blocked from that for which the drone provides coverage, a different coverage area and shape are calculated (see Fig. 7). In this newly redefined area, the centre of gravity may shift significantly and can thus cause the drone to move towards a better position.

The BISON algorithm: is discussed in detail in e.g., ([46, 47]), here we will restrict ourselves to a high level explanation of how the approach works.

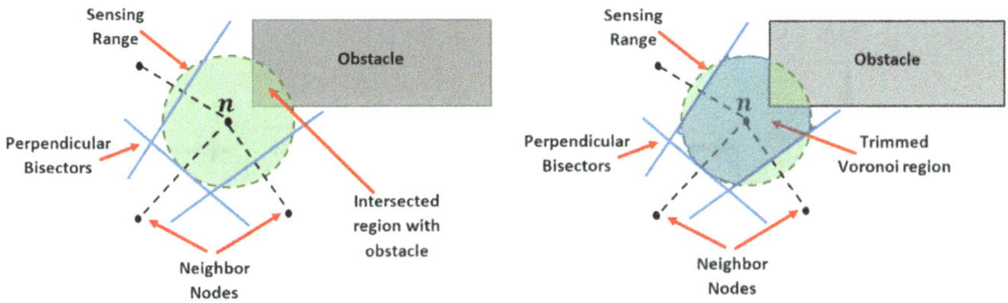

Fig. 7 Visual explanation of how the BISON approach, illustrated in Figs. 5 and 6, handles obstacles in the environment. As shown in detail in Fig. 6 the (original BISON) approach essentially relies continuously moving towards is centre of gravity, and obstacles can impact the centre's location significantly (right panel)

BISON will iteratively insert drones into an environment; or, more appropriately phrased: in a swarm where all drones are running BISON, the deployment of drones happens iteratively and is controlled by the application. A number of triggering conditions determine whether/when an additional device is added to the swarm. Initially a minimum number of drones is deployed. In Fig. 8, this number is set to 3, but this value can be determined by the operator based on either information or expectations about the environment or on the specifications of the drones or the type of sensing they are expected to do. After this, the algorithm (executed by every drone) enters a loop.

Drones will, in this order, **(1)** check their neighbours and report any disconnected neighbour back to the operator through the network. If the node is disconnected from the controller itself one of its peers will notice a disconnection and report it. Upon receiving this, the controller will initiate the booting up and deployment of an additional drone. Since drones are deployed iteratively, it does not matter how many drones are disconnected as new drones are released until all known drones are re-connected to the network again.

Once the check on the neighbours is done, any drone will **(2)** infer the position of its neighbours, calculate the perpendicular bisectors, calculate the resulting Voronoi cell as well as the cell's centre of gravity (centroid) and then move towards it. This process is illustrated in Fig. 6. If there are any obstacles sensed in the coverage field of the drone this is included in the calculation of the Voronoi region (see Fig. 7). Drones will then **(3)** move towards their (new) Voronoi centroid, calculate their movement distance and report that back to the controller where this value (or some estimate) will be used to eventually terminate the process: once the aggregated node movement in the system drops below a certain threshold Val_{th} the WSN deployment is considered to have stabilised and the algorithm terminates.

Termination criteria: there are a number of criteria to terminate the algorithm. The first, the *stability* criteria, measures the movement of the nodes in the swarm and terminates the deployment process when this value drops below a certain value. The reasoning is that the process underlying BISON will cause the nodes to be in

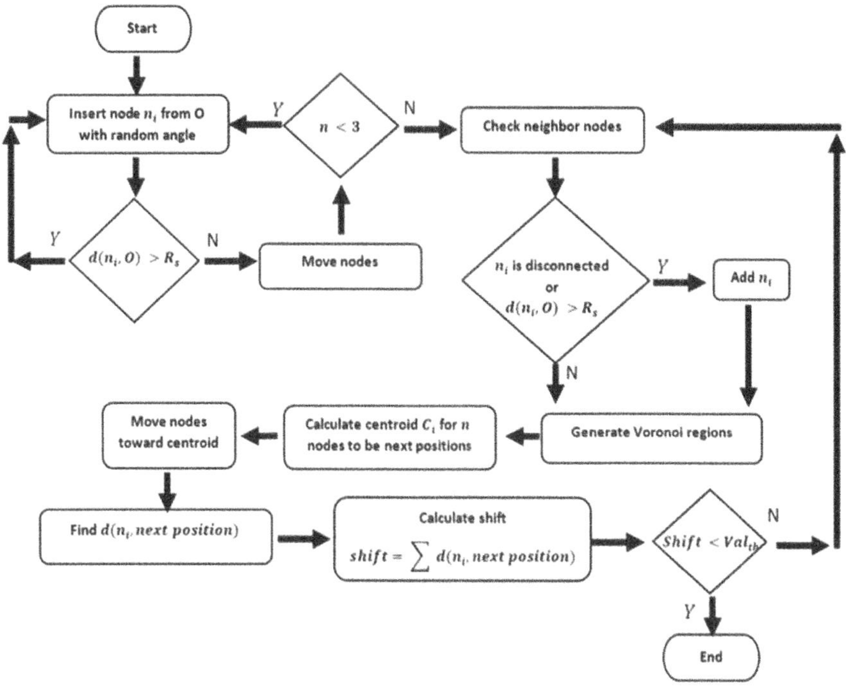

Fig. 8 A flowchart for the original BISON algorithm. Figure 9 provides the pseudo-code, for a detailed discussion of the algorithm we refer to [46, 47]

continuous movement, always reacting to changes in their neighbours coverage and position. Initially this will be substantial, as the nodes move into an unexplored environment. However, as the space is filled with nodes, their movements will grow smaller and smaller until, eventually, they will no longer move further into the space but only oscillate back and forth in reaction to their neighbours doing the same. When it reaches such a state, the network is basically cycling through solutions to the deployment problem that differ very little among themselves.

Instead of implementing complex assessment procedures to infer when this happens we can simply look at the aggregated node movement over some time window. In our implementation, this threshold value was set relative to the sensing range, namely to $\frac{R_S}{100}$.

The second criteria is, simply stated: stop *if there has not been any improvement in coverage in n iterations*. This is captured by summing up the individual coverage polygons and comparing them to the sum from the last iteration. If there is no gain, a counter is increased; if there is some gain the counter is reset. Both can be seen in Fig. 9 (the pseudo code for BISON).

Fig. 9 Pseudo-code for the original BISON algorithm (cf. the flowchart in Fig. 8 or [46, 47] for the details on the algorithm

```
Initialization;
Start from origin point (O);
Inject node randomly;
Val^th ← R_S/100;
Define Cov_Max;
count ← 0;
while (stability > Val^th || count > 15) do
    if (exist disconnected node || node away from O)
     then
        inject new node;
    end
    for each node n_i do
        Add Gaussian noise to neighbor nodes;
        Generate Voronoi region;
        Calculate Voronoi Centroid C_i;
        move n_i towards C_i;
        shift_i ← d(n_i, C_i);
    end
    stability ← ∑^N shift_i;
    if PAC ≥ Cov_Max then
        count = count +1;
        Cov_Max ← PAC;
    end
end
```

4.4.2 Augmenting the Original BISON Approach with Genetic Algorithms

As stated above, the deployment process is continuous with each node effectively reacting to changes in its neighbouring nodes' positions. In the beginning of the deployment, this change will be large and there will be relatively fast and seemingly directed dispersal of the drones into the environment. However, in the presence of obstacles or when there are many nodes in a small area, this process will slow down. Furthermore, the *shuffling* into the best location may take increasingly long with decreasing gains (this is one of the motivations behind the termination criteria). To overcome this and to speed up exploration of the environment, we developed a genetic algorithm (GA) augmentation of BISON, where node movement is not based in the Voronoi centroid of the drone but which instead uses a location identified by a GA (cf. Fig. 11).

Every time a node re-evaluates its position and tries to move, the GA is executed to identify a number of suitable candidate locations to move to (Fig. 10 shows the flow chart for the GA, Fig. 11 illustrates the finding of candidate solutions other than the centroid). This enables the drones to discover additional suitable locations and effectively combines the uniformity and convergence properties of the Voronoi-based approach with the enhancement in discovery rate and lifetime offered by the GA.
The GA: as shown in Fig. 10, the GA has three stages:

1. **Initialisation**: when started, the GA randomly generates *pop_size* locations as an initial population (of locations) inside the respective Voronoi region. These are evaluated using an objective function $\text{obj}_{fun}(n_i')$. Of those, the best $\frac{pop\_size}{2}$ solutions are kept as parents for the next generation.

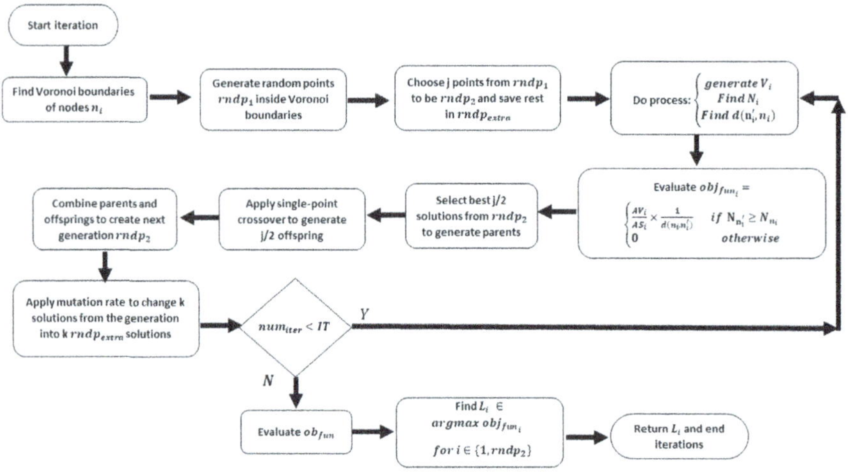

Fig. 10 A flowchart for the BISON-GA algorithm. Figures 12 and 13 provide the pseudo-code for the two variations of the GA that were used: the *Fixed Nodes* and the *Conditional* variation, respectively. For details of the algorithms see [44, 46]

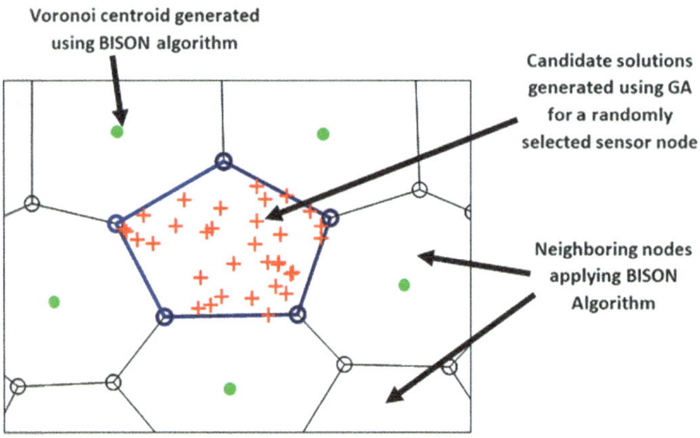

Fig. 11 Instead of moving a drone towards its Voronoi centroid, a GA can be used to generate a number of candidate locations to consider

2. **Genetic Operators**: the two standard operators, *cross-over* and *mutation* are used. Mutation is determined by the *mutation rate* **m**, the crossover step is implemented as a single-point crossover where two parent solutions (i.e., locations) are chosen at random and their y-axis values are switched. This generates two new offspring and is repeated $\frac{pop\_size}{2 \times 2}$ times to ensure that the parent- and the offspring-generation are of the same size. To foster diversity of, and explore new locations for, the candidate solutions, mutation is then applied to the offspring generation.

3. **Selection**: the best $\frac{pop\_size}{2}$ from both, the parent- as well as the offspring-generation, are kept and become the next generation. Of those, the best member is chosen to become the new position for the node to move to.

Steps 2 and 3 are repeated for # *iterations* before the best candidate solution is chosen; cf. Fig. 2 for the parameter space exploration. *The objective function for the GA*: the objective function $\text{obj}_{fun}(n_i')$ for each candidate solution n_i' (used in the flowchart presented in Fig. 10):

$$\text{obj}_{fun}(n_i') = \begin{cases} \frac{A_{V_i}}{A_{S_i}} \times \frac{1}{d(n_i, n_i')} & \text{if } N_{n_i'} \geq N_{n_i} \\ 0 & \text{otherwise} \end{cases} \tag{11}$$

with

- A_{V_i} the node's Voronoi area;
- A_{S_i} the node's sensing range area;
- $d(n_i, n_i')$ the distance between candidate (n_i') and existing (n_i) location;
- $N_{n_i'}$ the number of neighbours at the new location;
- N_{n_i} the number of neighbours at the existing location.

See Fig. 11 for an illustration of candidate solutions.
Two variations on BISON-GA: the use of GA is motivated by the need to distribute the nodes more effectively, without adding computational cost. Running the additional GA on top of the Voronoi approach incurs a computational cost. To mitigate this impact, two variations were conceived. Both aim to reduce the number of times the GA is used while maximising the benefit gained from doing so.

The first variation (cf. Fig. 12) limits the number of nodes that do use the algorithm to a fixed number of (randomly) pre-selected nodes which will then be able to make use of the GA during their deployment. Since neighbouring nodes affect each other, the impact of this ripples through the network and can benefit other drones as well, resulting in an overall improvement for the entire network. This approach was called

Fig. 12 Pseudo-code for the *Fixed Nodes* variation of the BISON-GA algorithm (cf. the flowchart in Fig. 10 or [44, 46] for the details of the algorithm

```
for each node n_i do
    Add Gaussian noise to neighbor nodes;
    Generate Voronoi region;
    if n_i ∈ GA_rnd then
        apply GA algorithm ;
        return next position L_i;
        move n_i towards L_i;
        shift_i ← d(n_i, L_i);
        else
            Calculate Voronoi Centroid C_i;
            move n_i towards C_i;
            shift_i ← d(n_i, C_i);
        end
    end
end
```

```
for each node n_i do
    Find number of neighbors N_i;
    Add Gaussian noise to neighbor nodes N_i;
    Generate Voronoi region;
    if for n_i (0 < N_i < 4 ) then
        apply GA algorithm ;
        return next position L_i;
        move n_i towards L_i;
        shift_i ← d(n_i, L_i);
        else if N_i ≥ 4 then
            Calculate Voronoi Centroid C_i;
            move n_i towards C_i;
            shift_i ← d(n_i, C_i);
            else
                n_i ←n_i;
                shift_i ← 0;
            end
        end
    end
end
```

Fig. 13 Pseudo-code for the *Conditional* variation of the BISON-GA algorithm (cf. the flowchart in Fig. 10 or [44, 46] for the details of the algorithm

Fixed Nodes, and was applied to a varying number of nodes: in the scenarios we tested we investigated performance for applying it to 1, 3, 7 and 10 nodes. See Sects. 4.6, 5.3 and 6.3 for the results.

A second approach (cf. Fig. 13) was designed using a different consideration: as stated above, the impact of cleverly chosen new candidate locations increases as the deployment evolves. This is due to the fact that initially drones are released into relatively empty spaces and are likely to move away from the injection point at great velocity, at least until they encounter the first obstacles. Once the space is getting filled, drones shift, from exploration to optimisation. In that later stage the movement of the drones is smaller. At that point is may be of increasing benefit to identify potential candidate locations other than the drone's centroids.

Therefore, the second variation of the GA approach is not based on specific nodes but instead on a condition which has to be met before the GA can be used (but is then used by all drones). Due to this, the second variation that was implemented and evaluated is referred to as *Conditional*, with the condition being a restriction on the number of neighbouring drones a drone has. In our case, this threshold for using the GA was set to (less or equal than) 3 (cf. Fig. 14).

4.5 Materials and Methods

The simulations were implemented in MATLAB R2017b. The parameter setting are listed in Table 3. The results presented in Table 2 are for independent evaluation of the three parameters *population size* (**p**), *# iterations* (**#**) and *mutation rate* (**m**). These choices for the respective missing parameters were initially guided in part by the existing values in literature (e.g., for the mutation rate) and in part by the design

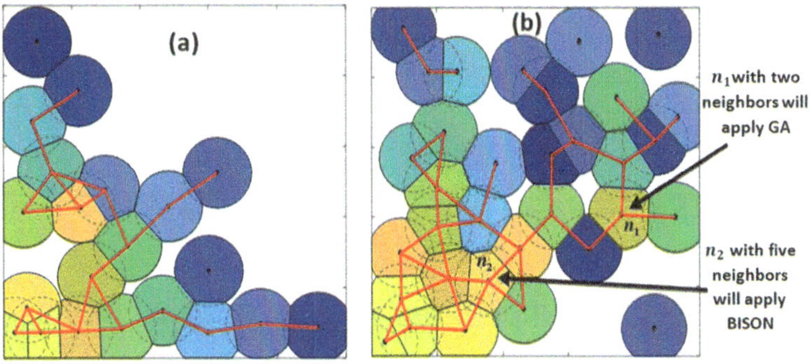

Fig. 14 Both, BISON-GA *Fixed Nodes* and BISON-GA *Conditional*, are using a genetic algorithm but they differ in to which nodes the GA is applied and, in principle to how many nodes they are applied

of the model problems we aim to solve (such as population size), but later confirmed by the results obtained.

The values outside the ranges shown in the table do not show promise of better overall performance (with respect to `PAC`, `CDT` and `steps`). However, within the evaluated ranges, changes impact performance similarly across algorithms. Since the parameters for specific real world scenarios will very likely be scenario dependent, we argue that the chosen ones suffice for our evaluation here and we hypothesis that they are good guesses for the initial iteration.

4.6 Results and Discussion

In this section we briefly discuss results comparing the algorithms. In the interest of brevity we only provide a few selected results and refer the reader to the two other results sections (Sect. 5.3 for results and comparison between results when the algorithms are applied to scenarios with communication noise; Sect. 6.3 for additional results using the metric of *drift* and *diffusion*). In addition, [44, 46, 47] provide additional and more detailed results.

We only compare the performances of BISON with its genetic algorithm enhanced siblings in two of the 4 environments, namely the obstacle-free environment and an environment with pillars or other signal blocking/scattering objects (left and right panels in both Figs. 15 and 16, respectively). Both of these environments are illustrated in Fig. 2 on page 10.

Table 2 Results for the parameter-space exploration for obstacle-free environments, using the BISON-GA fixed nodes algorithm

		BISON-GA 1 nodes			BISON-GA 3 nodes			BISON-GA 7 nodes			BISON-GA 10 nodes		
		PAC (%)	CDT (m)	st	PAC (%)	CDT (m)	st	PAC (%)	CDT (m)	st	PAC (%)	CDT (m)	st
→	**p** = 10	93.68	15.8	426	94.00	50.03	467	82.68	81.15	307	85.56	71.16	341
	±	1.2	2.69	48.7	3.4	5.1	68	2.2	12.1	53	3.13	26	63
	p = 15	95.78	22.7	526	84.64	41.06	330	85.42	108.5	440	80.27	147.6	454
	±	1.4	3.27	54.1	2.3	4.2	52	3.1	14.8	62	2.05	32	71
	p = 20	96.87	23.62	545	90.42	56.82	467	80.84	81.17	293	80.9	121.6	316
	±	1.9	4.06	56.3	2.9	5.5	63	1.6	12.1	48	2.21	28	56
→	# = 15	96.87	23.14	545	90.42	56.82	467	80.84	81.17	293	80.9	121.6	316
	±	1.7	2.5	49	4.5	14.6	77	2.58	15.9	28	0.61	27	16
	# = 30	93.8	16.15	418	80.92	23.75	273	87.98	99.74	3987	82.15	59.69	237
	±	1.2	1.6	42	4.1	9.6	58	3.47	17.8	36	1.11	13	14
	# = 50	98.2	18.9	600	92.67	56.31	532	83.3	54.65	351	81.41	63.99	301
	±	1.7	1.8	70	4.7	14.4	103	2.92	13.7	31	1.24	16	15
→	m = 0.05	96.87	23.67	545	90.42	56.82	467	80.84	81.17	293	80.9	121.6	3.16
	±	1.8	1.5	71	1.2	3.65	19	1.06	9.6	30	1.16	17	34
	m = 0.1	93.6	20.61	429	92.95	52.66	489	83.6	71.4	272	83.59	129.8	359
	±	1.5	1.2	66	1.3	2.9	23	2.04	7.4	28	1.94	19	37
	m = 0.2	92.69	20.09	399	93.14	55.99	530	81.58	97.84	351	81.06	87.98	421
	±	1.3	1.2	54	1.4	3.61	26	1.49	10.4	33	1.57	14	39

Shown are: the percentage of the area covered, PAC, the cumulative distance travelled by all nodes, CDT as well as the number of steps taken, **st**, for four different numbers of fixed nodes (1, 3, 7, 10). For each parameter *pop_size* (**p**), *# iterations* (**#**) and *mutation rate* (**m**) three values were investigated in isolation. The values used for the performance evaluation are indicated by arrows

Table 3 The parameter settings used for the performance evaluation

A	The environment to be explored	Set to 100×100 *length units*
R_S	The sensing range of all drones	$R_S = 10$ *length units*
$Valth$	For the regular BISON (cf. Fig. 9)	$R_S \times \frac{1}{100}$
$Valth$	For the BISON-GA	$R_S \times \frac{1}{25}$
counter	For the regular BISON (cf. Fig. 9)	15

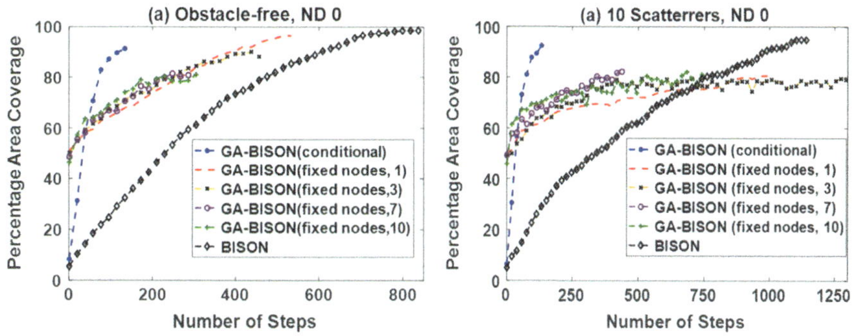

Fig. 15 Comparing algorithm performance on the basis of the total area covered (PAC). Shown are the performances for the different variations of the BISON algorithm with number of steps on the x-axis and the achieved coverage (in %) in the y-axis; for (left) for environments without obstacles as well as (right) for environments with signal scattering objects such as pillars, both shown in Fig. 2. Compare these results to those in Fig. 24 to see the impact of adding noise to the communication channels

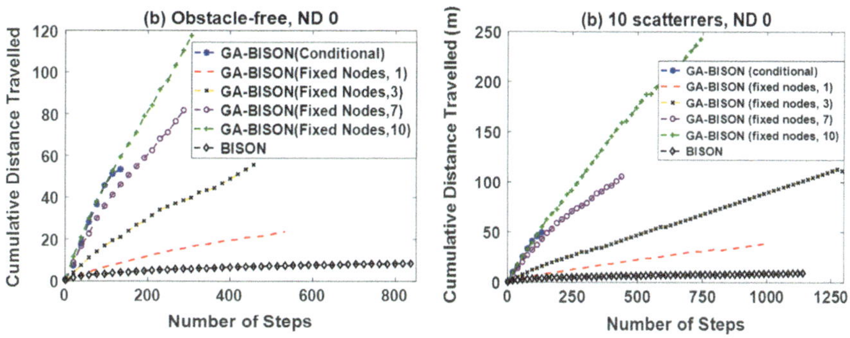

Fig. 16 Comparing algorithm performance when subjected to noise deviation (ND, reported are the results for ND = 0.05) on the basis of the total area covered (CDT). Shown are the performances for the different variations of the BISON algorithm with number of steps on the x-axis and the cumulative distance travelled (in distance units, here *meter*) in the y-axis; for (left) for environments without obstacles as well as (right) for environments with signal scattering objects such as pillars, both shown in Fig. 2. Compare these results to those in Fig. 25 to see the results with noise

4.6.1 Comparing the Area Coverage (PAC) achieved by the algorithms

With regard to the total area covered by the network by the time the algorithm terminates, we can see in Fig. 15 that the original BISON either outperforms all GA augmented variations or matches their performance.

However, in the case of the empty and obstacle-free environment (left panel) it does so at the cost of a much longer deployment time. In the presence of obstacles, the conventional BISON outperforms the fixed node GA variation but the conditional GA variation achieves similar coverage much faster.

4.6.2 Comparing the Cumulative Distance Travelled (CDT)

When comparing the performance of the algorithms and considering the cumulative distances travelled by all nodes in the network (see Fig. 16) we again see that the original BISON requires significantly longer to terminate, but that it does so with a fraction of the movement required for the other algorithms. Since movement is a major cost factor this suggests that the original BISON approach is more energy efficient while matching or outperforming the other algorithms with regard to coverage, but that this does come at the cost of deployment time.

Summary

Under a simplified view of the world, BISON outperforms the other algorithms albeit at the cost of deployment time. This indicates that the choice of algorithm could be driven by the need for timely coverage. It may also be worth observing that the main energy cost for a rotary wing drone (e.g., quadcopter) is incurred by the need to fight gravity to remain airborne and not by movement in the horizontal plane. Accordingly, minimising deployment time, even at the cost of longer cumulative distance travelled, is likely to improve overall energy efficiency substantially. However, these results rely on a significant simplification, namely that communication between nodes as well as measurements regarding the environment are perfect. This is unrealistic and can be disadvantageous when the system is deployed in the real world. In the next section, we enrich our simulations with communication noise, with counter-intuitive results.

5 The World Is a Messy Place (Full of Uncertainty and Noise)

Simulations and models are simplified representations of the real world. One fundamental aspect of the real world is the existence of noise and uncertainties, both of which can distort the outcome of our calculations. Investigating the impact of noise on our approaches is therefore a necessary part of the process, assuming that we are indeed serious about working towards an engineering solution that allows us to eventually deploy our drone swarm.

5.1 The Impact of Noise

In Fig. 6 we illustrated how BISON works. Specifically, in panel b, a drone identifies and determines the position of its neighbours. The precision of this measurements directly impacts node movement and thereby the overall performance of the approach. In the previous section we discussed the performance of BISON and BISON-GA, however, the discussed data was collected from simulations where the position of neighbouring drones was calculated perfectly. We now report the investigations of the algorithms that focused on performance under communication noise.

Subjecting the swarm of drones to signal distortion and noise means that the individual drones will calculate incorrect positions for their neighbours (cf. Fig. 17) and therefore will also calculate an incorrect Voronoi centroid for themselves. This means that their movement is not as intended.

This entails that the nodes are not moving towards the true Voronoi centroid. As shown in Fig. 18 this can lead to incomplete coverage as well as continued (and unnecessary) consumption of energy, depending on how much the noise causes the nodes to move around. In addition, noise can impact the usefulness of termination criteria: in a noisy environments signal distortions can cause nodes to keep moving back and forth.

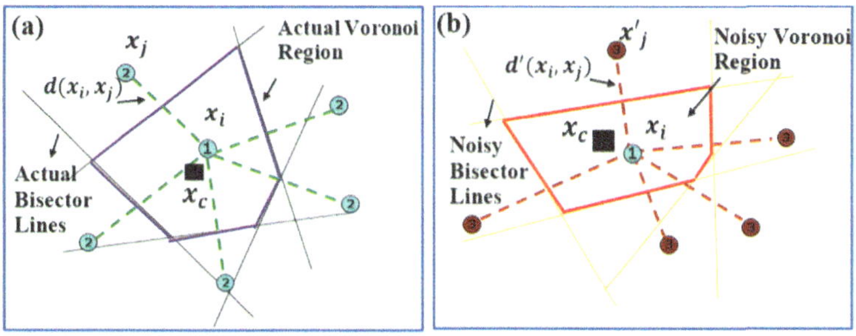

Fig. 17 Any type of wireless communication is subject to noise and distortion. Given the importance of the position of the neighbouring drones, incorrect measurements of those can significantly impact the accuracy of a node's actions and thus on the overall performance of the swarm. Compare the above to Fig. 6 (where the working of the BISON algorithm is illustrated); panel a depicts the actual state of affairs as shown in Fig. 6 while panel b shows some erroneous locations as well as the resulting Voronoi region. The actual location of nodes is shown in blue while erroneously calculated locations are depicted in red

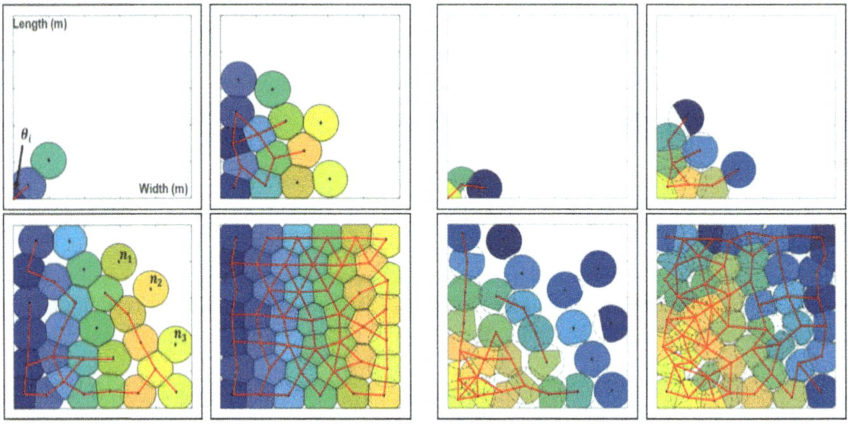

Fig. 18 The impact of communication noise (right panel) on the behaviour of the swarm as well as the changes in performance. When comparing the final results (bottom, right sub-panels in both panels) we see a significant coverage loss due to the miscalculations

Recall that we use the measure of aggregated node movement as an estimate for when the coverage has stabilized. Under noise, nodes may continue to move and thereby unnecessarily extend the runtime of the algorithm. To address this, the second termination criteria is used. We refer to Fig. 9 (the pseudo-code of the BISON algorithm) where a counter measures the number of times that the swarm improves on a pre-set minimum target coverage (PAC). This ensures that the algorithm will continue to loop until a minimum is reached while also terminating after a finite number of improvements over this performance, so as to avoid lengthy run times with ever so minimal improvements. This can be understood intuitively: the less clear the view on the world is (here: the location of other drones) the smaller the significance of tiny improvements (calculated using the imprecise data).

5.2 *Materials and Methods*

In the literature [78, 79], randomization has been used to implement signal distortion with the goal to replicate the effect of obstacles or imprecisions in the sensing equipment/process. In our work we chose to not do this but instead define a specific noise function. The reasons for this are twofold: (1) our goal is to investigate performance under varying noise levels so as to gain insights into the impact this has on performance, because (2) our work is intended to be (eventually) deployed in the real world, with real swarms and in real scenarios. Nothing in the real world is really random, and different scenarios will differ significantly in the specific characteristics of the environment. While random noise may allow us to gain some insights, in order to tailor the simulation to specific use cases when addressing the engineering aspects of our approach we need a more detailed understanding of *which level and type* of noise has what impact. Therefore, we used *Additive White Gaussian Noise* (AWGN), the most simplistic model of Gaussian distribution with zero mean, which we added to the signal with uniform power across the transmission channel. We modelled this as statistical noise with a probability density function $P(x)$ (normal distribution), formally defined as follows:

$$P(x) = \frac{1}{\sigma\sqrt{2\pi}} e^{\frac{(x-\mu)^2}{2\sigma^2}} \tag{12}$$

with the parameters μ and σ representing the mean and standard deviation of the random variable x, respectively. In this model, increasing the variance σ^2 (which in our case is given by $\sigma^2 = \frac{N_o}{2}$, with N_o the *noise power*) has a growing detrimental effect on signal accuracy. The effect of σ is, however, only felt within the communication range R_C. In other words, while visibility of neighbouring nodes (line-of-sight, LOS) may be maintained, their derived locations may be subject to error. Three different noise-levels were considered in our simulations: $\sigma = 0.01$ (low), $\sigma = 0.05$ (medium) and $\sigma = 0.1$ (high).

5.3 Results and Discussion

In this results section we first consider results under various different noise levels for all environment types (except the empty environment) as well as for all variations on these types discussed in Sect. 3.3 and illustrated in Figs. 2 and 4 on pages 10 and 11, respectively. We compare performance using ADT, the average distance travelled by all nodes in the network.

In the interest of brevity we report here only the results for the original BISON algorithm as the general impact of noise is consistent across algorithms [44]. Based on the results shown in Figs. 19, 20 and 21, we then fix the noise deviation to the value 0.05. As before, the interested reader is referred to ([44, 46, 47]) for a more detailed performance comparison and discussion.

5.3.1 Comparing the Impact of Noise Using ADT

With regard to environments with pillars and other signal scattering objects, we would assume that increasing the number of pillars results in an increase in *difficulty* (so to speak). In Fig. 19 we can see that (with respect to the average distance a single node is travelling from the injection point to its final location) the number of pillars hardly impacts performance. Increasing the number of objects to move around only slows the drones down. However, when adding/increasing the noise deviation this is significantly reduced across the board (while preserving the impact of more pillars). Counter-intuitively, communication noise speeds up the performance of the

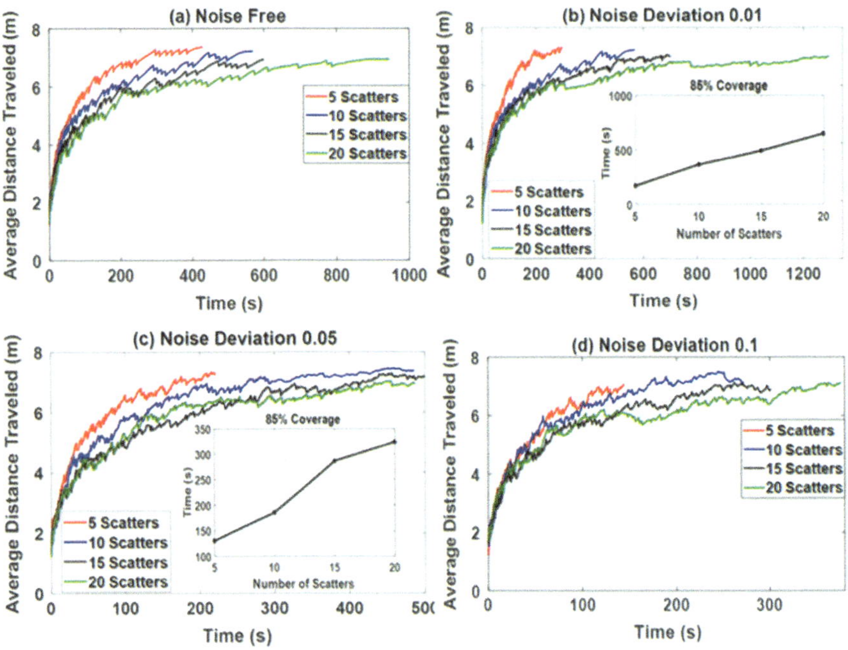

Fig. 19 Performance comparison of the original BISON algorithm in environments with 5, 10, 15 and 20 pillars. Reported is the average distance travelled (ADT) by each node. Across all 4 panels, this value (y-axis) seems to be consistent for independent of the number of objects with the performance differing only in the time required for the algorithm to terminate (x-axis). Note that the scale of the x-axis differs significantly between the panels, with a clear trend towards speedy convergence with increasing noise levels

algorithm. Panels b and d also show a comparison of the time taken to reach 85% PAC where we again see a significant improvement in deployment speed as noise increases.

For the environment types with walls/rooms (Fig. 20) and hallways/cracks (Fig. 21) we report the performance differences per individual variation of the respective scenario. These are discussed here to motivate using a noise deviation of 0.05 for the remaining results in this chapter.

5.3.2 Comparing the Area Coverage (PAC) Achieved by the Algorithms

With regard to the final WSN coverage, we first briefly discuss one of the 4 scenarios separately because of the inherent result-distorting properties it has: the scenario with hallways or cracks. We remind the reader that the overall size of the environment was fixed in our simulations, meaning that any obstacles added effectively reduce the maximum area that can possibly be covered.

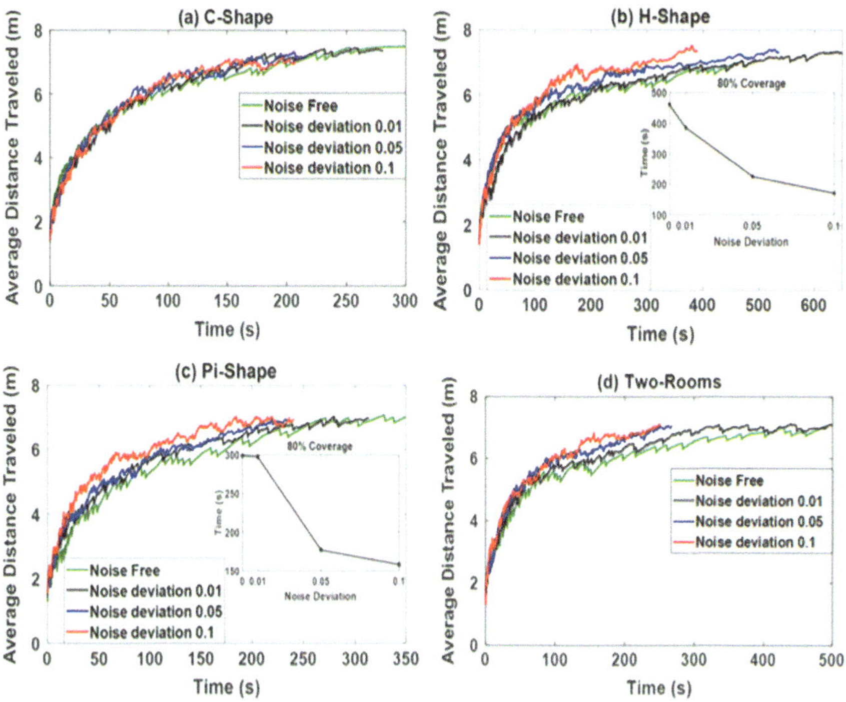

Fig. 20 Performance comparison of the original BISON algorithm in the 4 environments with walls (cf. left panel in Fig. 4 on page 11). A noise deviation value of 0.05 was used

This does not apply to the empty environment, and in case of pillars the difference between 5 and 20 pillars is not large enough to distort the outcome (in terms of PAC). In case of the walled scenarios, the main differences to pillars is the location and orientation of the walls.

For the scenario where we investigate elongated corridors of varying width we encounter the issue that for the narrowest of these the resulting crack is not wide enough to permit a drone to enter (see Fig. 22). In these scenarios, one quarter of the overall space is empty with the remaining three quarters constituting hallways of varying sizes. Bear this in mind when comparing the PAC results in Fig. 23.

In the previous section we compared the different algorithm types (Figs. 15 and 25). The following two figures report results from parallel investigations, differing only in that noise deviation was added. Since we above identified an appropriate noise deviation of 0.05, Figs. 24 (PAC) and 25 (CDT) report only the result obtained for this value. We refer to [44, 46, 47] for more results.

When we compare the results under noise (Fig. 24) to those previously discussed without noise (Fig. 15, page 29) we see that adding noise results in faster deployment times, sometimes with noticeable performance increase with regard to PAC. This is

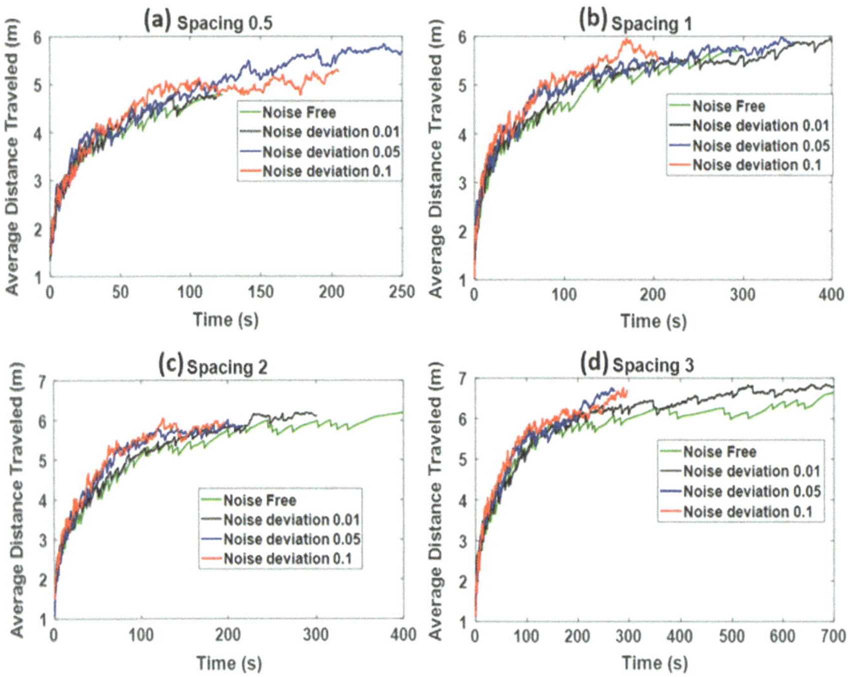

Fig. 21 Performance comparison of the original BISON algorithm in the 4 environments with hallways or cracks (cf. right panel in Fig. 4 on page 11). Noise deviation = 0.05

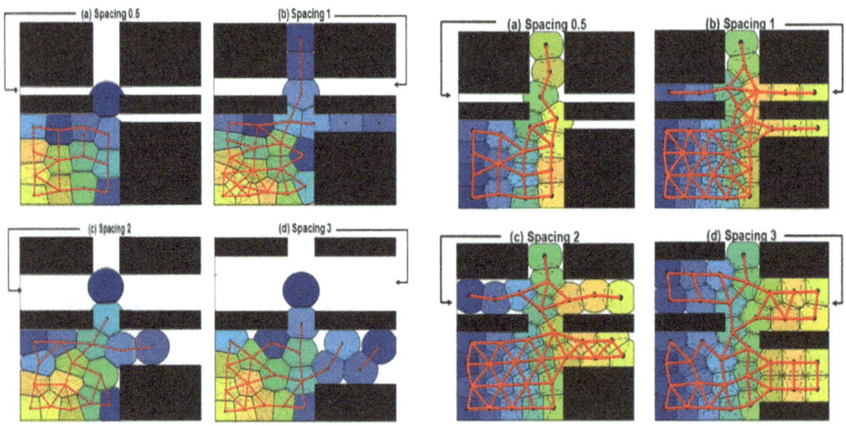

Fig. 22 Deployment (left) and a solution for a fully deployed WSN (right) in the 4 variations of the scenario with elongated corridors, representing hallways or cracks. Drones cannot enter the narrowest passage while, for the intermediate width, the drone's coverage is such that it fills it wall-to-wall. Due to this, the measure of PAC does not differ much between the variations of this scenario type (with the exception of variation a)

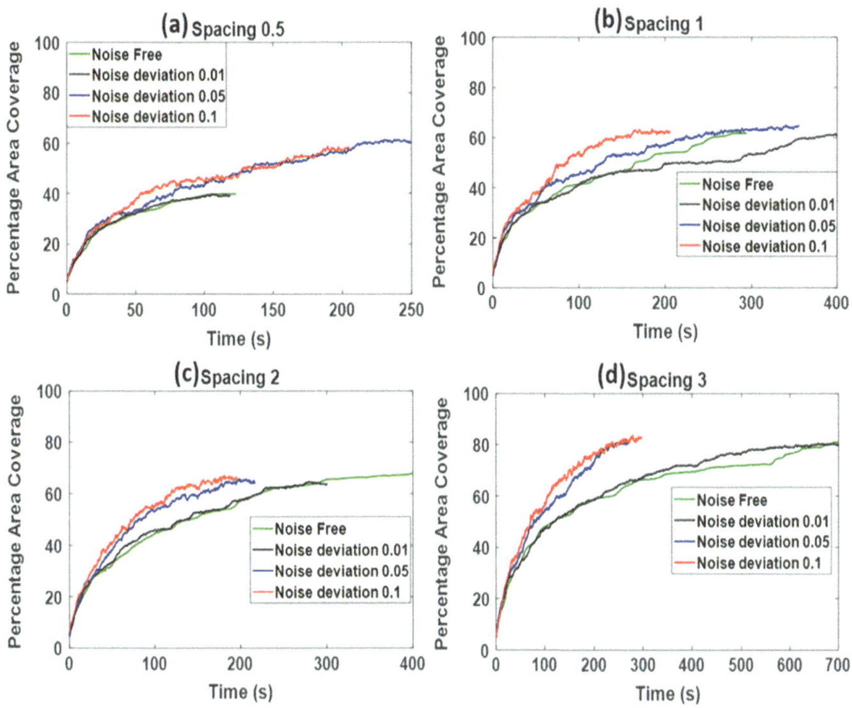

Fig. 23 Results for the area coverage (y-axis) achieved by the original BISON in the 4 variations of the hallway scenario type (cf. Fig. 22). With the exception of the thinnest hallway, panel a, the ultimately achieved PAC differs little between the different noise levels. However, as before, the deployment time seems to improve for increased noise

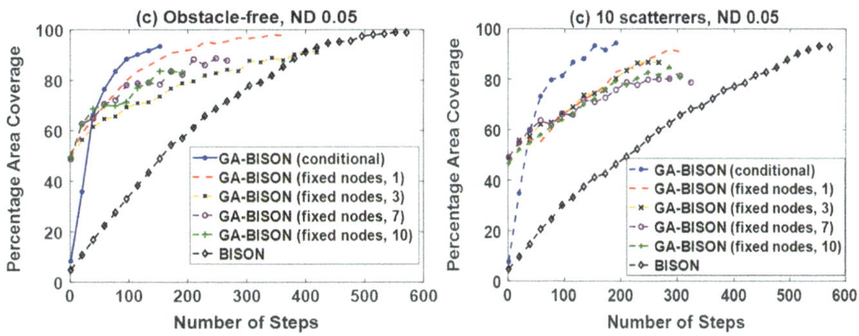

Fig. 24 Comparing algorithm performance when subjected to noise deviation (ND, reported are the results for ND = 0.05) on the basis of the total area covered (PAC). Shown are the performances for the different variations of the BISON algorithm with number of steps on the x-axis and the achieved coverage (in %) in the y-axis; for (left) for environments without obstacles as well as (right) for environments with signal scattering objects such as pillars, both shown in Fig. 2. See Fig. 15 for results without noise

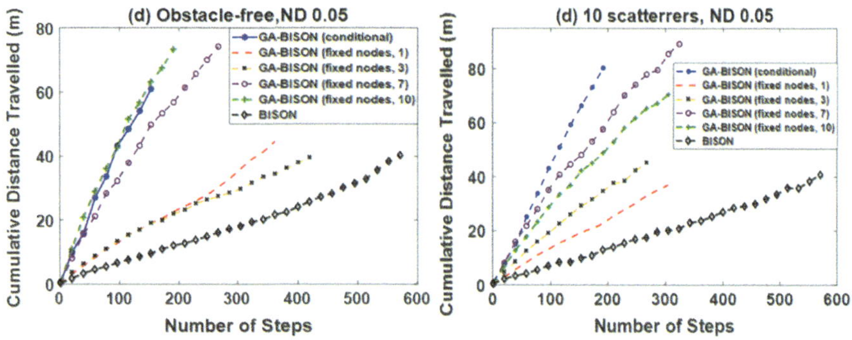

Fig. 25 Comparing algorithm performance on the basis of the total area covered (CDT). Shown are the performances for the different variations of the BISON algorithm with number of steps on the x-axis and the cumulative distance travelled (in distance units, here *meter*) in the y-axis; for (left) for environments without obstacles as well as (right) for environments with signal scattering objects such as pillars, both shown in Fig. 2. See Fig. 16 to compare these with the results without simulated communication noise

immediately visible when comparing the range of the x-axis, which for both panels in Fig. 24 is significantly shorter.

5.3.3 Comparing the Cumulative Distance Travelled (CDT)

The performance improvement witnessed for PAC is repeated for CDT. When comparing the results of the performance evaluation shown in Fig. 16 on page 28, we again notice improved performance both in the deployment time (x-axis) as well as the CDT, plotted on the y-axis. As above, the improvement is immediately obvious when comparing the range of the axis, which is shorter in all cases for the results under noise.

Summary

When comparing the performance of the algorithms we notice that, counter-intuitively, all algorithms perform better when the communication is subjected to noise. As noise is an inherent property of the real world, this is a good thing. As it turns out, by adding uncertainty to the system we enable the swarm to *stray from the direct path* so to say, which results in sub-optimal movement but also in a form of exploration. As we discuss in [44, 46, 47], this leads to a faster diffusion into the environment and can result in better solutions.

6 Swarm Behaviour: Drift and Diffusion

In the previous sections we have provided the motivation for our work, defined some more or less realistic environments within which we wanted to test our approaches and provided some performance measures to compare outcomes and algorithms. This work has previously been published and has been validated by peer review. In this previous paper we have also reported on the measure of the *drift-* and *diffusion-*coefficient, as used by [153] which provides a means to measure how drones disperse into an environment [157].

The argument we make in this chapter can be grounded in two claims:

1. Swarms of cyber-physical systems are becoming a reality, and
2. when such swarms are operating in the real world they will exhibit certain properties and behaviours of biological swarms.

Based on these two claims, we argue that an Internet of Drones, or any large enough collective of cyber-physical devices for that matter, will have to be studied similarly to how groups/flocks/schools/packs or swarms (etc) in nature are studied by biologists. We will discuss this in more detail in the next section but before we do we first motivate the claim that there is, indeed, merit in this type of investigation. To this end, we report on the performance evaluation of our algorithms which we conducted (in addition to what is reported in the previous two sections) using the concepts of drift and diffusion, from the drift-diffusion model, well known to biologists [138].

6.1 Measuring Group Behaviour in Biology: Drift and Diffusion

Our results in Sect. 5 clearly show the benefit of noise on the performance of our algorithms with regard to the resulting final WSN coverage and the rapid increase thereof during swarm deployment. The most likely quantitative explanation is that in the right circumstances, the right kind of noise leads to self-organized directional switching that, in the context of our performance measures, can be considered helpful. This is comparable to the positive impact of noise in biological systems [153].

To investigate the behaviour of the swarm we use the drift-diffusion model (DDM), a popular model for cognitive decision making under forced choices from biology [119]. To do so we consider the velocity of nodes in the network (independently of their direction) and argue that this one-dimensional view can already suffice to shed some light on how the group behaviour changes in the presence of noise. This does not allow us any claims about *why* this is the case, nor does it enable us to capitalize on this effect (other than by suggesting that signal noise is not, in itself, a disadvantage).

During the evaluation of the BISON algorithms we were not just interested in the distances travelled by the individual nodes but also in the collective behaviour of the swarm. In line with [153] we will use the *average* node-velocity to characterize *node behaviour* during the deployment of the WSN.

In our simulation, individual node velocity is determined by the calculation of the Voronoi centroid combined with the node's inference of its location (i.e., the *to* and *from* of the node's movement). The way we determine movement defines the overall collective deployment behaviour but considering the distance (as we do with ADT and CDT) only gives us insight into the performance of the entire system and does not tell us anything about the performance of the nodes in the context of the behaviour of the other nodes.

In line with [153] we avoid using a theoretical complex model/equation to determine node-velocity; instead we use empirical data (generated by our simulations) to estimate the diffusion (D) and the drift (F) coefficients of this equation. The former will capture the mean rate of change of average node velocity, while the latter quantifies the evolution thereof.

6.2 Materials and Methods

The coefficients for diffusion (D) and drift (F) are defined as follows [153]:

$$D\big(v(t)\big) = \frac{1}{2}\left\langle \frac{\big(v(t+\delta t) - v(t)\big)^2}{\delta t} \right\rangle \tag{13}$$

$$F\big(v(t)\big) = \left\langle \frac{v(t+\delta t) - v(t)}{\delta t} \right\rangle \tag{14}$$

with

$$v(t) = \frac{1}{n}\sum_{i=1}^{n} v_i(t) \tag{15}$$

where $v(t)$ is the average velocity of all the available nodes n at time t, $D\big(v(t)\big)$ is the diffusion coefficient, $F\big(v(t)\big)$ the drift coefficient and δt the chosen time step value.

6.3 Results and Discussion

In the interest of brevity we again restrict our results to a subset of those available and refer the reader to ([44, 46, 47]) for more results and discussion. Here we restrict our discussion to two scenarios, namely the empty environment and the one featuring the largest hallways/cracks (shown in Fig. 22, panel d).

The results shown in Figs. 26 and 27 are both showing the impact of noise on the behaviour of the swarm, plotted as *drift* and *diffusion*: Fig. 26 shows these metrics for the obstacle-free environment while Fig. 27 does so for the scenario with the widest simulated hallways. In both we provide the outcomes without noise (top row,

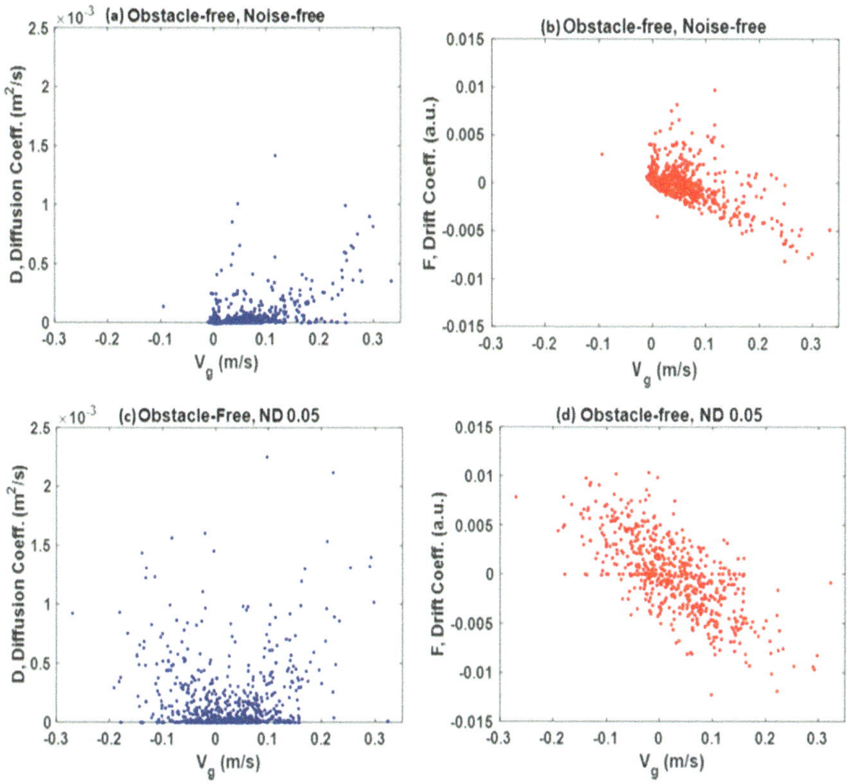

Fig. 26 A comparison of the drift and diffusion observed for the nodes of the WSN when deploying into the obstacle-free environment (see Fig. 27 for the environment with hallways). Despite being on the same scale, there is a noticeable difference between the noise-free (top panels) and the noisy environments (bottom panels)

diffusion plotted in blue, drift in red) and (bottom row) after noise was added. Clearly (and in accordance with the results shown in previous sections) the swarm behaviour changes when signals are distorted. The most likely quantitative explanation is that in environments/conditions where the communication is subjected to noise there is an increase in directional switching. This can be considered helpful in accomplishing some WSN tasks where e.g., exploration can provide an advantage.

We refrain from further discussions on the impact of noise, as this is not the intention of this chapter. Instead, we continue in the next section with a discussion on the need to study swarms of cyber-physical systems using tools and models used by biologists to study natural swarms, flocks or schools. Once this case is made, we then reflect on this and offer words of caution.

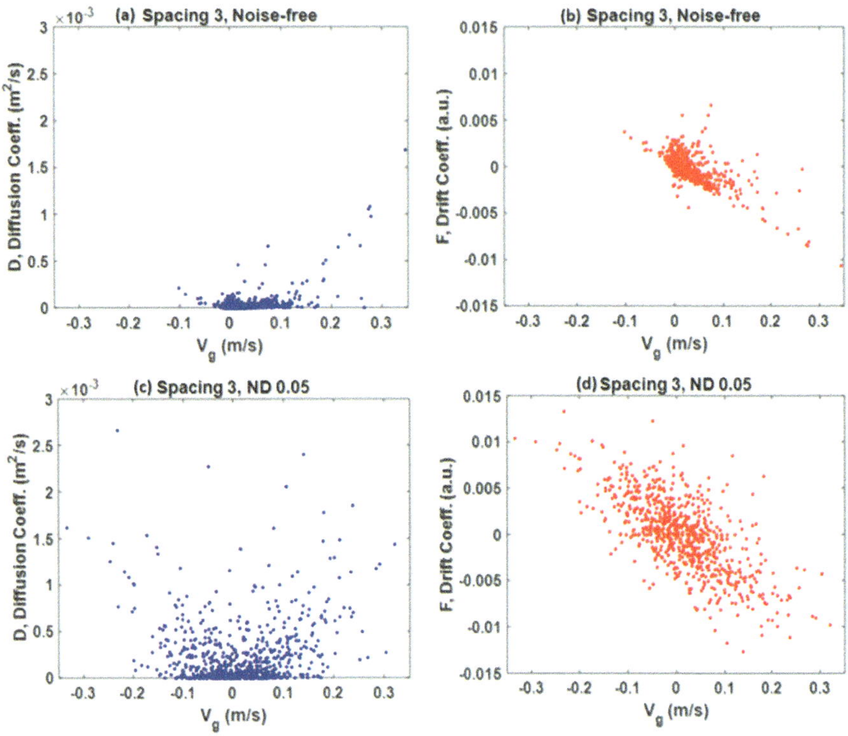

Fig. 27 A comparison of the drift and diffusion observed for the nodes of the WSN when deploying into the environment with large hallways (compare to Fig. 26). Despite being on the same scale, there is a noticeable difference between the noise-free (top panels) and the noisy environments (bottom panels)

7 Conclusion

*7.1 Towards an Inter*NET *of Drones: Swarming Behaviour*

Swarm Intelligence has attracted the attention of researchers for decades [94]. In biology, mathematical models have been proposed to explain collective behaviour in groups of animals, most famously for social insects such as ants [41] or termites [16] or for packs of mammals such as e.g., wolves [28, 49, 92]. For groups of sufficient size, hierarchical orders and structures within the group have been investigated [93, 107]. As swarms of cyber-physical systems (such as drones) progressively become a reality, the field needs to follow them into the real world: while simpler models and theories may suffice for simulations and controlled experiments (often ignoring real world aspects such as communication noise, friction, resistance and other distorting influences) there is an increasing need to study robot swarms "in the wild" in a similar manner to studying groups of animals. To this end, Bonnet et al. [21] propose a *collective decision coefficient* to measure coordination in groups of animals (and robots). Knebel et al. [76] report on heterogeneity in the context of swarm robotics [72, 139].

In the long run, similar to social theory [34] for living beings, the study of collectives of drones will need a unified approach capable of describing both stability and change in such quasi social systems. As the fields of AI and robotics evolve, there is increasing need for a theoretical foundation linking the behaviour of individual units to some desired organisational behaviour at the group level and, ultimately, in a society of devices spanning multiple swarms (similar to the so-called *system of systems*, we envision a *collective/swarm of swarms*).

7.1.1 Collective Behaviour in Groups of Individuals

When considering the finer details, no two animals in nature are ever exactly alike [69] and the same can be said for cyber-physical systems, simply because of variations inherent to the hardware (such as sensors and actuators which will always differ minutely in precision, cf. [120]). Furthermore, any real-world task that requires categorical decision making by the individual and on the basis of perception has to somehow navigate noisy, distorted and ambiguous signals [119]. The drift-diffusion model (DDM) [153] has been used to model sequential choices made on the basis of uncertain or incomplete information [99]. In the previous section we have discussed using drift and diffusion to measure some aspects of group behaviour. The point is not to propose these as the definitive measures to qualify collective behaviour, but to demonstrate the use of a methodology developed by biologists for a similar purpose. Our message is simply that the field needs to draw on work done in biology (or physics, or chemistry, for example) to advance swarm robotics further.

While it is perfectly fine to use performance measures such as the ones proposed in Sect. 3.4 to investigate the performance of a computational system with regard to its intended application, it falls painfully short of investigating the collective/swarm/group of cyber-physical systems in itself (i.e., as a group and not an application). There is an increasing amount of research directed at the studies of swarms (e.g., [9, 135]) and on the computational models that result in behaviour similar to that observed in groups of animals in nature (e.g., [92]). But these studies are mainly concerned with producing behaviour that appears to mimic that of groups of animal [153] and lack the means to investigate the relationships that potentially exist between individual- and group-level patterns and behaviours [108].

7.1.2 Focusing on the Direction of Motion

When studying groups of individuals in nature, collective movement is among the most striking phenomena investigated [153]. This may be driven by the visually stunning experience that such displays can create (consider watching a large flock of birds or marvelling at the seemingly organised chaos of an ant hill in full swing), but on a more technical level even such apparently simple coordination can pose a huge challenge. After all, the discrimination of movement in one's surrounding requires a number of complex operations such as e.g., an understanding of relative and absolute points of reference. Furthermore, understanding the direction of movement based on noisy signals is an example of sensory discrimination under uncertainty [138].

7.1.3 Directions of Research Other Than DDM

The drift-diffusion model, arguably the canonical computational model for the cognitive process that drives forced-choice decision making [119], has performed well for various tasks [146] but there is research questioning some of its assumptions such as e.g., that reaction time is a function of signal to noise ratio [138]. There are many aspects in decision making in the real world that cannot be described by the DDM, such as e.g., the benefit of being able to combine temporally distant signals (i.e., remembering what was good the last time when we were in a similar situation) [99]. In [51] the authors propose what they claim to be an "*equivalent Bayesian model*" which (unlike the DDM) links information directly in the stimulus to the decision process.

7.1.4 Implications for Other Aspects of the Internet of Drones

Among biologists, the use of mathematical models has become a recognised research tool for the study of collective behaviour [37]. Such models can be applied to the dynamics underlying the formation of group behaviours. The study of directional changes in the movement of autonomous agents capable of rational decision making is certainly of interest to biologists [153] and, we argue, to the relatively young field of computational swarm intelligence. Individuals in a cooperative collective pass information between each other, and do so over noisy and inherently unreliable channels. In nature, this information can make the difference between survival or death (e.g., when the information passed is in reference to an external threat). It has been shown that the effectiveness (with regard to success) of this passing of messages can be a function of the density of the swarm [137]. This alone suggests that there is benefit to be gained from investigating swarms of cyber-physical systems using tools from biology.

Under a simplified view, a swarm is a decentralised sensing and acting organism with localised information storage, processing capability as well as individual characteristics. Just as the survival chances of a group in nature can be increased though better adaptation to, and understanding of, the environment, so is there a clear potential for gain in swarms of drones which inherently act so that their individual orientations, velocities etc. drive an emergent behaviour of the collective. Many interesting discoveries with regard to the role of collective phenomena in animal and human behaviours have been made and there is no reason to think that collectives of machines cannot also benefit from them in similar ways [143]. We are standing on the brink of an extremely exciting time, when machines will reach human and possibly above-human levels of intelligence and when collectives of machines start interacting autonomously in meaningful ways, chances are that this will give rise to artificial collective- or swarm-intelligence.

7.2 A Word of Caution

Whenever talking about swarming, flocking, recruitment, division of labour or, indeed, any kind of social behaviour, it is paramount to keep in mind that these collective phenomena exist for a purpose. This is certainly the case for biological examples, where the corresponding behavioural traits (and associated physiological functions) have evolved through selective pressure because of the advantages that they confer to those individuals exhibiting them. We argue that the same considerations should apply to cyber-physical systems, the designers of which are attempting to leverage nature-inspired distributed intelligence to achieve a certain goal. We feel obliged to mention that this has not always been the case and that, particularly with regard to swarming, the ability for a group of individual units to coordinate their movement through collective decision [39] has occasionally been regarded as its own reward [9], notwithstanding its usefulness in a practical application scenario. Biologists commonly accept that flocking in birds or schooling in fish has evolved as a means to mitigate the adverse effects (e.g. risk of collision) of maintaining a very high population density, which is otherwise beneficial, for instance because it constitutes a good defence strategy against predators. This begs the following question: in which actual use-case does packing together a large number of drones in a small volume bring sufficiently significant advantages to justify fitting them with the sensors, actuators and decision-making capabilities required for coordinated flying? We cannot think of many, which is not to say that nature-inspired design of cyber-physical collectives is a futile endeavour, only that copying biological swarms for its own sake has no intrinsic value. Accordingly, we advocate a "looser" approach to biomimicry, a design methodology that, whilst retaining the desirable properties of the biological model (such as the low or absent reliance on centralised control mechanisms that supports flexibility and robustness in the face of unpredictable or adverse circumstances), does not focus on reproducing any given naturally occurring phenomenon, however impressive it may be. In the context of the *Internet of Drones*, this could mean that the swarm exhibits collective dynamics that, although highly efficient with respect to its intended function, do not resemble anything found in nature, because the selective pressures that apply are entirely different. For the sake of example, let us consider the case in which the objective is to achieve homogeneous distribution of an arbitrary number of units over an area of equally arbitrary size and shape. From exploration and surveillance to providing reliable network coverage, it is straightforward to imagine a wide range of practical applications for such an ability. Conversely, it hard to think of a situation in which the ability to fulfil this objective would be of any use to a biological swarm. On the contrary, recruitment mechanisms and coordinated movement typically exist to maintain cohesion and focus the activity of the collective on a given location (e.g. food source or threat) because, in the biological world, this is the kind of behaviour rewarded by natural selection. The ability to establish and maintain a distribution pattern that simultaneously maximises coverage and preserves line-of-sight between first neighbours is of no use to social insects (the main source of inspiration for cyber-physical swarms), despite being an emergent property of other, usually competitive biological systems (e.g. regular hexagonal lattice formed by stickleback nests). So a drone swarm using co-operative social interactions to achieve the common goal of distributing the collective resources homogeneously has no clear biological equivalent, even though its behavioural repertoire may be comprised of algorithmic "primitives" found in nature.

8 Abbreviations and Notations

The abbreviations used in this chapter are listed in Table 4.

Table 4 **ReWrite** Well these are the abbreviations we use throughout the document

Abbreviation	Meaning
TNO	Netherlands Organization for Applied Scientific Research
VTT	Technical Research Centre of Finland
IoT	**I**nternet **o**f **T**hings
IoD	**I**nternet **o**f **D**rones
IoDT	**I**nternet **o**f **D**rone **T**hings
SAR	**S**earch **A**nd **R**escue
UAV	**U**nmanned **A**erial **V**ehicles
MRS	**M**ulti-**R**obot **S**ystem
HEL	**H**igh **E**nergy **L**aser
AWGN	**A**dditive **W**hite **G**aussian **N**oise
GA	**G**enetic **A**lgorithm
PSO	**P**article **S**warm **O**ptimization
ACO	**A**nt **C**olony **O**ptimization
BISON	**B**io-**I**nspired **S**elf-**O**rganizing **N**etwork
N2N	**N**ode to **N**ode
WLAN	**W**ireless **L**ocal **A**rea **N**etwork
WSN	**W**ireless **S**ensor **N**etwork
mWSN	**m**obile **W**ireless **S**ensor **N**etwork
LOS	**L**ine **o**f **S**ight
PAC	**P**ercentage **A**rea **C**overage
ADT	**A**verage **D**istance **T**ravelled
CDT	**C**umulative **D**istance **T**ravelled
D	**D**rift
F	**D**iffusion
DDM	**D**rift-**D**iffusion **M**odel
E_{total}	the combined energy that a node has access to
E_M	the energy cost incurred through movement
E_S	the energy cost incurred by operating the node's sensing equipment
E_{other}	any other energy cost included in E_{total} but not in E_M or E_S

Acknowledgements The authors acknowledge support from the UAE ICTFund grant "Bio-inspired Self-organizing Services" as well as the valuable input from their colleague Dymitr Ruta (EBTIC). AFI acknowledges motivating discussion with Prof. V. Kumar (Univ. of Pennsylvania).

References

1. Abbasi, M., Bin Abd Latiff, M.S., Chizari, H.: Bioinspired evolutionary algorithm based for improving network coverage in wireless sensor networks. Sci. World J. **2014**, 839486 (2014). https://doi.org/10.1155/2014/839486
2. Abo-Zahhad, M., Sabor, N., Sasaki, S., Ahmed, S.M.: A centralized immune-Voronoi deployment algorithm for coverage maximization and energy conservation in mobile wireless sensor networks. Inf. Fusion **30**, 36–51 (2016). https://doi.org/10.1016/j.inffus.2015.11.005
3. Adnan, M.A., Razzaque, M.A., Ahmed, I., Isnin, I.F.: Bio-mimic optimization strategies in wireless sensor networks: a survey. Sensors **14**(1), 299–345 (2014). https://doi.org/10.3390/s140100299
4. Aftab, F., Khan, A., Zhang, Z.: Bio-inspired clustering scheme for internet of drones application in industrial wireless sensor network. Int. J. Distrib. Sens. Netw. **15**, 155014771988990 (2019). https://doi.org/10.1177/1550147719889900
5. Al-Kaff, A., Madridano, A., Campos, S., García, F., Martín, D., de la Escalera, A.: Emergency support unmanned aerial vehicle for forest fire surveillance. Electronics **9**(2) (2020). https://doi.org/10.3390/electronics9020260
6. Al-Naji, A.A., Perera, A., Mohammed, S., Chahl, J.: Life signs detector using a drone in disaster zones. Remote Sens. **11**, 2441 (2019). https://doi.org/10.3390/rs11202441
7. Almeida, M., Hildmann, H., Solmazc, G.: Distributed UAV-swarm-based real-time geomatic data collection under dynamically changing resolution requirements. In: UAV-g 2017—ISPRS Archives of the Photogrammetry, Remote Sensing and Spatial Information Sciences. Bonn, Germany (2017)
8. Atia, D.Y., Ruta, D., Poon, K., Ouali, A., Isakovic, A.F.: Cost effective, scalable design of indoor distributed antenna systems based on particle swarm optimization and prufer strings. In: 2016 IEEE Congress on Evolutionary Computation (CEC), pp. 4159–4166 (2016)
9. Awasthi, S., Balusamy, B., Porkodi, V.: Artificial Intelligence Supervised Swarm UAVs for Reconnaissance, pp. 375–388 (2020). https://doi.org/10.1007/978-981-15-5827-6_33
10. Aziz, N.A.A., Mohemmed, A.W., Zhang, M.: Particle swarm optimization for coverage maximization and energy conservation in wireless sensor networks. In: Di Chio, C., Brabazon, A., Di Caro, G.A., Ebner, M., Farooq, M., Fink, A., Grahl, J., Greenfield, G., Machado, P., O'Neill, M., Tarantino, E., Urquhart, N. (eds.) Applications of Evolutionary Computation, pp. 51–60. Springer, Berlin, Heidelberg (2010)
11. Aziz, N.A.B.A., Mohemmed, A.W., Sagar, B.S.D.: Particle swarm optimization and Voronoi diagram for wireless sensor networks coverage optimization. In: 2007 International Conference on Intelligent and Advanced Systems, pp. 961–965 (2007)
12. Bamann, C., Henkel, P.: Visual-inertial odometry with sparse map constraints for planetary swarm exploration. In: 2019 IEEE International Conference on Industrial Cyber Physical Systems (ICPS), pp. 290–295 (2019)
13. Banimelhem, O., Mowafi, M.Y., Aljoby, W.A.Y.: Genetic algorithm based node deployment in hybrid wireless sensor networks. Commun. Netw. **05**, 273–279 (2013)
14. Bartolini, N., Calamoneri, T., La Porta, T.F., Silvestri, S.: Autonomous deployment of heterogeneous mobile sensors. IEEE Trans. Mob. Comput. **10**(6), 753–766 (2011)
15. Bartolini, N., Ciavarella, S., Silvestri, S., Porta, T.L.: On the vulnerabilities of Voronoi-based approaches to mobile sensor deployment. IEEE Trans. Mob. Comput. **15**(12), 3114–3128 (2016)

16. Beckers, R., Holland, O., Deneubourg, J.L.: From local actions to global tasks: stigmergy and collective robots. In: Proceedings of the Workshop on Artificial Life, pp. 181–189. MIT Press, Cambridge, MA (1994)
17. Bhargava, A., Verma, S.: Kate: Kalman trust estimator for internet of drones. Comput. Commun. (2020). https://doi.org/10.1016/j.comcom.2020.04.027
18. Bonabeau, E., Dorigo, M., Theraulaz, G.: Inspiration for optimization from social insect behaviour. Nature **406**(6791), 39–42 (2000). https://doi.org/10.1038/35017500
19. Bonabeau, E., Sobkowski, A., Theraulaz, G., Deneubourg, J.L.: Adaptive task allocation inspired by a model of division of labor in social insects. In: Biocomputing and Emergent Computation: Proceedings of BCEC97, pp. 36–45 (1997)
20. Bonabeau, E., Theraulaz, G., Deneubourg, J.L., Aron, S., Camazine, S.: Self-organization in social insects. Trends Ecol. Evol. **12**(5), 188–193 (1997)
21. Bonnet, F., Mills, R., Szopek, M., Schönwetter-Fuchs, S., Halloy, J., Bogdan, S., Correia, L., Mondada, F., Schmickl, T.: Robots mediating interactions between animals for interspecies collective behaviors. Sci. Robot. **4**(28) (2019). https://doi.org/10.1126/scirobotics.aau7897
22. Borreguero, D., Velasco, O., Valente, J.: Experimental design of a mobile landing platform to assist aerial surveys in fluvial environments. Appl. Sci. **9**(1), 38 (2018). https://doi.org/10.3390/app9010038
23. Boubeta-Puig, J., Moguel, E., Sánchez-Figueroa, F., Hernández, J., Preciado, J.C.: An autonomous UAV architecture for remote sensing and intelligent decision-making. IEEE Internet Comput. **22**(3), 6–15 (2018). https://doi.org/10.1109/MIC.2018.032501511
24. Bridgwater, T., Winfield, A., Pipe, T.: Reactive virtual forces for heterogeneous and homogeneous swarm exploration and mapping. In: Conference Towards Autonomous Robotic Systems, pp. 247–261 (2017). https://doi.org/10.1007/978-3-319-64107-2_20
25. van den Broek, A.C., Dekker, R.J.: Geospatial intelligence about urban areas using SAR. In: Ehlers, M., Michel, U. (eds.) Remote Sensing for Environmental Monitoring, GIS Applications, and Geology VII, vol. 6749, pp. 199–210. International Society for Optics and Photonics, SPIE (2007). https://doi.org/10.1117/12.738486
26. van den Broek, B., van der Velde, J., van den Baar, M., Nijsten, L., van Heijster, R.: Automatic threat evaluation for border security and surveillance. In: Counterterrorism, Crime Fighting, Forensics, and Surveillance Technologies III, vol. 11166, pp. 113–122. Int. Society for Optics and Photonics, SPIE (2019). https://doi.org/10.1117/12.2532308
27. Bruinsma, O.H.: An analysis of building behaviour of the termite Macrotermes subhyalinus (Rambur). Ph.D. thesis, Wageningen University (1979). http://edepot.wur.nl/202106
28. Cafazzo, S., Marshall-Pescini, S., Essler, J.L., Virányi, Z., Kotrschal, K., Range, F.: In wolves, play behaviour reflects the partners' affiliative and dominance relationship. Anim. Behav. **141**, 137–150 (2018). https://doi.org/10.1016/j.anbehav.2018.04.017
29. Camazine, S., Deneubourg, J., Franks, N., Sneyd, J., Bonabeau, E., Theraula, G.: Self-organization in Biological Systems. Princeton Studies in Complexity. Princeton University Press, Princeton (2003)
30. Camazine, S., Deneubourg, J.L., Franks, N.R., Sneyd, J., Theraulaz, G., Bonabeau, E.: Self-organization in Biological Systems. Princeton University Press, Princeton (2001)
31. Chang, Y.S.: An enhanced rerouting cost estimation algorithm towards internet of drone. J. Supercomput. (2020). https://doi.org/10.1007/s11227-020-03243-9
32. Chen, M., Wang, H., Chang, C.Y., Wei, X.: SIDR: a swarm intelligence-based damage-resilient mechanism for UAV swarm networks. IEEE Access **8**, 77089–77105 (2020). https://doi.org/10.1109/ACCESS.2020.2989614
33. Chien, S.: Plenary talk: automated detection and tracking of plumes at 67p/Churyumov-Gerasimenko in osiris/rosetta image sequences: summary report. In: 14th International Symposium on Artificial Intelligence, Robotics and Automation in Space (i-SAIRAS) (2018)
34. Coleman, J.: Foundations of Social Theory. Belknap Series. Belknap Press of Harvard University Press (1994). https://books.google.nl/books?id=a4Dl8tiX4b8C
35. Conesa-Muñoz, J., Valente, J., Del Cerro, J., Barrientos, A., Ribeiro, A.: A multi-robot sense-act approach to lead to a proper acting in environmental incidents. Sensors **16**(8) (2016). https://doi.org/10.3390/s16081269

36. Corson, F., Couturier, L., Rouault, H., Mazouni, K., Schweisguth, F.: Self-organized notch dynamics generate stereotyped sensory organ patterns in drosophila. Science **356**(6337) (2017). https://doi.org/10.1126/science.aai7407
37. Couzin, I.D., Krause, J.: Self-organization and collective behavior in vertebrates. Adv. Study Behav. **32**, 1–75 (2003). https://doi.org/10.1016/S0065-3454(03)01001-5
38. Cremer, F., Schutte, K., Schavemaker, J., den Breejen, E.: A comparison of decision-level sensor-fusion methods for anti-personnel landmine detection. Inf. Fusion **2**, 187–208 (2001). https://doi.org/10.1016/S1566-2535(01)00034-3
39. Dai, F., Chen, M., Wei, X., Wang, H.: Swarm intelligence-inspired autonomous flocking control in UAV networks. IEEE Access **7**, 61786–61796 (2019). https://doi.org/10.1109/ACCESS.2019.2916004
40. Deneubourg, J.L.: Application de l'ordre par fluctuations a la description de certaines étapes de la construction du nid chez les termites. Insect. Soc. **24**(2), 117–130 (1977)
41. Deneubourg, J.L., Aron, S., Goss, S., Pasteels, J.M.: The self-organizing exploratory pattern of the argentine ant. J. Insect Behav. **3**, 159–168 (1990)
42. Dorigo, M.: Optimization, learning and natural algorithms. Ph.D. thesis, Politecnico di Milano, Milan, Italy (1992)
43. Du, Q., Faber, V., Gunzburger, M.: Centroidal Voronoi tessellations: applications and algorithms. SIAM Rev. **41**(4), 637–676 (1999). https://doi.org/10.1137/S0036144599352836
44. Eledlebi, K., Hildmann, H., Ruta, D., Isakovic, A.F.: A hybrid voronoi tessellation/genetic algorithm approach for the deployment of drone-based nodes of a self-organizing wireless sensor network (WSN) in unknown and GPS denied environments. Drones **4**(3) (2020). https://doi.org/10.3390/drones4030033
45. Eledlebi, K., Ruta, D., Hildmann, H., Saffre, F., Hammadi, Y.A., Isakovic, A.F.: Coverage and energy analysis of mobile sensor nodes in obstructed noisy indoor environment: a voronoi approach, in IEEE Transactions on Mobile Computing (2020). https://doi.org/10.1109/TMC.2020.3046184, https://ieeexplore.ieee.org/document/9300245
46. Eledlebi, K., Ruta, D., Saffre, F., Al-Hammadi, Y., Isakovic, A.F.: Autonomous deployment of mobile sensors network in an unknown indoor environment with obstacles. In: Proceedings of the Genetic and Evolutionary Computation Conference Companion, GECCO '18, pp. 280–281. Association for Computing Machinery, New York, NY, USA (2018). https://doi.org/10.1145/3205651.3205725
47. Eledlebi, K., Ruta, D., Saffre, F., Al-Hammadi, Y., Isakovic, A.F.: A model for self-deployment of autonomous mobile sensor network in an unknown indoor environment. In: Zhou, Y., Kunz, T. (eds.) Ad Hoc Networks, pp. 208–215. Springer, Cham (2018)
48. Eledlebi, K.K.: Bio-inspired self organizing networks (BISON) algorithm for blanket coverage in unknown indoor environments. Doctoral dissertation, Khalifa University of Science and Technology, Abu Dhabi, UAE (2019)
49. Escobedo, R., Muro, C., Spector, L., Coppinger, R.P.: Group size, individual role differentiation and effectiveness of cooperation in a homogeneous group of hunters. J. R. Soc. Interface **11**(95), 20140204 (2014). https://doi.org/10.1098/rsif.2014.0204
50. Fabra, F., Zamora, W., Reyes, P., Sanguesa, J., Calafate, C., Cano, J.C., Manzoni, P.: MUSCOP: mission-based UAV swarm coordination protocol. IEEE Access **8**, 72498–72511 (2020). https://doi.org/10.1109/ACCESS.2020.2987983
51. Fard, P.R., Park, H., Warkentin, A., Kiebel, S.J., Bitzer, S.: A Bayesian reformulation of the extended drift-diffusion model in perceptual decision making. Front. Comput. Neurosci. **11**, 29 (2017). https://doi.org/10.3389/fncom.2017.00029
52. Farsi, M., Elhosseini, M.A., Badawy, M., Ali, H.A., Eldin, H.Z.: Deployment techniques in wireless sensor networks, coverage and connectivity: a survey. IEEE Access **7**, 28940–28954 (2019)
53. Fernandes, C., Ramos, V., Rosa, A.: Varying the population size of artificial foraging swarms on time varying landscapes (2005)
54. Gharibi, M., Boutaba, R., Waslander, S.L.: Internet of drones. IEEE Access **4**, 1148–1162 (2016)

55. Ghosh, N., Banerjee, I., Samanta, T.: Energy efficient coverage of static sensor nodes deciding on mobile sink movements using game theory. In: 2014 Applications and Innovations in Mobile Computing (AIMoC), pp. 118–125 (2014)
56. Glick, B.S.: Let there be order. Nat. Cell Biol. **9**(2), 130–132 (2007). https://doi.org/10.1038/ncb0207-130
57. Goyal, A., Kumar, N., Dua, A., Kumar, N., Rodrigues, J., Jayakody, D.N.: An efficient scheme for path planning in internet of drones, pp. 1–7 (2019). https://doi.org/10.1109/GLOBECOM38437.2019.9014305
58. Gundry, S., Zou, J., Sahin, C.S., Kusyk, J., Uyar, M.U.: Autonomous and fault tolerant vehicular self deployment mechanisms in MANETs. In: 2013 IEEE International Conference on Technologies for Homeland Security (HST), pp. 595–600 (2013)
59. Hall, R.: An internet of drones. IEEE Internet Comput. **20**, 68–73 (2016). https://doi.org/10.1109/MIC.2016.59
60. Halloy, J., Sempo, G., Caprari, G., Rivault, C., Asadpour, M., Tâche, F., Saïd, I., Durier, V., Canonge, S., Amé, J.M., Detrain, C., Correll, N., Martinoli, A., Mondada, F., Siegwart, R., Deneubourg, J.L.: Social integration of robots into groups of cockroaches to control self-organized choices. Science **318**(5853), 1155–1158 (2007). https://doi.org/10.1126/science.1144259
61. Hildmann, H., Kovacs, E.: Review: Using unmanned aerial vehicles (UAVs) as mobile sensing platforms (MSPs) for disaster response, civil security and public safety. Drones **3**(3), 59 (2019). https://doi.org/10.3390/drones3030059
62. Hildmann, H., Kovacs, E., Saffre, F., Isakovic, A.F.: Nature-inspired drone swarming for real-time aerial data-collection under dynamic operational constraints. Drones **3**(3), 71 (2019). https://doi.org/10.3390/drones3030071
63. Hildmann, H., Martin, M.: Adaptive scheduling in dynamic environments. In: 2014 Federated Conference on Computer Science and Information Systems, vol. 2, pp. 1331–1336. IEEE (2014). https://doi.org/10.15439/2014F357
64. Hildmann, H., Martin, M.: Resource allocation and scheduling based on emergent behaviours in multi-agent scenarios. In: International Conference on Operations Research and Enterprise Systems, pp. 140–147. Insticc, Scitepress, Lisbon, Portugal (2015)
65. Hildmann, H., Nicolas, S., Saffre, F.: A bio-inspired resource-saving approach to dynamic client-server association. IEEE Intell. Syst. **27**(6), 17–25 (2012)
66. Holland, J.: Emergence: From Chaos to Order. Popular Science/Oxford University Press (2000)
67. Hussein, A., Al-Kaff, A., de la Escalera, A., Armingol, J.M.: Autonomous indoor navigation of low-cost quadcopters. In: 2015 IEEE International Conference on Service Operations and Logistics, and Informatics (SOLI), pp. 133–138 (2015)
68. Imai, H., Iri, M., Murota, K.: Voronoi diagram in the Laguerre geometry and its applications. SIAM J. Comput. **14**, 93–105 (1985)
69. Jentink, F.A.: On a new antelope, Cephalophus Coxi, from North-Western Rhodesia. Notes Leyden Museum **28**, 117–119 (1906). http://www.biodiversitylibrary.org/part/150988
70. Kantaros, Y., Thanou, M., Tzes, A.: Distributed coverage control for concave areas by a heterogeneous robot-swarm with visibility sensing constraints. Automatica **53**, 195–207 (2015). https://doi.org/10.1016/j.automatica.2014.12.034
71. Kaur, S., Uppal, R.S.: Dynamic deployment of homogeneous sensor nodes using genetic algorithm with maximum coverage. In: 2015 2nd International Conference on Computing for Sustainable Global Development (INDIACom), pp. 470–475 (2015)
72. Kengyel, D., Hamann, H., Zahadat, P., Radspieler, G., Wotawa, F., Schmickl, T.: Potential of heterogeneity in collective behaviors: a case study on heterogeneous swarms. In: Chen, Q., Torroni, P., Villata, S., Hsu, J., Omicini, A. (eds.) PRIMA 2015: Principles and Practice of Multi-agent Systems, pp. 201–217. Springer, Cham (2015)
73. Khamis, A., Hussein, A., Elmogy, A.: Multi-robot Task Allocation: A Review of the State-of-the-Art, pp. 31–51. Springer, Cham (2015). https://doi.org/10.1007/978-3-319-18299-5_2

74. Khelil, A., Beghdad, R.: Esa: an efficient self-deployment algorithm for coverage in wireless sensor networks. Procedia Comput. Sci. **98**, 40–47 (2016). https://doi.org/10.1016/j.procs.2016.09.009. The 7th International Conference on Emerging Ubiquitous Systems and Pervasive Networks (EUSPN 2016)/The 6th International Conference on Current and Future Trends of Information and Communication Technologies in Healthcare (ICTH-2016)/Affiliated Workshops
75. Klein, R.: Voronoi Diagrams and Delaunay Triangulations, pp. 2340–2344. Springer, New York, NY (2016). https://doi.org/10.1007/978-1-4939-2864-4_507
76. Knebel, D., Ayali, A., Guershon, M., Ariel, G.: Intra- versus intergroup variance in collective behavior. Sci. Adv. **5**(1) (2019). https://doi.org/10.1126/sciadv.aav0695
77. Kumar, A., Muhammad, B.: On how internet of drones is going to revolutionise the technology application and business paradigms. In: 21st International Symposium on Wireless Personal Multimedia Communications (WPMC), pp. 405–410 (2018). https://doi.org/10.1109/WPMC.2018.8713052
78. Kusyk, J., Zou, J., Gundry, S., Sahin, C., Uyar, M.: Metrics for performance evaluation of self-positioning autonomous manet nodes. In: 2012 35th IEEE Sarnoff Symposium, pp. 1–5 (2012)
79. Kusyk, J., Zou, J., Gundry, S., Sahin, C., Uyar, M.: Performance metrics for self-positioning autonomous MANET nodes. J. Cybersecur. Mob. **2**, 151–173 (2013). https://doi.org/10.13052/jcsm2245-1439.223
80. Lee, C.Y.: Cooperative drone positioning measuring in internet-of-drones. In: IEEE 17th Annual Consumer Communications & Networking Conference (CCNC), pp. 1–3 (2020). https://doi.org/10.1109/CCNC46108.2020.9045111
81. Lee, H., Kim, Y., Han, Y., Park, C.Y.: Centroid-based movement assisted sensor deployment schemes in wireless sensor networks. In: 2009 IEEE 70th Vehicular Technology Conference Fall, pp. 1–5 (2009)
82. Long, T., Ozger, M., Çetinkaya, O., Akan, O.: Energy Neutral Internet of Drones (2018). https://doi.org/10.17863/CAM.21199
83. Lv, Z.: The security of internet of drones. Comput. Commun. **148** (2019). https://doi.org/10.1016/j.comcom.2019.09.018
84. Mahboubi, H., Aghdam, A.G.: Distributed deployment algorithms for coverage improvement in a network of wireless mobile sensors: relocation by virtual force. IEEE Trans. Control Netw. Syst. **4**(4), 736–748 (2017)
85. Mahboubi, H., Moezzi, K., Aghdam, A.G., Sayrafian-Pour, K.: Distributed sensor coordination algorithms for efficient coverage in a network of heterogeneous mobile sensors. IEEE Trans. Autom. Control **62**(11), 5954–5961 (2017)
86. Maraiya, K., Kant, K., Gupta, N.: Application based study on wireless sensor network. Int. J. Comput. Appl. **21**, 9–15 (2011)
87. van der Mark, W., Heuvel, J., den Breejen, E., Groen, F.: Camera based motion tracking for data fusion in a landmine detection system **1**, 20–22 (2003). https://doi.org/10.1109/IMTC.2003.1208269
88. Mason, R.: Feasibility of Laser Power Transmission to a High-altitude Unmanned Aerial Vehicle. Project Air Force report, RAND (2011)
89. McGuire, K.N., De Wagter, C., Tuyls, K., Kappen, H.J., de Croon, G.C.H.E.: Minimal navigation solution for a swarm of tiny flying robots to explore an unknown environment. Sci. Robot. **4**(35) (2019). https://doi.org/10.1126/scirobotics.aaw9710
90. Muñoz, P., R-Moreno, M., Barrero, D., Ropero, F.: Mobar: a hierarchical action-oriented autonomous control architecture. J. Intell. Robot. Syst. (2018). https://doi.org/10.1007/s10846-018-0810-z
91. Muñoz, P., R-Moreno, M., Castaño, B.: 3Dana: a path planning algorithm for surface robotics. Eng. Appl. Artif. Intell. **60**, 175–192 (2017). https://doi.org/10.1016/j.engappai.2017.02.010
92. Muro, C., Escobedo, R., Spector, L., Coppinger, R.: Wolf-pack (Canis lupus) hunting strategies emerge from simple rules in computational simulations. Behav. Process. **88**(3), 192–197 (2011). https://doi.org/10.1016/j.beproc.2011.09.006

93. Nagy, M., Ákos, Z., Biro, D., Vicsek, T.: Hierarchical group dynamics in pigeon flocks. Nature **464**(7290), 890–893 (2010). https://doi.org/10.1038/nature08891
94. Nayyar, A., Le, D., Nguyen, N.: Advances in Swarm Intelligence for Optimizing Problems in Computer Science. CRC Press, Boca Raton (2018). https://books.google.nl/books?id=BidxDwAAQBAJ
95. Nayyar, A., Nguyen, B.L., Nhu, N.: The Internet of Drone Things (IoDT): Future Envision of Smart Drones, pp. 563–580 (2020). https://doi.org/10.1007/978-981-15-0029-9_45
96. Nayyar, A., Singh, R.: Ant colony optimization—computational swarm intelligence technique. In: 2016 3rd International Conference on Computing for Sustainable Global Development (INDIACom), pp. 1493–1499 (2016)
97. Nayyar, A., Singh, R.: Ant colony optimization (ACO) based routing protocols for wireless sensor networks (WSN): a survey. Int. J. Adv. Comput. Sci. Appl. **8** (2017). https://doi.org/10.14569/IJACSA.2017.080220
98. Nepusz, T., Vicsek, T.: Controlling edge dynamics in complex networks. Nat. Phys. **8**(7), 568–573 (2012). https://doi.org/10.1038/nphys2327
99. Nguyen, K.P., Josić, K., Kilpatrick, Z.P.: Optimizing sequential decisions in the drift-diffusion model. bioRxiv (2018). https://doi.org/10.1101/344028
100. Nicolis, G., Prigogine, I.: Self-organization in Nonequilibrium Systems: From Dissipative Structures to Order Through Fluctuations. Wiley, New York (1977)
101. Norouzi, A., Zaim, A.: Genetic algorithm application in optimization of wireless sensor networks. Sci. World J. **2014** (2014). https://doi.org/10.1155/2014/286575
102. Olsson, L., Jerneck, A., Thoren, H., Persson, J., O'Byrne, D.: Why resilience is unappealing to social science: theoretical and empirical investigations of the scientific use of resilience. Sci. Adv. **1**(4) (2015). https://doi.org/10.1126/sciadv.1400217
103. Onuki, A.: Interface instability induced by an electric field in fluids. Phys. A: Stat. Mech. Appl. **217**(1), 38–52 (1995). https://doi.org/10.1016/0378-4371(94)0002
104. Osman, M., Hussein, A., Al-Kaff, A., García, F., Cao, D.: A novel online approach for drift covariance estimation of odometries used in intelligent vehicle localization. Sensors **19**(23) (2019). https://doi.org/10.3390/s19235178
105. Ostrom, E.: A diagnostic approach for going beyond panaceas. Proc. Natl. Acad. Sci. USA **104**(39), 15181–15187 (2007). https://doi.org/10.1073/pnas.0702288104. 7353[PII]
106. Ostrom, E.: A general framework for analyzing sustainability of social-ecological systems. Science **325**(5939), 419–422 (2009). https://doi.org/10.1126/science.1172133
107. Palla, G., Derényi, I., Farkas, I., Vicsek, T.: Uncovering the overlapping community structure of complex networks in nature and society. Nature **435**(7043), 814–818 (2005). https://doi.org/10.1038/nature03607
108. Parrish, J.K., Viscido, S.V., Grünbaum, D.: Self-organized fish schools: an examination of emergent properties. Biol. Bull. **202**(3), 296–305 (2002). https://doi.org/10.2307/1543482. PMID: 12087003
109. Pietrabissa, A., Liberati, F., Oddi, G.: A distributed algorithm for ad-hoc network partitioning based on Voronoi tessellation. Ad Hoc Netw. **46**, 37–47 (2016). https://doi.org/10.1016/j.adhoc.2016.03.008
110. Pike, G.: Legal issues: the internet of drones. SSRN Electron. J. (2015). https://doi.org/10.2139/ssrn.2963623
111. Priyadarshi, R., Gupta, B., Anurag, A.: Deployment techniques in wireless sensor networks: a survey, classification, challenges, and future research issues. J. Supercomput. **76**(9), 7333–7373 (2020). https://doi.org/10.1007/s11227-020-03166-5
112. Puri, V., Nayyar, A., Raja, L.: Agriculture drones: a modern breakthrough in precision agriculture. J. Stat. Manag. Syst. **20**(4), 507–518 (2017). https://doi.org/10.1080/09720510.2017.1395171
113. Qu, Y., Georgakopoulos, S.V.: A centralized algorithm for prolonging the lifetime of wireless sensor networks using particle swarm optimization. In: WAMICON 2012 IEEE Wireless Microwave Technology Conference, pp. 1–6 (2012)

114. Rahmani, N., Nematy, F., Rahmani, A.M., Hosseinzadeh, M.: Node placement for maximum coverage based on voronoi diagram using genetic algorithm in wireless sensor networks (2011)
115. Raman, S., Raina, G., Hildmann, H., Saffre, F.: Ant-colony based heuristics to minimize power and delay in the internet. In: IEEE International Conference on Green Computing and Communications 2013. Beijing, PR China (2013)
116. Ramirez-Atencia, C., R-Moreno, M., Camacho, D.: Handling swarm of UAVs based on evolutionary multi-objective optimization. Prog. AI **6** (2017). https://doi.org/10.1007/s13748-017-0123-7
117. Rehman, A., Paul, A., Ahmad, A., Jeon, G.: A novel class based searching algorithm in small world internet of drone network. Comput. Commun. (2020). https://doi.org/10.1016/j.comcom.2020.03.040
118. Ropero, F., Muñoz, P., R-Moreno, M.: TERRA: a path planning algorithm for cooperative UGV-UAV exploration. Eng. Appl. Artif. Intell. **78**, 260–272 (2019). https://doi.org/10.1016/j.engappai.2018.11.008
119. Roxin, A.: Drift-diffusion models for multiple-alternative forced-choice decision making. J. Math. Neurosci. **9**(1), 5 (2019). https://doi.org/10.1186/s13408-019-0073-4
120. Rubenstein, M., Cornejo, A., Nagpal, R.: Programmable self-assembly in a thousand-robot swarm. Science **345**(6198), 795–799 (2014). https://doi.org/10.1126/science.1254295
121. Saffre, F., Halloy, J., Shackleton, M., Deneubourg, J.L.: Self-organized service orchestration through collective differentiation. IEEE Trans. Syst. Man Cybern. Part B: Cybern. **36**(6), 1237–1246 (2006). https://doi.org/10.1109/TSMCB.2006.873214
122. Saffre, F., Hildmann, H., Deneubourg, J.L.: Can individual heterogeneity influence self-organised patterns in the termite nest construction model? Swarm Intell. **12**, 101–110 (2017)
123. Saffre, F., Simaitis, A.: Host selection through collective decision. ACM Trans. Auton. Adapt. Syst. **7**(1), 4:1–4:16 (2012). https://doi.org/10.1145/2168260.2168264
124. Sasai, Y.: Cytosystems dynamics in self-organization of tissue architecture. Nature **493**(7432), 318–326 (2013). https://doi.org/10.1038/nature11859
125. Schavemaker, J., Cremer, F., Schutte, K., den Breejen, E.: Infrared processing and sensor fusion for anti-personnel land-mine detection (2001)
126. Schutte, K., Cremer, F., den Breejen, E., Schavemaker, J., Benoist, K.: Anti-personnel landmine detection using depth fusion, pp. 1–4 (2001). https://doi.org/10.1109/EUMA.2001.338976
127. Senouci, M.R., Mellouk, A., Asnoune, K., Bouhidel, F.Y.: Movement-assisted sensor deployment algorithms: a survey and taxonomy. IEEE Commun. Surv. Tutor. **17**(4), 2493–2510 (2015)
128. Sharma, B., Srivastava, G., Lin, J.: A bidirectional congestion control transport protocol for the internet of drones. Comput. Commun. **153** (2020). https://doi.org/10.1016/j.comcom.2020.01.072
129. Sharma, N., Magarini, M., Alam, M.: Internet of Drones Enabled Smart Cities, pp. 107–133 (2019). https://doi.org/10.4018/978-1-7998-1253-1.ch006
130. Sheltami, T., Mahmoud, A., Alafari, K., Shakshuki, E.: Self-organizing sensor networks: coverage problem. In: 2012 26th Biennial Symposium on Communications (QBSC), pp. 91–96 (2012)
131. Soleymani, T., Trianni, V., Bonani, M., Mondada, F., Dorigo, M.: Bio-inspired construction with mobile robots and compliant pockets. Robot. Auton. Syst. **74**, 340–350 (2015). https://doi.org/10.1016/j.robot.2015.07.018
132. Sosa San Frutos, R., Al Kaff, A., Hussein, A., Madridano, Á., Martín, D., de la Escalera, A.: Ros-based architecture for multiple unmanned vehicles (UXVS) formation. In: Moreno-Díaz, R., Pichler, F., Quesada-Arencibia, A. (eds.) Computer Aided Systems Theory—EUROCAST 2019, pp. 11–19. Springer, Cham (2020)
133. Staudinger, E., Shutin, D., Manss, C., Viseras, A., Zhang, S.: Swarm technologies for future space exploration missions. In: 14th International Symposium on Artificial Intelligence, Robotics and Automation in Space (I-SAIRAS) (2018). https://elib.dlr.de/120345/

134. Stergiopoulos, Y., Tzes, A.: Voronoi-based coverage optimization for mobile networks with limited sensing range—a directional search approach, pp. 2642–2647 (2009). https://doi.org/10.1109/ACC.2009.5160709
135. Stolfi, D.H., Brust, M., Danoy, G., Bouvry, P.: A cooperative coevolutionary approach to maximise surveillance coverage of UAV swarms, pp. 1–6 (2020). https://doi.org/10.1109/CCNC46108.2020.9045643
136. Strogatz, S.: Sync: How Order Emerges from Chaos in the Universe, Nature, and Daily Life. Hachette Books (2012)
137. Sumpter, D., Buhl, J., Biro, D., Couzin, I.: Information transfer in moving animal groups. Theory Biosci. **127**(2), 177–186 (2008). https://doi.org/10.1007/s12064-008-0040-1
138. Sun, P., Landy, M.S.: A two-stage process model of sensory discrimination: an alternative to drift-diffusion. J. Neurosci. **36**(44), 11259–11274 (2016). https://doi.org/10.1523/JNEUROSCI.1367-16.2016
139. Szwaykowska, K., Romero, L.M., Schwartz, I.B.: Collective motions of heterogeneous swarms. IEEE Trans. Autom. Sci. Eng. **12**(3), 810–818 (2015)
140. Valente, J., Almeida, R., Kooistra, L.: A comprehensive study of the potential application of flying ethylene-sensitive sensors for ripeness detection in apple orchards. Sensors **19**(2) (2019). https://doi.org/10.3390/s19020372
141. Valente, J., Roldán, J., Garzón, M., Barrientos, A.: Towards airborne thermography via low-cost thermopile infrared sensors. Drones **3**(1) (2019). https://doi.org/10.3390/drones3010030
142. Valente, J., Sanz, D., Barrientos, A., del Cerro, J., Ribeiro, A., Rossi, C.: An air-ground wireless sensor network for crop monitoring. Sensors **11**(6), 6088–6108 (2011). https://doi.org/10.3390/s110606088
143. Vicsek, T.: A question of scale. Nature **411**(6836), 142 (2001). https://doi.org/10.1038/35078161
144. Vicsek, T.: Complexity: the bigger picture. Nature **418**(6894), 131 (2002). https://doi.org/10.1038/418131a
145. Vieira, M.A.M., Vieira, L.F.M., Ruiz, L.B., Loureiro, A.A.F., Fernandes, A.O., Nogueira, J.M.S.: Scheduling nodes in wireless sensor networks: a Voronoi approach. In: 28th Annual IEEE International Conference on Local Computer Networks, 2003. LCN '03. Proceedings, pp. 423–429 (2003)
146. Wald, A., Wolfowitz, J.: Optimum character of the sequential probability ratio test. Ann. Math. Stat. **19**(3), 326–339 (1948). http://www.jstor.org/stable/2235638
147. Wang, D., Xie, B., Agrawal, D.P.: Coverage and lifetime optimization of wireless sensor networks with Gaussian distribution. IEEE Trans. Mob. Comput. **7**(12), 1444–1458 (2008)
148. Wang, G., Cao, G., La Porta, T.F.: Movement-assisted sensor deployment. IEEE Trans. Mob. Comput. **5**(6), 640–652 (2006)
149. Watteyne, T.: Energy-efficient self-organization for wireless sensor networks. Ph.D. thesis, INSA de Lyon (2008)
150. Werfel, J., Petersen, K., Nagpal, R.: Designing collective behavior in a termite-inspired robot construction team. Science **343**(6172), 754–758 (2014). https://doi.org/10.1126/science.1245842
151. Yao, J., Ansari, N.: QoS-aware power control in internet of drones for data collection service. IEEE Trans. Veh. Technol. **PP**, 1 (2019). https://doi.org/10.1109/TVT.2019.2915270
152. Yao, J., Ansari, N.: Online task allocation and flying control in fog-aided internet of drones. IEEE Trans. Veh. Technol. **PP**, 1 (2020). https://doi.org/10.1109/TVT.2020.2982172
153. Yates, C.A., Erban, R., Escudero, C., Couzin, I.D., Buhl, J., Kevrekidis, I.G., Maini, P.K., Sumpter, D.J.T.: Inherent noise can facilitate coherence in collective swarm motion. Proc. Natl. Acad. Sci. **106**(14), 5464–5469 (2009). https://www.pnas.org/content/106/14/5464
154. Wang, Y.-C., Hu, C.-C., Tseng, Y.-C.: Efficient deployment algorithms for ensuring coverage and connectivity of wireless sensor networks. In: First International Conference on Wireless Internet (WICON'05), pp. 114–121 (2005)
155. Zafeiris, A., Vicsek, T.: Group performance is maximized by hierarchical competence distribution. Nat. Commun. **4**(1), 2484 (2013). https://doi.org/10.1038/ncomms3484

156. Zhang, H., Hou, J.: Maintaining sensing coverage and connectivity in large sensor networks. Ad Hoc Sens. Wirel. Netw. **1** (2004). https://doi.org/10.1201/9780203323687
157. Zheng-Jie, W., Wei, L.: A solution to cooperative area coverage surveillance for a swarm of MAVs. Int. J. Adv. Robot. Syst. **10**(12), 398 (2013). https://doi.org/10.5772/56801
158. Zou, J., Gundry, S., Kusyk, J., Sahin, C.S., Uyar, M.U.: Bio-inspired and Voronoi-based algorithms for self-positioning of autonomous vehicles in noisy environments. In: Proceedings of the 8th International Conference on Bioinspired Information and Communications Technologies, BICT '14, pp. 17–22. ICST (Institute for Computer Sciences, Social-Informatics and Telecommunications Engineering), Brussels, BEL (2014). https://doi.org/10.4108/icst.bict.2014.257917
159. Zou, J., Kusyk, J., Uyar, M.U., Gundry, S., Sahin, C.S.: Bio-inspired and Voronoi-based algorithms for self-positioning autonomous mobile nodes. In: MILCOM 2012—2012 IEEE Military Communications Conference, pp. 1–6 (2012)

Underwater Drones for Acoustic Sensor Network

Meeta Gupta, Adwitiya Sinha, and Shikha Singhal

Abstract Apparently, two-thirds of our planet is covered by water bodies, of which oceanic coverage deeply impacts terrestrial life. Oceans form a majestic resource of most of the amazing marine creatures and natural resources, especially mineral oil. This makes underwater monitoring one of the most demanding areas to explore. For monitoring the oceanic layer, a specialized form of underwater vehicle network is employed using underwater drones for acoustic monitoring. The underwater drones are autonomously operated and remotely tracked using onshore and offshore base stations. The underwater research draws eminence from a wide variety of application-based research, including collaborative undersea exploration, naval surveillance and other oceanographic mining and engineering tasks. Underwater communication confronts unique challenges that involve severe bandwidth constraints, high bit error rates, variable propagation delays and limited spatial correlation. Such constraints mandate the use of sound waves for information transfer that further requires complex signal processing at receiving ends. Owing to such complex capabilities, acoustic sensors are quite expensive and therefore, sparsely deployed. Moreover, oceanic tides and drifts results in high vulnerability for acoustic devices towards failure due to fouling and corrosion underwater. Our research is focused on presenting surveys of architecture and applications of underwater drones for acoustic sensor networks abided by significant characteristics and challenges.

Keywords Acoustic sensors · Underwater vehicle · Underwater drone · Autonomous vehicles · Oceanic challenges · Acoustic architecture · Underwater communication

1 Introduction

Tremendous increase in demand for pervasive applications of sensor networks in diverse research fields and huge progress made in microelectronics technology has

M. Gupta · A. Sinha (✉) · S. Singhal
Department of Computer Science and Engineering, Jaypee Institute of Information Technology, Noida, Uttar Pradesh, India
e-mail: mailtoadwitiya@gmail.com

R. Krishnamurthi et al. (eds.), *Development and Future of Internet of Drones (IoD): Insights, Trends and Road Ahead*, Studies in Systems, Decision and Control 332,
https://doi.org/10.1007/978-3-030-63339-4_2

greatly promoted underwater sensor networks. The acoustic sensor network has a wide range of applications, supporting water level detection, oceanographic pollution monitoring, oceanic data collection, disaster prevention, offshore resource exploration, tactical navigation and surveillance, water quality analysis, fisheries application, etc. Traditional approach employed for ocean-bottom and column monitoring, involved temporary installation of oceanographic acoustic sensor nodes under water and recovering back the instruments after recording relevant data. Such approach often results in huge latency in receiving and communicating the sensed information to on-shore sinks. Moreover, acoustic sensors are small and delicate electronic devices that establish aural links to communicate with each other. Also, these nodes are battery operated; hence possess highly constrained processing, transmitting and storage capabilities.

Therefore, if some form of failure occurs during the acoustic communication, chances of data loss would eventually become very high. Resultantly, traditional real time oceanic monitoring confronts challenges due to absence of backbone network, frequent network failure and restricted sensor hardware [1, 2]. In order to overcome these issues, innovative design and deployment strategies are developed for better performance of underwater acoustic networks [6]. Underwater acoustic network is used in both oceanic and small aqueous water bodies like rivers, lakes, sea, etc. The acoustic networks are a combination of sensor nodes and Autonomous Underwater Vehicles (AUVs), also called underwater drones. This submerged network of heterogeneous nodes and mobile drones are further connected to surface stations and the backbone networks, like internet, RF links, etc. to execute monitoring tasks over the oceanic region. This systematic configuration of the network creates an interactive environment for scientists to gather real time information in collaborative monitoring missions [1]. For making this application feasible, a two-way acoustic communication link (up/down-link) is required between sensor nodes and AUV underwater.

The underwater environment and specific deployment strategies pose unique challenges for acoustic sensor networks. The basic architecture of an acoustic sensor is given in Fig. 1. Some of the significant dissimilarities between terrestrial and underwater wireless sensor networks (UWSNs) are summarized as following:

i. Cost—The acoustic nodes deployed underwater are much more costly than terrestrial nodes. The complex structure of transceivers built-in within acoustic nodes, especially required for hardware protection in underwater, magnifies the cost.
ii. Deployment—Terrestrial sensor networks support highly dense deployment of nodes while in underwater applications, node deployment is generally sparser. This is because of the cost factor and challenges involved in the underwater deployment process.
iii. Power—The power required in communication of acoustic signals underwater is comparatively more than in case of terrestrial sensor networks. The reason behind is larger distances between channels, continuous water drift and complex signal-level processing at receiving end.

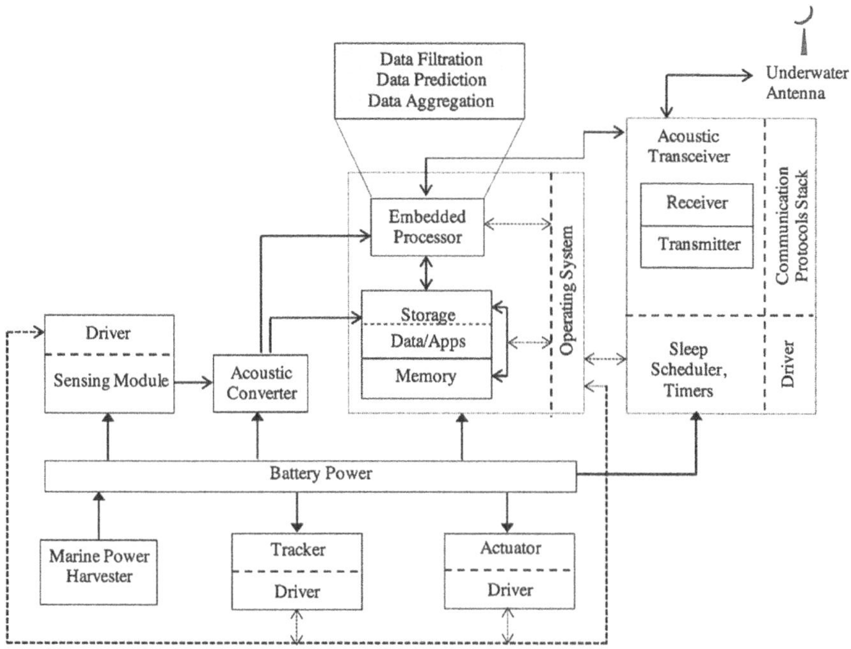

Fig. 1 Acoustic sensor node architecture

iv. Memory—Unlike terrestrial sensors, underwater sensors have irregular buffer capability due to which acoustic sensors require frequent data caching.

Underwater acoustic networks as compared with terrestrial sensor networks face distinctive challenges in terms of limited bandwidth, energy consumption, temporary handoffs, frequent connectivity loss, high bit-rate error, large deviation in propagation delay, etc. [1, 3, 4]. Therefore, the communication and routing protocols developed for terrestrial scenarios cannot be directly utilized for underwater operations. Hence, underwater acoustic standards and protocol suites should be designed specifically.

2 Dynamics of Drone Modeling

Due to the addition of hydrodynamic mass, gravity and buoyancy forces, damping, lifting forces, thruster forces, coriolis and centripetal forces, and disturbances in environment, the AUV's dynamics is closely related and highly non-linear [18, 19]. It is important to have a coordinate system in order to determine the AUV's dynamic behavior in a fluid environment. For the development of AUV's 6 Degrees of Freedom (6-DoF) are used to represent the motion of AUV in six independent coordinates [21] as shown in Fig. 2.

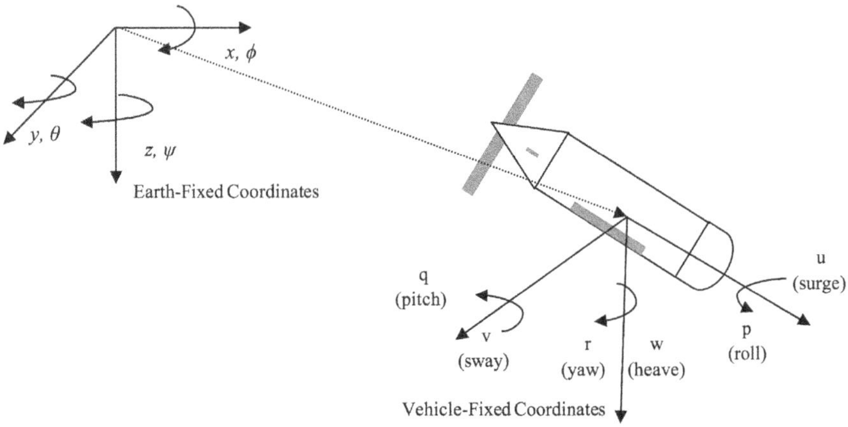

Fig. 2 The coordinates system of AUV

The 6-DoF components are named as Surge, Sway, Heave, Roll, Pitch and Yaw. The position and angular components in 6-DoF are denoted by $(x, y, z, \phi, \theta, \psi)$. The 6-DoF velocity components denoted in the order are (u, v, w, p, q, r). The force denoted in the order (X, Y, Z, K, M, N) are attributable to 6-DoF. All these symbols used for 6-DoF are given in Table 1.

The mathematical model of AUV's consist of kinematic part and dynamic part. Two coordinates frame are used to represent kinematics of AUV's:

(a) Earth-Fixed frame (E-frame): This frame is connected to the globe the x-axis points toward the north, the y-axis toward the east and the z-axis toward the middle of the globe. Often known as the inertial framework, guidance and navigation are carried out in this framework
(b) Body-fixed frame (B-frame): This frame is connected to the AUV, In B-Frame the x-axis connects to the forward direction, the y-axis connects to the right of the AUV and the z-axis connects vertically downwards. This frame is used to express velocity of the vehicle.

Table 1 Six symbols of DoF

DoF	Motion and direction	Position and angles in E-frame	Velocity in B-frame	Force and moments
1	Surge	x	U	X
2	Sway	y	V	Y
3	Heave	z	W	Z
4	Roll	ϕ	P	K
5	Pitch	θ	Q	M
6	Yaw	ψ	R	N

These two reference frames determine the orientation, velocity and acceleration of the vehicle with respect to earth and vehicle coordinates and helps in navigation of the AUV. The motion of AUV [20] is described be the following vectors as given below

(i) Position vector in E-frame $\eta_1=[x, y, z]^T$
(ii) Euler vector in E-frame $\eta_2=[\phi, \theta, \psi]^T$
(iii) Linear velocity vector in B-frame $v_1 = [u, v, w]^T$
(iv) Velocity of angles in B-frame $v_2 = [p, q, r]^T$
(v) Forces vector $\tau_1 = [X, Y, Z]^T$
(vi) Moments vector $\tau_2 = [K, M, N]^T$.

The velocity parameters of the AUV are determined from the B-frame and the velocity in the E-frame are denoted by $\dot{\eta}_1$ and $\dot{\eta}_2$ and are determined using transformation Eqs. (1)–(8) given below. These are known as kinematics equation of motion.

$$\dot{\eta}_1 = \left[\dot{x}, \dot{y}, \dot{z}\right]^T \tag{1}$$

$$\dot{\eta}_2 = \left[\dot{\phi}, \dot{\theta}, \dot{\psi}\right]^T \tag{2}$$

$$\begin{aligned}\dot{x} = u(\cos\theta\cos\psi) + v(\sin\phi\sin\theta\cos\psi - \cos\phi\sin\psi) \\ + w(\cos\phi\sin\theta\cos\psi - \sin\phi\sin\psi)\end{aligned} \tag{3}$$

$$\begin{aligned}\dot{y} = u(\cos\theta\sin\psi) + v(\sin\phi\sin\theta\sin\psi - \cos\phi\cos\psi) \\ + w(\cos\phi\sin\theta\sin\psi - \sin\phi\sin\psi)\end{aligned} \tag{4}$$

$$-\dot{z} = u(\sin\theta) + v(-\sin\phi\cos\theta) + w(-\cos\phi\cos\theta) \tag{5}$$

$$\dot{\phi} = p + q\sin\phi\tan\theta + r\cos\phi\tan\theta \tag{6}$$

$$\dot{\theta} = q\cos\phi - r\sin\phi \tag{7}$$

$$\dot{\psi} = \frac{q\sin\phi + r\cos\phi}{\cos\theta} \tag{8}$$

The dynamics equation of motion and force applied on AUV in 6-DoF is given by τ (vector of control inputs) by Eq. (9)

$$\tau = M(\nu)\nu + C(\nu)\nu + D(\nu)\nu + g(\eta) \tag{9}$$

Here: $M(\nu) = 6 \times 6$ inertia matrix containing mass and rigid body

$C(\nu) = 6 \times 6$ Coriolis and centripetal forces matrix
$D(\nu) = 6 \times 6$ linear and nonlinear hydrodynamic damping force matrix
$g(\eta) = 6 \times 1$ vector for gravitational and buoyancy forces
$\nu = [u, v, w, p, q, r]^{\mathrm{T}}$.
The 6×1 vector for control input forces in B-frame is given by Eq. (10)

$$\tau = [X, Y, Z, K, M, N]^{\mathrm{T}} \tag{10}$$

The 6-DoF equations of forces according to Newton-Euler formulation are given below in Eqs. (11)–(16)

$$X = m\left[(u_r - v_r r + w_r q) - x_G\left(q^2 + r^2\right) + y_G(pq - r) + z_G(pr + q)\right] \tag{11}$$

$$Y = m\left[(v_r - w_r p + u_r r) - y_G\left(r^2 + p^2\right) + z_G(qr - p) + x_G(qp + r)\right] \tag{12}$$

$$Z = m\left[(w_r - u_r q + v_r p) - z_G\left(p^2 + q^2\right) + x_G(rp - q) + y_G(rq + p)\right] \tag{13}$$

$$K = I_x\dot{p} + \left(I_Z - I_y\right)qr + m[y_G(\dot{w}_r + pv_r - qu_r) - z_G(\dot{v}_r + ru_r - pw_r)] \tag{14}$$

$$M = I_y\dot{q} + (I_x - I_z)rp + m[z_G(u_r + w_r q - v_r r) - x_G(w_r + pv_r - u_r q)] \tag{15}$$

$$N = I_z\dot{r} + \left(I_y - I_z\right)pq + m[x_G(v_r + u_r r - pw_r) - y_G(u_r + qw_r - v_r r)] \tag{16}$$

In the above equations of force applied on vehicle, the variables $I_x, I_y, and I_z$ are the rotational inertia about x, y and z axes, m is the mass of the vehicle. $x_G, y_G, and z_G$ are center of gravity locations, $u_r, v_r, and w_r$ are the translational velocities related to sway, surge and heave degrees of freedom.

3 Underwater Acoustic Architecture

Architectures suggested by researchers for underwater acoustic scenarios, mainly include static two- and three-dimensional underwater networks and mobile architectures using acoustic underwater vehicles or AUVs [14, 15].

3.1 *Static Two-Dimensional UW-ANs*

This architecture consists of a group of sensors, which are anchored to the ocean bed. The topology is not susceptible to change after deployment (Fig. 3). This topological

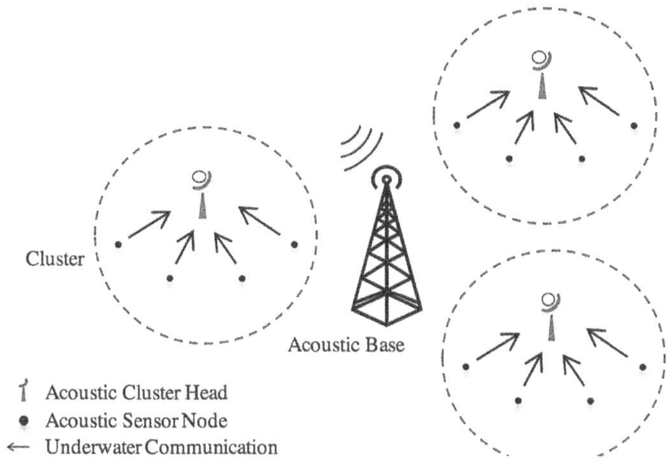

Fig. 3 Static two-dimensional UW-ANs

arrangement could be grid, tree, cluster, bus, star, etc. Applications which support this type of architecture, include environmental monitoring of underwater plantation, water mineral-content detection, observation of underwater plates in tectonics, etc. [1, 5].

There can be one or more underwater sinks (also known as base stations) available, which interconnect a group of sensor nodes, thereby forming clusters at the ocean bed. They are vested with the responsibility of forwarding data from sensors, which are moored at ocean bed, to stations at the surface level. Underwater sinks are assigned dual links to support two-way acoustic communication links. The first link is designated to communicate with sensor nodes in cluster to accomplish following tasks:

i. To send commands, queries and network-relevant configurations to the acoustic sensor nodes (i.e. underwater sinks to sensor nodes)
ii. To collect sensed data from sensor nodes for further relaying and processing (i.e. sensor nodes to underwater sinks).

The second link is used to send aggregated sensory information to the surface station. The surface station is configured to handle multiple communication links with underwater sinks. Within the underwater clustered boundary, the sensor nodes send their gathered data directly in one hop to its corresponding sink. This data can be made available to other clusters in multi-hop paths via intermediate nodes. However, for single hop transmission, network throughput is likely to get reduced due to increased interference through high power transmission. While, in multi hop transmission, short-distance communication results in considerable energy savings and, thus network capacity is increased. As a drawback, multi-hop forwarding endorses complex functionalities in packet routing, thereby increasing the relaying overhead and latency incurred in communication due to long traversals towards underwater sink.

3.2 Static Three-Dimensional UW-ANs

In static three-dimensional Under Water Acoustic Network (UW-AN), the nodes are deployed at different depths from objects floating over the water surface (ships, etc.), as illustrated by Fig. 4. The applications that implement this type of architecture, mainly involve surveillance or monitoring of different oceanic phenomena, for instance—water quality, pollution, chemicals in water, etc. [5].

The acoustic nodes are deprived of the actuation unit that imparts self-driven mobility. This arrangement allows collection of three-dimensional environmental data that was previously infeasible to scrutinize, as in ocean bottom sensing. In static 3D UW-AN, sensor nodes are anchored from the floating objects above the water surface with long-length wires to adjust the depth of sensors. According to another type of 3D arrangement, sensors are linked to the bottom of an ocean bed through floating buoys. The floating buoys are fixed to the ocean bed and can be bloated by air pumps, thereby lifting the sensor nodes upward. The length of the buoys further controls the height of nodes towards the ocean surface.

The functioning of a static 3D acoustic network is severely challenged with ensuring sensing and communication coverage. Therefore, for the purpose of addressing such issues hybrid underwater networking is proposed to achieve high performance 3-D monitoring.

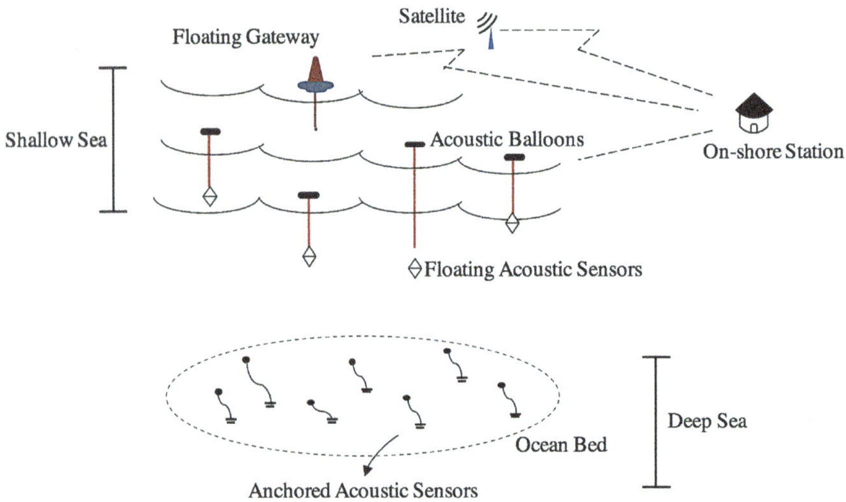

Fig. 4 Static three-dimensional UW-ANs

3.3 Three-Dimensional Hybrid UW-ANs

The hybrid network is composed of several sensors that are stocked over Autonomous Underwater Vehicles (AUVs). These autonomous vehicles provide additional support for gathering information and improving network capacity (Fig. 5). They have a high energy reserve and powerful actuation that drives them independently under water. AUVs often act as routers, or cluster heads for performing network installation, system reconfiguration, data gathering and processing. The network topology can be adjusted to meet specific requirements of any application. This architecture also ensures reliability in case of node failures. Successful implementation of such architecture would require some coordination algorithms like adaptive sampling and self-configuration [1, 5, 7].

i. Adaptive Sampling—This phenomenon is particularly used to control AUVs regarding data storage and transmission for further useful processing at off-shore sinks. The application of this approach is pioneering in 3D monitoring missions. For example, in some given area where a high sampling rate is required, the bulk of sensor nodes can be manageably amplified to perform continuous and effective sampling.
ii. Self-configuration—This allows usage of certain controls strategies to detect node failure and damaged channels. A request is sent to AUVs for re-installation and maintenance of mal-functioned sensor nodes. In case maintenance of a failed

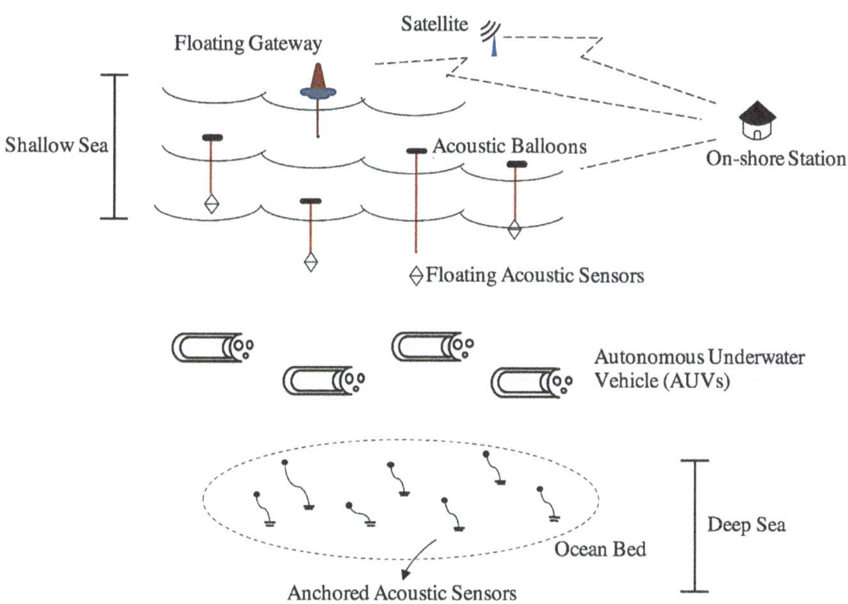

Fig. 5 Three-dimensional hybrid UW-ANs

node is worthless, requests can be triggered to deploy new sensor nodes in the concerned region for hassle-free operations.

In UW-ANs hybrid architecture poses additional challenges due to the presence of autonomous vehicles. Constraints like cost, time, delay and energy consumption may vary because of the type of AUVs deployed according to specific application requirements [11]. Moreover, battery storage and speed of the underwater vehicles also needs to be optimally set for increasing the network performance. Finally, managing decisions regarding network reconfiguration and topological control in the event of node failure becomes very crucial for successful functioning of acoustic networks. Overall, the ultimate goal of the AUVs is to manage and maintain underwater sensor networks by relying more on local interactions rather than onshore communication.

4 Control and Navigation of Drone Vehicle

The sovereignty architecture of an AUV has three core structures [18, 19]. The guidance system manages the creating the route of the vehicle; the control system applies and calculates the suitable manoeuvring forces to the vehicle; and the navigation system provides an estimate of the current position of the vehicle.

Apart from completing their own task, these three systems, also need to work cooperatively to enable a vehicle to fulfil its goals reliably. Due to the presence of complex non-linear forces such as hydrodynamic mass, lifting, drag, centripetal, gravity, Coriolis, buoyancy and thruster forces acting on the vehicle, the dynamics of underwater vehicles are nevertheless closely coupled and highly nonlinear. Engineering issues connected to the high density, non-uniform and un-structured marine setting and therefore the non-linear response of vehicles create it tough to attain a high degree of autonomy. The management must be sensible enough to get info from the system and build its own control mechanisms with no human intervention.

Control system of an AUV has two sub systems namely: Control law and control allocation. The control law subsystem generates the generalized forces in six degrees of freedom using the present and desired states. The control allocation subsystem distributes this generalized force among vehicle actuators. The Control system is modular, which allows the two subsystems to be programmed and enforced individually, however each need to operate collaboratively to manage the vehicle accurately.

4.1 Control Law Subsystem

The Control system receives inputs like desired trajectory from the guidance system and current state estimates of position and velocity from navigation system, which are internally received by control law subsystem. The intermediate signals from control law subsystem are the six degree of freedom generalized forces which are passed to control allocation subsystem. The output of control allocation subsystem are control signals to accurately control the AUV. The various control law design techniques are used to control AUV's.

4.1.1 Feedback Linearization

This control technique solves many practical non-linear control problems. It transforms a non-linear machine into a linear machine by changing control inputs and variables. Afterwards Proportional Derivative (PD) controller stabilizes the system. Due to parameters uncertainty and disturbances, this technique does not always guarantee powerful and robust system.

4.1.2 PID Control

PID controller is linear. PID schemes are not very successful for the management of nonlinear AUV dynamics of models working in heavy wave and current disturbances in unknown environments. The PID controller is composed of three parameters, namely the proportional, integral and derivative. Reaction to the current error is determined by the proportional value. The integral value is based on the sum of recent errors defining the answer. The derivative value determines the reaction rate at which the error went up. PID methods are suitable for non-complex AUV's operating in environments without any external disturbances. PID method generates an error signal relating the current state and the required state.

$$u(t) = x(t) - y(t) \tag{17}$$

here $u(t)$ is the error signal, $x(t)$ is the AUV desired state, and y (t) is the AUV current state, and this error signal is modified to apply a corrective action to the AUV denoted by $\tau(t)$, the output of PID controller.

$$\tau(t) = K_p u(t) + K_I \int_0^f u(\lambda) d\lambda + K_D \frac{d}{dt} u(t) \tag{18}$$

where K_p = Proportional value of the error
K_I = factor of the integral value of the error
K_D = factor of the derivative value of the error.

4.1.3 Sliding Mode Control (SMC)

This is a scheme that uses a specific flipping term to bear the consequences of factors that were not considered during the process of designing the controller. The Control law for SMC is defined by an Eq. (19) having sufficiently large value of $T(\dot{\eta}, \eta)$ and result in η converging to zero.

$$\tau = -T(\dot{\eta}, \eta)sign(s), T(\dot{\eta}, \eta) > 0 \tag{19}$$

If η is substituted by the gap between the present and desired AUV states, then the reference trajectory can be followed by the above control law equation. The SMC provides a robust control scheme even if there are uncertainties in parameters. There are two types of SMC schemes namely coupled SMC and uncoupled SMC.

(i) Uncoupled SMC: An assumption is made in uncoupled SMC that no coupling exist between various degrees of freedom and simple manoeuvring is used. In this scheme implementation of the controller is easier.
(ii) Coupled SMC: It is an unconventional way of designing a control law is used, in which coupling among the degrees of freedom is retained. In a combined SMC body frame, only guidance and navigation data are transformed from an E-frame to a B-Frame. The body frame already has latest velocity and acceleration data hence no rotations are needed for implementing controller. This scheme of control is more computationally efficient.

4.1.4 Linear Quadratic Gaussian Controller (LQG)

This controller works on linear system and assumes a Gaussian noise and cost function is taken to be quadratic. LQG controller can be implemented in both linear and nonlinear controller. But linear altitude control produces chattered control signals which results in skidding of vehicle. Whereas non-linear control reduces chattering and skidding of vehicle. LQG control can be extended with the help of Kalman Filter for improving the performance.

4.1.5 Fuzzy Logic Control

It is based on rules, which are stored in natural language using If-then format and can be easily understood by non-specialist person also. Fuzzy logic control is slow to implement and can avoid complex hydrodynamic modelling of the vehicle. This type of controller should be used when mathematical model of the vehicle is not available.

4.1.6 Adaptive Control

It is a non-linear control system. This type of control is useful in dynamic and time varying systems, having uncertainties in the ocean. In this controller, the system adapts to the changing dynamics by itself, by first setting all the control gains initially to zero. This type of control system gives more accurate control of the 6-DoF by using velocity and position estimated generated by on-board sensor navigation system.

4.2 Control Allocation Subsystem

The undertaking of the control allocation subsystem is to make the proper signs for all actuators for the power to be applied to the vehicle generalized from the control strategy. Since various actuators can apply power to a particular DoF, the control circulation is liable for the most gainful use of the perfect capacity to the vehicle for each accessible actuator. Power utilization is of specific significance for every free vehicle, since it is a basic factor in deciding the length of the mission. It is the duty of control distribution to discover balance between all the actuators for applying both the preferred powers to the vehicle, simultaneously limiting the power expended. This negotiating act helps the vehicle to preserve the maneuverability offered by all actuators while allowing for as long as possible mission duration. The force F from total n actuators is applied to the vehicle and is denoted by Eq. (20).

$$F = CKv \tag{20}$$

Here C is the actuator configuration represented as a matrix of size 6 × n and where K is a square and matrix of size n x n representing coefficient of force and v is the control input of size n × 1. The various control allocation schemes are illustrated in the following section [22].

4.2.1 Non-optimal Scheme

This is one of the simplest and implementation efficient scheme for control allocation which requires only one matrix multiplication given by Eq. (21)

$$v = (CK)^{-1}F \tag{21}$$

The drawback of this technique is that is does not minimizes power consumption. Hence a more intelligent and optimization based control allocation scheme is needed which minimizes power consumption.

4.2.2 Quadratic Programming

This scheme resolves the constraints of said non-optimal scheme by using an optimization scheme for constrained and unconstrained actuators inputs. There are two types of quadratic programming control allocation optimization schemes namely linear and non-linear quadratic programming optimization schemes. In both the schemes weighing matrix are introduced, allowing actuators to use control surfaces over tunnel thrusters and hence the actuators can apply generalized force generated by control law and minimize the power consumption. Both linear and non-linear quadratic programming control allocation schemes gives two solutions explicit (online) solution and iterative(offline) solutions. Iterative solutions supports actuator reconfiguration due to failure by applying multiple iterations, whereas the explicit solution needs to be recalculated again. The linear quadratic control allocation scheme assumes non rotatable actuators while non-linear quadratic programming control allocation scheme supports rotatable actuators. These schemes have large computational requirement.

4.2.3 Two-Stage Scheme

In this control allocation scheme is broken down in two sub-problems. The first sub-problem tackles control allocation to primary rotor and surface area and the second sub problem solves control allocation to tunnel rotors. Thus, the control surfaces for low power consumption are used to the maximum extent possible, whereas the higher power consumption tunnel thrusters are only used as necessary to provide their extra manoeuvrability. This scheme works like a two-stage non optimal scheme. But this scheme is less computation intensive than the quadratic programming yet allowing to use control surfaces more than tunnel thrusters.

4.3 Navigation of AUV's

The navigation system helps in determining the closest approximation of the Autonomous Underwater Vehicle's current state, regardless of what information is shared by the sensor. For the safe operation and repair of an AUV, good navigation knowledge is crucial. The location from which the data has been retrieved must be correctly identified for the data obtained by an AUV to be of interest. Global Positioning System (GPS) is easily accessible for vehicles in land and air, it is used to continuously give accurate information regarding the position of the vehicle to the navigation system. But GPS is largely impractical for underwater vehicles as GPS signals cannot easily propagate through water. Thus based on the observations from entoperipheral and epiperipheral sensors, the navigation system finds out the best approximation of the vehicles current state. The GPS system is used only when available. Some type of sensor fusion technique is used, such as particle filtering, to get the best estimate of the existing operational state, and enable a rectification method to be used when GPS is available. A single navigation system cannot address all the demands of AUV operations. As a result, a number of navigation methods have been developed each satisfying certain categories. The different navigation techniques have been grouped in four general categories [23, 24].

4.3.1 Inertial Navigation System (INS)

INS is mainly used on larger, more costly AUVs, although continuing technology aims to make this capability cost-effective for smaller cars. The navigation suite typically consists of an INS, a velocity sensor, and a receiver of the GPS. Activating the navigation system requires the GPS, and when the AUV is on the surface for location determination. To determine the modified location, the vehicle's accelerations are combined twice in time in inertial navigation systems. For high-quality commercial grade INS systems the drift rates are as high as few kilometers per hour. For operation near the seabed, the Doppler Velocity Sonar (DVS) sensors are used to measure the vehicle's speed in comparison to the ground. A well-integrated INS-DVS system can provide navigation accuracy greater than 0.1% of travelled distance. The problem with relying solely on inertial navigation is that location error increases unbounded because the distance travelled by the vehicle increases. The growth rate would depend on the currents of the ocean, the vessel's speed, and the accuracy of the dead reckoning sensors.

4.3.2 Acoustic Navigation

Acoustic signals can easily perpetuate underwater and can be used as "signposts" to direct an AUV's movement without resurfacing. Two types of techniques mainly used are ultra-short baseline (USBL) and long baseline (LBL). A number of transponders are deployed in and surveyed in LBL navigation systems. The vehicle transmits an acoustic wave, which each signpost returns as it receives. The location is calculated by calculating the time of travel between the vehicle and each signpost, geometry of array of signposts and calculating the local sound speed of the profile. Using this information, the relative distances between the signposts array and vehicle can be measured. While LBL navigation provides the best navigation accuracy of any underwater navigation system, it requires professional operators and can entail significant costs. In USBL system, the vehicle can calculate both range and angle of signpost using a multi-element receiver array. The tolerance from the vehicle to signpost can be calculated by calculating the difference between arrival times of a single sonar ping between two or more hydrophones. LBL and USBL techniques can report error from several sources. Primary errors can be divided in two main types: the imagined array geometry error, and the expected sound speed of the profile error.

4.3.3 Geophysical Navigation

If an reliable environmental a priori map is available, the use of estimates of geophysical parameters, like magnetic field, gravitational anomaly, and bathymetry, is one approach to global position estimation. These methods are based on the correspondence of sensor data with an environmental graph. The vehicle does manoeuvring by matching collected sensed data with an a priori chart of data stored. The two main issues in an n-dimensional map or sensor data set are: firstly creation of a priori map is difficult and cost is high and secondly finding a crest in the n-dimensional correlation surface is quite complex. For actuality, there might not be a high-quality map of the operational area of interest. Geophysical techniques were limited to bathymetric methods, largely due to the need for precise a priori maps. The number of feasible applications of AUV can increase if AUV can create a map of unknown area and then use it for navigation.

4.3.4 Optical Navigation

Optical technologies are a significant choice for providing environmental information. Such devices may be applied either with a camera or the optical sensor array. Optical navigation system have limited scope due to restricted transmission of light through water. There are two examples of optical systems: first example in which the AUV has to detect and follow active landmarks within a structure and second example in which AUV has to recognize the patterns created with active marks. The optical navigation can be implemented using array of sensors or using set of cameras. The performance of the optical detectors are assessed for its ability to estimate the location, orientation and forward velocity of AUVs in relation to a fixed underwater light source. Accuracy of optical navigation systems is less as compared to geophysical and acoustic navigation.

5 Design Issues of Acoustic Drones

Some of the specific design issues and constraints involved in deploying and maintenance of acoustic sensor networks for efficient underwater functioning, are enlisted as follows [7].

i. Power and Load Balancing—The acoustic vehicles are deployed with the vision to impart underwater operations and promote longer network life. Though the AUVs have much larger rechargeable power reserves as compared to homogeneous acoustic sensors, their source of power is restricted. Moreover, frequent recharging would require relocation of vehicles to the off-shore charging station, which may eventually lead to disruption of network service under water. Therefore, the available energy needs to be utilized optimally so as to carry out complex processing, routing and other control tasks. Also, search load (i.e. the span of area covered by AUVs) should be uniformly distributed among all other deployed underwater vehicles for achieving better efficiency and workload balancing in the network.
ii. Fault Tolerance—The AUVs should be configured in a manner so as to achieve fault tolerance against mechanical failure of any nearby vehicle. The failed AUV should be replaced with highest priority so that the disrupted segment

can become operative without affecting other segments of the network. Hence, AUVs should be capable of predicting its failure and get recovered or replaced themselves for maintaining network performance.

iii. Data Processing and Command Dissemination—During search operations, data is collected by different AUVs at discrete time-steps, thus generating bulk of samples over a large period of time [8]. This necessitates aggregation of the gathered data before transmitting the collective summary to the nearest offshore sink for further analysis. The sinks and other AUVs send control signals i.e., requests and commands to other AUVs in the network to schedule data collection and communication services.

iv. Limited Sensing and Communications Range—The communication range of participating AUVs are much smaller as compared to the huge monitoring area underwater. For signaling and communications, the acoustic networks may operate in two modes. One mode for sending control signals at a higher range, requiring a lower bit-rate. Another mode can be set for transmitting data with higher bit-rate performed at a shorter range.

v. Directional Underwater Acoustic Search—The directional search is highly beneficial in case of recovery and search operations. AUVs must survey the location of target to the direction of target movement

6 Underwater Network Challenges

Significant challenges that deeply impact the performance of underwater networks of acoustic sensors are highlighted as follows.

i. Energy Consumption—Sensors nodes are mainly distributed with the aim to sense an area and record relevant data for a long network lifespan. The sensor network lifetime is directly proportional to residual battery life in the individual sensors. Therefore, it requires efficient scheduling of duty cycles of sensors in a manner such that all nodes simultaneously in active mode could be avoided to save the sensor battery. Another important factor that affects the scheduling of sensor nodes includes high mobility, ocean drift and uncertainty among networks due to environmental noise and frequent link failures.

ii. Synchronization—Another challenging issue dealt with underwater sensor networks is time-synchronization. There are several RF-based synchronization techniques proposed for terrestrial wireless sensor networks [9]. However, unlike terrestrial sensor networks, underwater WSNs use acoustics or sound waves for data transmission. The underwater propagation speed of sound is much lower (i.e.) as compared to the speed of radio waves in terrestrial applications. The synchronization issue is mainly raised due to clock drifting between sensor nodes. Research in [10] experimentally showed that the drift rate was 50 ppm after a span of 30 days, which was found to create a clock difference of 130 s. According to a study conducted in [12] suggested a clock synchronization mechanism called Network Time Protocol (NTP). Also, Code Division Multiple

Access (CDMA) can be applied to prevent the loss due to drifts in clock or lack of synchronization.

iii. Routing—One of the biggest challenges in UW-ANs is communication and routing. Path loss is generally encountered due to attenuation, random noise, high propagation delays, etc., which can greatly degrade the performance of communication quality underwater. Moreover, an underwater acoustic network is not capable of sending and receiving data simultaneously, hence supporting half-duplex and asymmetric communication. Moreover, data rate for underwater applications is much lower to effectively support carrier-sense based transmission schemes. Overall, routing poses a critical issue in UW-ANs, regarding efficient storage and retrieval of the data from sensor nodes that minimizes waiting time and propagation delay [13].

iv. Localization—Another major issue in underwater applications involves localization. Placement strategies become quite essential for anchoring sensors at desired geo-locations while preventing them from displacement due to ocean drift and current. The arrival time or RSSI (Received Signal Strength Indication) techniques that are usually employed for terrestrial applications for computing exact location of sensor nodes, cannot be directly utilized for underwater applications. This is due to the fact that in UW-ANs, variable sound speed and non-negligible mobility of acoustic sensors set unique challenges in localization. Moreover, other localization techniques using GPS, fail to perform under water. Research conducted in [12, 14] innovated range-based methods along with several range-free mechanisms for application specific requirements.

Some of the characteristics of underwater acoustic channels, like extensive delays incurred in propagation, bandwidth constrained, Doppler shift, frequent variable phase and amplitude fluctuations and multipath interference; requires critical concern while developing protocols and standards [16, 17]. This includes ensuring high accuracy in protocol performance, less processing and communication overhead and improved scalability.

7 Emerging Applications of Underwater Drones

Underwater communication with acoustic sensors confronts distinct challenges, specifically due to inherent mobility and large propagation delays. Acoustic sensor-based application greatly assists in the exploration of ocean and deep waters. This broadly includes seismic monitoring, natural resources excavation, warfare surveillance and unmanned operations along with generic environmental issues, involving tracking underwater climatic conditions, aquatic life supervision, variation of chemical content of water, etc. [25–35].

i. Seismic Monitoring—This application is highly resonant, especially around earthquake prevalent zones. Higher dimensional seismic surveys are often employed, in which acoustic devices closely coordinate with each other and collaboratively collect location-relevant data. Additionally, in-network processing is carried out by sensors before transmitting to the offshore stations via satellite for complex-level processing.

ii. Ocean Sampling Networks—Sensors nodes and underwater vehicular nodes are engaged in compact and collaborative adaptive sampling of the three-dimensional ocean environment. Some experimental study includes Monterey-Bay field project, which incorporates the benefits of AUVs with new ocean architectural models. This allows new possibilities to explore in observing the environment and enables us to predict good characteristics of oceanic environments.

iii. Environmental Monitoring—The wireless acoustic networks underwater are highly specialized in performing pollution-level monitoring, tracking ocean/wind current and analysis of aquatic life of fishes, micro-organisms, etc. Pollution monitoring mainly involves recording the presence of essential chemical contents that negatively impacts biological life forms underwater. Also, ocean currents are tracked in order to enable weather forecasting, climate change detection and tracking influence of human actions on aquatic ecosystems.

iv. Oceanic Explorations—The acoustic sensor networks are also used to detect underwater oilfields and other natural reservoirs. Such exploration also assists in determining effective paths for installing undersea cables required for communication, underwater accommodation (e.g. aquatic resorts and labs) and geomagnetic event detection (e.g. earthquake, volcanic eruptions, etc.).

v. Disaster Prevention—Underwater networking helps to observe seismic activities from remote locations to provide tsunami warnings. In addition, for smaller aqueous bodies (e.g. lakes, canals, wetlands, etc.), change in water level is dynamically recorded to forecast possibilities of flood. Information collected locally can be sent to nearby coastal or inshore stations for deriving predictions of natural disaster.

vi. Marine Warfare—Acoustic surveillance is often targeted using underwater sensor networks. Unmanned autonomous vehicles are deployed with the task of tracking enemy ships and submarines for raising alert signals to host security portals. For such applications, UW-ANs predominantly offers critical and continuous inspection with wide coverage of geographical area underwater.

Therefore, deployment of acoustic networks in a large variety of crucial underwater applications justifies the rapid technical growth achieved in acoustic communication and aquatic sensing technologies.

8 Conclusion and Discussion

For monitoring the oceanic layer, a specialized form of underwater vehicle network is employed using underwater drones for acoustic monitoring. This chapter reviews literature related to UN-AN's for AUV. The design of the acoustic sensor node and the differences between surface and marine sensor systems have been highlighted. The equations of kinematics and dynamics of the AUV's modeling based of 6-DoF have been presented. The architectures of under-water acoustic networks such as two stage, three stage and hybrid architectures are discussed and among these the hybrid architecture is most difficult of implement due to the presence of autonomous vehicles.

The three core subsystems of AUV's namely guidance, navigation and control work together to control and manage the AUV. The guidance system finds the trajectory of the vehicle; the navigation system provides an estimate of the current status of the vehicle; and the control system measures and applies the proper forces to manoeuvre the vehicle. The different control laws such as PID, SMC and fuzzy control are discussed along with feedback linearization technique which is then followed by control allocation algorithms such as non-optimal, quadratic programming schemes are presented. The aim of control subsystem is to minimize the power consumption of the AUV. The AUV navigation techniques primarily used are INS and acoustic navigation. Geophysical navigation seems to be more promising hence needs to be explored more. The design issues of AUV such as power estimation, load balancing, fault tolerance etc. are mentioned, followed by challenges such as energy consumption, synchronization and routing and localization are discussed. In the last emerging applications of AUV's are presented for further exploration and research.

References

1. Akyildiz, I.F., Pompili, D., Melodia, T.: Underwater acoustic sensor networks: research challenges. Ad Hoc Netw. **3**, 257–279 (2005)
2. Sozer, E.M., Stojanovic, M., Proakis, J.G.: Underwater acoustic networks. IEEE J. Oceanic Eng. **25**(1), 72–83 (2000)
3. Proakis, G., Rice, J.A., Sozer, E.M., Stojanovic, M.: Shallow water acoustic networks. In: Proakis, J.G. (ed.) Encyclopaedia of Telecommunications. Wiley and Sons, Hoboken (2003)
4. Proakis, J.G., Sozer, E.M., Rice, J.A., Stojanovic, M.: Shallow water acoustic networks. IEEE Commun. Mag. **39**, 114–119 (2001)
5. Lin, W., Li, D., Tan, Y., Chen, J., Sun, T.: Architecture of underwater acoustic sensor networks: a survey. In: First International Conference on Intelligent Networks and Intelligent Systems, pp. 155–159
6. Pompili, D., Melodia, T., Akyildiz, I.F.: Deployment analysis in underwater acoustic wireless sensor networks. In: Proceedings of ACM International Workshop on Underwater Networks, Los Angeles, CA
7. Cayirci, Erdal, Tezcan, Hakan, Dogan, Yasar, Coskun, Vedat: Wireless sensor networks for underwater surveillance systems. Ad Hoc Netw. **4**(4), 431–446 (2006)
8. Yoon, Seokhoon, Qiao, Chunming: Cooperative search and survey using autonomous underwater vehicles (AUVs). IEEE Trans. Parallel Distrib. Syst. **22**(3), 364–379 (2011)
9. Elson, J., Girod, L., Estrin, E.: Fine-grained network time synchronization using reference broadcasts. In: Fifth Symposium on Operating Systems Design and Implementation, Boston, MA, pp. 147–163
10. Heidemann, J., Wei, Y., Wills, J., Syed, A., Yuan, L.: Research challenges and applications for underwater sensor networking. In: IEEE Conference on Wireless Communications and Networking, pp. 228–235

11. Li, K., Shen, C.C., Chen, G.: Energy-constrained bi-objective data muling in underwater wireless sensor networks. In: 7th IEEE International Conference on Mobile Ad-hoc and Sensor Systems (MASS), pp. 332–341
12. Syed, A.A., Heidemann, J.: Time synchronization for high latency acoustic networks. In: 25th IEEE International Conference on Computer Communications, pp. 1–12
13. Partan, J., Kurose, J., Levine, B.N.: A survey of practical issues in underwater networks. In: 1st ACM International Workshop on Underwater networks, New York, pp. 17–24
14. Erol-Kantarci, M., Mouftah, H.T., Oktug, S.: A survey of architectures and localization techniques for underwater acoustic sensor networks. IEEE Commun. Surv. Tutor. **13**(3), 487–502 (2011)
15. Gkikopouli, A., Nikolakopoulos, G., Manesis, S.: A survey on underwater wireless sensor networks and applications. In: 20th Mediterranean Conference on Control & Automation (MED), Barcelona, Spain, July 3–6 2012, pp. 1147–1154
16. Anguita, D., Brizzolara, D., Parodi, G.: Building an underwater wireless sensor network based on optical communication: research challenges and current results. In: International Conference on Sensor Technologies and Applications, pp. 476–479
17. Chen, K., Ma, M., Cheng, E., Yuan, F., Wei, S.: A survey on MAC protocols for underwater wireless sensor networks. IEEE Commun. Surv. Tutor. **16**(3), 1433–1447 (2014)
18. Vervoort, J.H.A.M.: Modeling and Control of an Unmanned Underwater Vehicle. Master Traineesh. Rep, pp. 5–15 (2009)
19. Cruz, N. (ed.): Autonomous Underwater Vehicles. BoD–Books on Demand (2011)
20. Vahid, S., Javanmard, K.: Modeling and control of autonomous underwater vehicle (AUV) in heading and depth attitude via PPD controller with state feedback. Int. J. Coastal Offshore Eng. **4**, 11–18 (2016)
21. Dinc, M., Hajiyev, C.: Autonomous underwater vehicle dynamics. In: Autonomous Vehicles, p. 81
22. Fossen, T.I., Johansen, T.A., Perez, T.: A Survey of Control Allocation Methods for Underwater Vehicles, Underwater Vehicles. In: Inzartsev, A.V. (Ed.), InTech (2009). https://doi.org/10.5772/6699
23. Leonard, J.J., Bennett, A.A., Smith, C.M., Jacob, H., Feder, S.: Autonomous Underwater Vehicle Navigation. MIT Marine Robotics Laboratory Technical Memorandum (1998)
24. González-García, J., Gómez-Espinosa, A., Cuan-Urquizo, E., García-Valdovinos, L.G., Salgado-Jiménez, T., Cabello, J.A.: Autonomous underwater vehicles: localization, navigation, and communication for collaborative missions. Appl. Sci. **10**(4), 1256 (2020)
25. Puri, V., Nayyar, A., Raja, L.: Agriculture drones: a modern breakthrough in precision agriculture. J. Stat. Manage. Syst. **20**(4), 507–518 (2017)
26. Ganesan, R., Raajini, X.M., Nayyar, A., Sanjeevikumar, P., Hossain, E., Ertas, A.H.: BOLD: bio-inspired optimized leader election for multiple drones. Sensors **20**(11), 3134 (2020)
27. Khan, N.A., Jhanjhi, N.Z., Brohi, S.N., Nayyar, A.: Emerging use of UAV's: secure communication protocol issues and challenges. In: Drones in Smart-Cities, pp. 37–55. Elsevier (2020)
28. Nayyar, A.: Flying adhoc network (FANETs): simulation based performance comparison of routing protocols: AODV, DSDV, DSR, OLSR, AOMDV and HWMP. In: 2018 International Conference on Advances in Big Data, Computing and Data Communication Systems (icABCD), pp. 1–9. IEEE
29. Nayyar, A., Nguyen, B.L., Nguyen, N.G.: The internet of drone things (IoDT): future envision of smart drones. In: First International Conference on Sustainable Technologies for Computational Intelligence, pp. 563–580. Springer, Singapore (2020)
30. Khan, N.A., Jhanjhi, N.Z., Brohi, S.N., Usmani, R.S.A., Nayyar, A.: Smart traffic monitoring system using unmanned aerial vehicles (UAVs). Comput. Commun. (2020)
31. Nayyar, A.: An encyclopedia coverage of compiler's, programmer's & simulator's for 8051, PIC, AVR, ARM, arduino embedded technologies. Int. J. Reconfigurable Embed. Syst. **5**(1) (2016)

32. Nayyar, A., Puri, V., Le, D.N.: Comprehensive analysis of routing protocols surrounding underwater sensor networks (UWSNs). In: Data Management, Analytics and Innovation, pp. 435–450. Springer, Singapore (2019)
33. Nayyar, A., Balas, V.E.: Analysis of simulation tools for underwater sensor networks (UWSNs). In: International Conference on Innovative Computing and Communications, pp. 165–180. Springer, Singapore (2019)
34. Nayyar, A., Ba, C.H., Duc, N.P.C., Binh, H.D.: Smart-IoUT 1.0: a smart aquatic monitoring network based on internet of underwater things (IoUT). In: International Conference on Industrial Networks and Intelligent Systems, pp. 191–207. Springer, Cham (2020)
35. John, S., Menon, V. G., Nayyar, A.: Simulation-based performance analysis of location-based opportunistic routing protocols in underwater sensor networks having communication voids. In: Data Management, Analytics and Innovation, pp. 697–711. Springer, Singapore

Smart Agriculture Using IoD: Insights, Trends and Road Ahead

N. Hema and Manish Sharma

Abstract The agriculture sector has the great scope for the improvement with the use of latest technological advancements and the decision support systems in order to have surplus crop productions in the market for the increasing population. The different agricultural activities in the fields can be performed and monitored with the use of smart devices, remote sensors, UAVs/Drones, IoT (Internet of Things) for data gathering and analysis to timely fulfil the demand of food in the market and support survival on the planet. IoT based drones come as boon to analyze data on the go and uniformly cover the crop area remotely. IoT enabled devices such as Unmanned Aerial Vehicles (UAVs) when integrated with different sensors and cameras can be applied in a variety of agricultural applications related to crops management for capturing sensor and image-based data. The newer technologies such as IoT and UAVs called as IoD (Internet of Drones) have a lot of potential to change the practices that have been performed in the area of agriculture. Such technologies will enable the farmers to make the decisions relatively in the lesser time to manage the crops in an effective manner by reducing the cost and increasing the yields. The presented chapter discusses various concepts related to IoT based drone technologies and the different types of sensors and the devices that can be used to collect the data from the farm lands. This chapter discusses overviews and the strategies to acquire the data with the UAVs using static or fixed type approaches. This chapter discusses about how IoD are used in the various applications of agriculture to achieve smart and precision farming. Emerging technologies and development of agricultural IoD with its challenges to adapt itself in harsh field environment for achieving accuracy are discussed.

Keywords Precision farming · Remote sensing · Digital image processing · Smart agricultural sensors · Internet of Drones · Agricultural applications · Agricultural services

N. Hema (✉) · M. Sharma
Jaypee Institute of Information Technology, Noida, India
e-mail: hema.n@jiit.ac.in

R. Krishnamurthi et al. (eds.), *Development and Future of Internet of Drones (IoD): Insights, Trends and Road Ahead*, Studies in Systems, Decision and Control 332,
https://doi.org/10.1007/978-3-030-63339-4_3

1 Introduction

Unmanned Aerial Vehicles (UAV or Drones) are becoming popular each day in terms of their academic, research, and commercial successes as they are being used for the purpose of observation and capturing information from the different geometrical perspectives in almost every application. Yao et al., discussed that UAV [1] are designed to use as general-purpose tools for remote sensing and information capture, but the associated methods for processing and analysis of data are completely independent in nature and specific to the application.

Puri et al. [2] mentioned different types of modern drones that are used specifically for smart and precision agriculture. Incorporation of Wi-Fi technology in drones brings a new dimension of drones called IoD (Internet of Drones). Such technologies are in form of First Person View (FPV), and that drones can be integrated with various large data capturing devices such as HD cameras like GoPro, Parrot, DJI, and many others to stream agricultural field real-time video of flight over smart phone or tablet or servers for various monitoring purpose. Currently, 85% of drone technology is mainly utilized for military and rest 15% by civilians for diverse applications but these usages are gradually changing over the time. Around 80% of the U.S. market for commercial drones is estimated to be for agriculture applications in near future as stated by Eriksson and Lundin [3].

There are varieties of sensors available that can be used with UAVs in plug-and-play fashion to perform the specific activities at the earliest and generate the insights for the problem to be solved. But there is a deficiency that can be seen in the systematic development and analysis of these systems with characteristics specific to the problem.

Drone technology with its huge potential is changing the pace of every industry in which it is entering. It is helping the professional to look at the problems from the different angles and design the sophisticated solutions that are much more reliable, robust, and efficient in terms of performance and delivering results. Drones business is impacting heavily many industries such as Infrastructure, Transport, Security, Entertainment and Media, Insurance, Telecommunication, Mining, and many more [4].

One such sector that still needs to be worked upon heavily is ‘Agriculture’. There are numerous activities in the agriculture field that are time consuming, inefficient, and inaccurate, which need to be optimized with the inclusion of the modern technologies. The traditional ways of farming need to meet every aspect of the modernization in order to get benefits and reap the best of the crop productions.

According to the surveys conducted and the forecasts, there is a part of the world population in the different countries, which is living extremely hungry and suffering, even with the current rate of food productions. The situation is becoming more alarming every year as the resources such as land and water to produce the food as per the demand and survive are getting scarcer. It is also estimated that the food production needs to be increased by 50% worldwide by 2050 in order to feed this large hungry population with the limited resources.

Those involved in the farming business need to understand and assess the situation with deep considerations to include the modern technology in the agriculture field. The inclusion of Information and Communication Technologies (ICT) to the agriculture for the betterment is not an option but it has become an evident necessity. The farming communities need to adapt the agriculture not only to the climate but also to the other challenges, as these factors play key roles in affecting the conditions for growing food. Modern techniques can help the community to grow crops with increased productions by generating the required information in no time to take actions and respond to the problems detected in the fields.

Drones can collect high resolution images and videos which can be used to run analysis and precisely detect and locate the problems with the crop, so that the appropriate measures can be taken to avoid any losses. Drone technologies can provide the valuable data which can influence policies and help in decision making, despite the presence of some limitations in the implementations and use. Drones are the solutions to the many problems in the agriculture sector, along with the support and extensive collaboration between governments, various industry giants, and major domain experts. This technology has the enormous potential that will give the agriculture sector an all-round, high end makeover in terms of laying out plans to set up the field and strategy by gathering real time data, and processing information on considering requirements for soil, planting, irrigation, crop spraying, crop monitoring, and health assessment in advance or during the agricultural cycle. Swarm of interconnected UAV are also used for real-time feedback for application like uniform pesticides and fertilization [1]. Environment monitoring is one of the important applications of drones especially in agricultural fields.

Figure 1 shows the architecture of Internet of drone for agriculture, how the various agricultural sensors equipped with drones are able to communicate with cloud services for real-time data analysis. Communication of drones to the cloud system is achieved through edge, mist and fog computing [5]. Next section discusses about the techniques used by drones for environment monitoring.

2 Remote Sensing and Digital Image Processing for Environment Monitoring by Drones

2.1 Remote Sensing

Remote sensing means any process which uses electromagnetic energy to collect the environmental information about the environment and its objects such as soil, vegetation cover, water resources, areas or phenomenon without having any physical contact with it. There can be several variations to the remote sensing in terms of the environment being sensed, energy forms, ways of measuring energy emissions [6]. Remote sensing is basically of two types, firstly the passive remote sensing in which the sensor does not have any radiation source of its own, but records the reflected

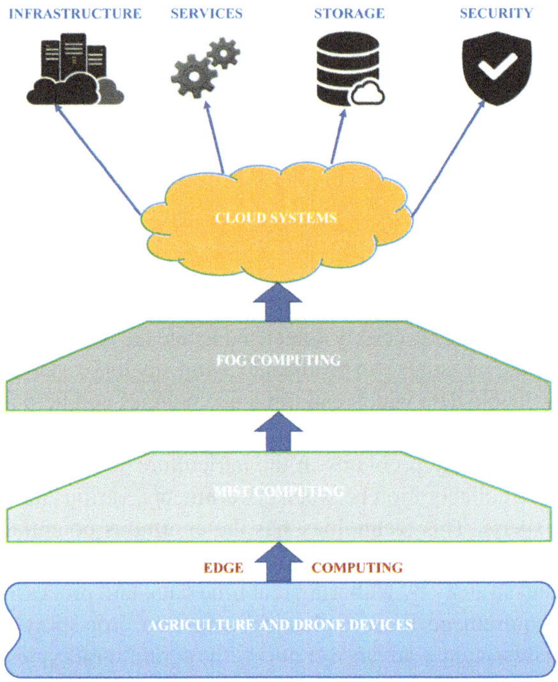

Fig. 1 Architecture of Internet of drone for agriculture

radiation from the surface of the object. Second is the active remote sensing, where the sensor has its own source to emit radiation and captures the reflectance returning to device [7].

With the advancements in the fields of the aviation and the space travel, along with the invention of the photography, made it possible for the remote sensing devices to capture the images from the air with balloons and the planes, and from space using satellites. Earlier photography was based on traditional methods but the modern photography uses digital techniques for satellite based remote sensing [8]. The advancements done in the areas of drones (or UAVs) have become more promising in terms of flexibility and performance with the added advantages to be applied in the field of agriculture and GIS for better understanding of the problems with a closer look at the fields. It enables on demand and quick deployment to monitor the health of the crops and take necessary control decisions and actions.

In remote sensing, the most used and popular spectral ranges are the visible light, near infrared, shortwave infrared, thermal infrared, and microwave bands. These wavelengths have the characteristics specific to the applications and the devices used in the different domains and specially in agriculture sector. Optical sensors can gauge the radiation levels emitted from the earth's surface in an appropriate manner when there is no or less disturbance caused by the presence of the variety of atmospheric elements such as water vapors, gases, particles, and etc.

The term Green Vegetation (GV) is associated with the amount of green content present in the vegetation. Vegetation types can be distinguished depending upon the large amount of infrared light that is reflected from the surface of the leaves, making this wavelength very suitable for the remote sensing applications in agriculture and GIS. A low degree of reflectance for near infrared or short-wave infrared light indicates the status on the health of the leaves being dry, ill, or suffering with water stress.

Vegetation Indices (VI) are defined as unit less radiometric measures which can be used to spectrally characterize the biophysical properties of the plants. Vegetation indices are given by ratios or defined as differences between two or more spectral bands in the VIS, NIR, SWIR regions. A number of vegetation indices are defined in order to capture information about every aspect of plants in relation to the environment, such as NDVI (Normalized Difference Vegetation Index), SAVI (Soil Adjusted Vegetation Index), MSI (Moisture Stress Index), LWCI (Leaf Water Content Index), and etc.

2.2 *Remote Sensing Platforms in Agriculture*

Depending upon the elevation from the earth's surface at which these systems are placed and the area that needs to be surveyed or monitored, remote sensing platforms can be classified into three major categories as:

- Spaceborne remote sensing
- Airborne remote sensing
- Ground based remote sensing.

These methods can be used to map the large or small areas in GIS and agricultural applications and is as shown in Fig. 2. A deeper consideration must be given to the spatial, spectral and temporal resolution for designing and evaluating a system to capture the images. These resolutions can help the professionals to understand the problem and design the systems specific to the requirements for delivering the results. Remote sensing technology applied in the agriculture fields can be used to forecast the crop productions, assess the requirement of nutrients by the plants, check water demands, perform pest detection and weed control [7].

2.2.1 Spaceborne Remote Sensing

There are various systems such as space shuttles, space stations, and satellites in this category that can help capture the information with cameras mounted on them. The most stable form of remote sensing is satellite based as they are orbiting the earth continuously. The satellites can be categorized based on their orbital geometry and timing as geostationary, polar, and sun synchronous. Sun synchronous satellites are used for remote sensing and can be used in agriculture sector to provide overall mapping of crop type, land cover, assessing general conditions of the crops, estimating crop yield in terms of acreage. The payload for remote sensing satellites

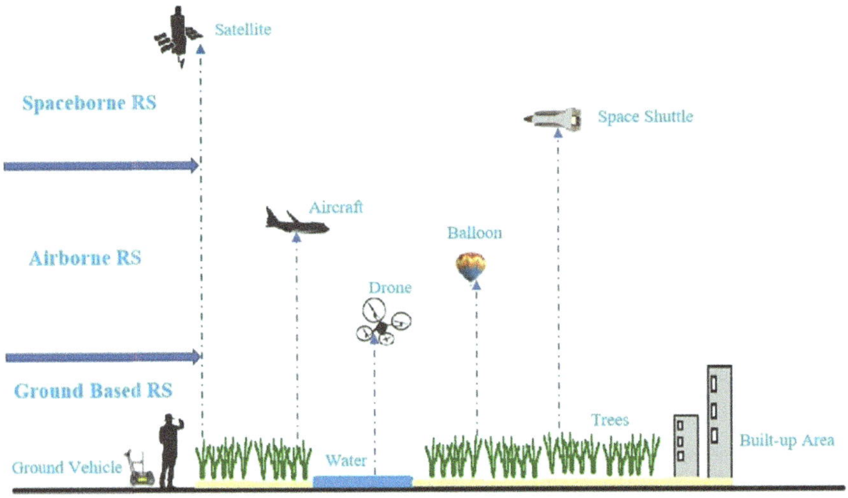

Fig. 2 Remote sensing platform

can include photographic systems, electro-optical sensors, microwave or LiDAR systems in addition to the different sensors incorporated along with data processing and transmission capabilities.

Most remote sensing satellites have been designed to transmit data to ground receiving stations located throughout the world. Satellite images basically covers a very large area but due to the limited resolution available, field specific information is not available for analysis of different problems. Besides this, there are other limitations such as the images captured by the satellites are affected by the varying weather conditions, satellites with high resolution imagery have a longer revisit time (1–16 days) making them not suitable for applications that need images at regular intervals, even with the capability to observe the problems with in the fields. VIS and IR bands of the electromagnetic spectrum are primarily used for identification of clouds, but can also be used to calculate spectral vegetation indices and thus to estimate the amount of biomass. Examples of remote sensing satellites include Quickbird, IKONOS, MODIS, TerraASTER, IRS, etc.

2.2.2 Airborne Remote Sensing

In earlier times, airborne systems were the only means of doing non-ground based remote sensing operations. It was done mainly with the balloons, which were very unstable and the flight path was not predictable. At present, airplanes and helicopters are the most common vehicles for performing aerial remote sensing activities. These have the capability to fly very high to capture the information in high resolution about a larger area and even closer to the ground for finer details. Airborne remote sensing technology employs both manned and unmanned aerial vehicles, but the advancements in the technology have more often replaced the manned systems with the unmanned ones. Unmanned Aerial Vehicles (UAVs or Drones) are majorly suited for collecting information about the field due to their characteristics such as low implementation cost, light weight, and low cruising airspeed.

Drone systems show their advantages in use as they can be deployed and controlled remotely to capture high resolution imagery with flexibility in time, quick and repeated number of operations in comparison to other means of remote sensing methods. These systems have the ability to fly close to the ground and perform the activity assigned in an area where it can be too hazardous or difficult to reach. More recently, the combination of rapid development of low cost and small UAVs, and the improvements of conventional sensors in terms of cost and size, has led to new and promising scenarios in environmental remote sensing and monitoring. This technology provides the images with spatial resolution in the range of 1–20 cm, and which can act as a bridge to fulfil the spatial resolution gap between the airborne systems in the range 0.2–2 m and the ground-based systems with less than 1 cm.

Drone technology equipped with light weight imaging systems are used for generating very high-resolution images for managing activities in site-specific crops, specialty crops and measuring canopy chemistry, total biomass, Vegetation greenness and health (Normalized Difference Vegetation Index (NDVI), Enhanced Vegetation Index (EVI)), and Vegetation biochemistry and heterogeneity. These data can be used to derive many key indicators of ecological health and changes in the ecosystems, including ecosystem structure, Digital Elevation Model (DEM), and Digital Surface Model (DSM).

2.2.3 Ground Based Remote Sensing

Ground based remote sensing platforms are also called as terrestrial platforms as they capture the information about the Earth's surface while being intact to it. This type of remote sensing can be useful to understand the various features over and under the surface of the earth. The ground-based platforms can be either fixed or moving depending upon the requirements. Examples in this category includes small handheld devices, tripods, booms, towers and cranes which provide better spatial, spectral, and temporal resolution in comparison to other remote sensing methods. Handheld devices and cranes can be used to collect information in a portable way while others are used mostly in fixed manner. These devices prove to be very useful in assessing the type of the plant stresses such as biotic and abiotic in case of the small fields as they provide a closer look at the plant's health, identifying species, and any other condition in the field.

The wide variety and availability of cheap electronics components and sensors along with the computers have given the entire process a boost to dramatically reduce the time in comparison to the manual methods. The data processing and representation capabilities have enhanced further when compared to the earlier times. With this technology, there are demerits in terms of the efficiency as the manpower is still involved and time requirement is still high even for small fields.

UAV uses airborne methods for remote sensing of environment. Data collected from ground and satellite based methods are later superimposed with drone data for accurate localization of area of interest. Satellite based remote sensing has limitation during cloudy condition, in such situation drone with internet connectivity plays an important role.

2.3 *Digital Image Processing*

The data captured by the various remote sensing devices is in the form of digital images i.e., digital data, which presents a main advantage in terms that it can be processed, analyzed, stored, and communicated with the help of computer systems. Digital image processing is concerned basically with four types of operations namely image restoration, image enhancement, image classification, image transformation [6].

Drone technology in combination with the advanced image analytics can help fulfill the gap in between the current agricultural productions and the future needs, and it can also help to understand the environment around us in a better way. The precise mapping of the areas suffering with the specific type of plant stress, can be addressed with application of the right amount of water, fertilizers, pesticides, or herbicides. Image analytics techniques can be used to assess the difference between the pre and post disaster states of the vegetation to correctly estimate the reduction in the yield to work upon insurance procedures. Drones and the imaging capabilities can be applied in agriculture from the soil condition mapping to the yield estimation, covering all major activities performed in the farms.

UAV technology along with image processing and analytics is in the constant state of continuous upgradation and producing systems that are promising to deliver the performance to correctly assess and estimate the conditions in the agriculture sector. These technologies have the potential to transform the agricultural sector in the coming years. The development in the areas of machine learning and deep learning will help in further automating the key activities of the sector and shifting the cognitive skills from humans to the trained algorithms. Besides this, other technologies like big data and cloud computing will further help in large scale processing and analytics on the information to provide real time insights with improved connectivity and interactions. All these technologies will ultimately lead to increased productions in the agricultural field and help in bringing down agricultural products prices.

3 Drone Technology and Developments

The development of the UAV assisted remote sensing systems are taking agricultural activities at all new levels in terms of the productions, precision, applications, etc. Drones have shown the ability to fly at low altitude for taking a closer look at the problem with the use of high-resolution cameras and help in early detection of any issues noticed in the fields. UAV technology has exhibited number of advantages in comparison to the other means of remote sensing in terms of low cost, increased flexibility, repeated on demand surveying and monitoring, and easy operations.

3.1 Drone Systems

The drone technology is advancing at a faster pace with the new components having low cost, robustness, and reliability. UAV systems are having their applications in almost every industry according to their type of design. There are different types of UAVs such as fixed-wing, rotary-wing, blimps, flapping wings, parafoil-wing [9]. Most commonly used drone systems are fixed-wing and rotary-wing [10]. Some of hybrid-wing are also designed using fixed-wing.

3.1.1 Fixed Wing

These unmanned planes are generally having rigid or non-movable wings supported on a fuselage and a tail that forms the propulsion system using a motor and a propeller. These require a medium for launching such as a run-way or a catapult launcher from the ground to fly in air. Fixed wing drones have the advantages as they can carry more payload, increased flight time, large coverage of areas with each flight. But fixed wing drones are more costly in comparison to the other types. These types of drones are used specially for pesticide spraying and dispatch of fertilizers. Examples of such fixed wing drones are Ag Eagle RAPID, Precision Hawk Lancaster and SenseFly eBee SQ Nixon [11].

3.1.2 Rotary Wing

This type of drones is also known as Vertical Take-Off and Landing (VTOL) or rotorcraft as it consists of rotary blades and the propulsion system. These have the advantage to stay in a position during flight for a very long time with the high ability to maneuver in any direction. However, there are limitations that this type of drones is unable to carry heavy payload and flight time is also very low on their part. But these are most used for different types of missions and very specially in agriculture sector for surveying and monitoring. The added advantage is that they have lower cost in comparison to the other types [10]. BLY-A DRONE, DJI MG-1 AGRAS are some examples of the rotary wing UAV which can lift minimum supplies for precision agriculture.

3.2 Components of an Agricultural Internet of Drone

A multi-copter UAV system consists of a number of electronic, electrical, and mechanical parts along with different types of sensors to form the complete, operational, and integrated unit. Depending upon the type of the application, specific components can be added to the system. All the parts of a copter system are vital for a smooth and safe flight, but there are certain components classified as essential ones for designing any multi-copter drone system. Different components of an unmanned aerial vehicle for agricultural purpose [12, 13] is as shown in Fig. 3 and its technical details are discussed as follows:

1. Frame

The frame can be considered as the backbone of the drone systems to which other components are placed on it. It is designed in such a way that it is suitable for the application requirements with its low weight for increased endurance, high tolerance for crashes, and ability to carry heavy payloads as per the needs. The design of the frame must be given importance, as it plays a key role to stabilize the system during flight. A variety of frames can be considered to design the mission application such as tri-copter, quadcopter, hexa-copter, etc. For agriculture purpose, mostly quadcopters and hexa-copters are used for monitoring and spraying activities. To reduce the weight of drone glass fiber frame are also used instead of steel or aluminum. Glass fiber quadcopter was built by Saheb and Babu [14] for agriculture purpose. Flame wheel is also popular due to its features such as relatively inexpensive, durability, spacious central plate, compact and easy to assemble. Foldable frames are also used to decrease the usage space.

2. Propellers

These are the components of a drone located on the sides of a copter which are responsible for lifting the drone in the air. Propellers are responsible for transforming rotary

Fig. 3 Drone components

motion into linear thrust. Propellers are having two important properties namely, pitch and the diameter. The multirotor drones such as quadcopter, hexacopter, and octocopter the propellers are arranged in pairs for rotating either clockwise or anti-clockwise to create a perfect balance. Propellers speed variation allows the drone to ascend, descend, hover or affect its pitch, roll and yaw. Propeller's motor can vary the speeds by changing the voltage supplied by an Electronic Speed Controller (ESC) module. Speed control is easier in shorter propellers and require less energy due to reduced inertia. However, for greater stability longer propellers are used which in turn needs to run in particular RPM and thus require powerful motors. Typically, heavy-lift drones will require longer propellers with smaller pitch [15].

Propellers are generally made of plastic, but high-quality parts are made of carbon fiber. This is also an area of innovation in designing better propellers to improve the flying experience and increase the flight time. To avoid any kind of damage to the propellers while operating, guards can also be added to the setup. Spraying Drone needs 30 Inch Hobby wing Carbon Plastic Propellers.

3. Motor

Motors play a very important role in designing any drone system, so it becomes necessary to consider the high-quality motors. The motors are generally of two types brushed DC motor and brushless DC motor to be used in the drone systems. The latest drones use brushless DC motors as they have advantages in terms of speed to torque characteristic, noise-free running, reliability, and high speed with longer operating life.

Whenever quadcopter motor selection needs to be done, we have to keep motor thrust and weight ratio in fix. It is generally said that the motors thrust should be twice as much as the total weight of the drone. Wider and taller motor produces larger torque and more powerful motor has larger RPM and this requires larger voltage. There is tradeoff between the higher KV motor and lower KV motor, the former has high rotation and less torque, and the latter has higher torque with less rotation. A more efficient motor can help save the battery power, which means power to perform more in a mission.

Frame size, propeller size and motor size are interrelated [16]. Lower frame size requires short propeller and lower power motor and vice versa. Typically, 150 mm frame requires 3-inch propeller, 1306 size motor, and 3000 kV speed and for larger frame size such as 450 mm requires 8, 9- or 10-inch propeller, 2212 size motor and larger 1000 kV motor speed. Selection of motor is based on the thrust, voltage consumption, efficiency and that of the weight of drone. For agricultural drones a highly efficient RPX series brushless DC motors are used.

4. Flight Controller

A flight controller can be considered as the brain of the drone system, as it is central to the functioning of the drone and performs the task of managing the speed of the motors to stabilize the entire system during the flight. It is able to perform calculations and operations depending upon the inputs given by the users as commands and the readings obtained from the sensors, and sends signals finally to the ESC (Electronic Speed Controller) unit for adjusting the speed of the motors. It can also perform operations like triggering cameras and other payloads, autopilot and fail safe.

To build autonomous drones, flight controller board has some selection criteria such as processor, open source firmware, autonomous functionality, affordability, FPV, microcontroller or Linux based environment, frame size, and popularity (basically for online resources and help).To develop autonomous drone from scratch, some of the popular flight controller boards that are available in the market are BeagleBone Blue, Navio2, Pixhawk, Pixhawk Cube, Naza Flight Controller and CC3D Revolution [17]. Most of these boards has 32-bit processor expect Navio2 which has 64-bit processor. Variants of TopXGun T1A flight controller are available in markets and are used for agricultural drone.

5. Power Source

The power source can be considered as the lifeline of the copter system, which is needed to run the different devices and the sensors connected to it. There can be different forms of the power source for a drone system such as solar power, batteries, petroleum, which are used as per the applications. The most successful form is the battery (Lithium Polymer), which takes a very less space to store, lightweight, low cost and very safe in comparison to other means of the power sources.

6. Wireless Communication

Unmanned Aerial Vehicles (UAVs) and Wireless Sensor Networks (WSNs) can be used as one of the solutions for the economic Precision Agriculture (PA) applications, such as aerial crop monitoring and smart spraying tasks. Wifi, mobile, WSN, zigbee, raspberry pi and various other communication technologies are used for the wireless communication. Generally, WSN based UAV's has the wireless receivers to capture the data from smart sensors that are installed at ground level with low range transmission. Further, this data will be transmitted on cloud through wireless and internet for analysis and action to be taken.

IoD main challenge is the efficient communication. Communication is required between inbuilt components or between drones or with main network. Ground based communication uses 2.4 GHz communication spectrum. However, long range communication is complex and expensive due to different terrains, dense trees and dense forest. For such real-time application, UAV vehicles are deploying Flying Adhoc Networks (FANETs) which is more reliable and economic [5, 18]. Connected UAVs are best coordinated using FANETs which are designed for rapidly changing environments.

7. Global Positioning System

Global Positioning System (GPS) provides location and time information anywhere on the Earth. Along with magnetometer, it can provide elevation and direction information also. Some of the latest drones have also added GLONASS standard to get the real time locations. GPS is required to localize the agricultural defective area on real-time basis [19]. Localization is required to gather information such as field contouring defects, production defects mapping, soil fertilizer defect mapping, and so on. GPS receivers or DGPS (differential GPS) is interfaced with drones which help to generate the visual maps of defective area of agriculture.

8. Geographic Information Systems

Geographic Information Systems (GIS) is used when we have to analyze the larger area for agricultural application [20]. Droughts, floods, swarms of insects and poor farming area analysis can be done using GIS software. Along with GPS, GIS is helpful in being able to map and project current and future fluctuations of weather information such as precipitation, temperature, wind direction and crop output, and more.

3.3 Imaging Cameras Used by Agricultural Drones

The remote sensing technology deals with the capturing of information about the objects from a distance in the form of images in different regions of the electromagnetic spectrum and followed by the interpretation at a later point of time. The imaging cameras cover a wide range of electromagnetic spectrum and are classified as RGB cameras, multispectral, hyperspectral, thermal, and LiDAR cameras. These cameras come in different sizes, with specific features, and varied costs. In order to benefit the various industries such as agriculture, some manufactures have designed the low-cost versions of their high-end cameras. It is estimated that an average volume of multispectral, thermal, and LiDAR imaging sensors is in gigabytes whereas for hyperspectral cameras, it is in terabytes. The section below discusses the details of the various imaging sensors [1] used by agricultural drones.

3.3.1 RGB Cameras

Drones equipped with normal RGB cameras are the most frequently used sensor for number of activities in precision agriculture. These sensors are having higher resolution and low cost in comparison to the other sensing devices, and are easy to operate and process data. The quality of the RGB cameras can be a key to the success of the application, which must consider the following factors such as camera lens, resolution, and quality of the image sensors, etc. The integrated cameras are easy to transport and operate but lack in terms of the customization whereas mounted mode have the flexibility to mount different RGB cameras specific to the application [21]. Many of the expert users work in different domains expecting the flexibility to change the cameras as per the requirements for achieving better results. Remote sensing applications [1] in agriculture field at large depend upon the RGB cameras for activities like detecting forest canopy, monitoring crop growth, and etc. These cameras have limitations to analyze the lot of vegetation features, so are used along with other types camera sensors. Examples belonging to this category are Sony A9, Nikon D850, etc.

3.3.2 Multispectral Cameras

Multispectral cameras are used extensively, as they have the capabilities to capture the images with high resolution in the red and near infrared band for calculating different vegetation indices [22]. Multispectral cameras are used for quantifying the state of the monitored vegetation in terms of chlorophyll content, water content in leaves, ground cover. NDVI, GNDVI, ENDVI are some of the commonly calculated vegetation indices in the near infrared band to support the analysis and interpretation of the vegetation health in the region. The benefits associated with the use of multispectral cameras come with high cost in comparison to the RGB cameras, so these are mostly used by the experts in the domains. Another limitation is that, these cameras have their own formats to deal with, but with the advancements in the ongoing research, the data to deal with are getting more generalized and common [23]. Cameras to capture multispectral images include multiSPEC4C, MCA-6 Tetracam, Quest Condor5 UAV, etc.

3.3.3 Light-Weight Hyperspectral Cameras

Hyperspectral cameras available these days have the capability to capture the images in the hundreds of narrow bands, but are very costly and have limitations in integrations with UAVs [24]. These cameras are less accessible due to these factors but capture the large volumes of information in different bands, that may be very beneficial for applications requiring to calculate vegetation indices. These systems employ complex pre-processing methods to capture the meaningful information from the images. By design, there are certain disadvantages and constraints with the use of hyperspectral cameras, but such limitations can be handled by making some adjustments to the drones, standardization among the format and processing capabilities. Hyperspectral cameras have their applications in agriculture for more precise classification and detection of the problems with much higher spatial and temporal resolutions in the fields to provide better understanding. Hyperspectral cameras available for drone based image capturing are Rikola hyperspectral camera, Resonon Pika NIR-640, etc.

3.3.4 Thermal Infrared Cameras

Being the passive sensors, thermal cameras are used to gauge the temperature of various surfaces and objects [25]. The way temperature measured using UAV borne thermal sensor is very different from as it is measured in the case of airborne or spaceborne thermal sensors [26], as the effect of atmosphere can be ignored, and readings are approximately accurate. Thermal imagery requires optimal weather conditions and no light is needed, making it suitable for carrying out operations in the night. Thermal cameras placed on the UAVs are able to capture the temperature levels on low flying altitude for humans, livestock detection, identification of sick animals, detecting fire centers, etc. Thermal scans data can also be combined with the data captured from the others sensor means to estimate the biophysical parameters of the crops in precision farming and water evaporation to plan irrigation. When designing an application that uses the thermal camera, one needs to consider using the RGB camera relative to the thermal cameras, for the purpose of identifying thermal camera shots. FLIR Vue Pro, senseFly thermoMAP, etc., are the cameras used for thermal imaging requirements.

3.3.5 LiDAR

In addition to the above types of sensors, laser scanners are also known as LiDAR (Light Detection And Ranging) which are used extensively for activities in environmental sciences. These systems are known to capture the high precision geometric data in terrestrial scanning for creating elevation maps where drainage is required or can be used to generate crop models in 3D to monitor water stress at different stages of the plant growth. There are certain examples of popular LiDAR data analysis software such as ArcGIS, AutoCAD or ENVI. These systems exhibit higher reliability and ability to capture the information, but with a very high cost of operations in comparison to other means of sensing, makes it inaccessible to use. RGB and LiDAR systems together have the capability to correctly perform measurements and interpretations [1]. Narváez et al., [27] also discuss about LiDAR and thermal images fusion for crop analysis. To perform imaging in this category, RIEGL VUX—40, Livox Mid-40 cameras are available.

Figure 4 shows the general hierarchy of applications, services and imaging sensors in drones for smart agriculture. In next section, various services provided by agricultural drones are discussed in detail.

4 Applications of Drones in Agriculture

Drones or UAV systems are gaining popularity with the increased number and variety of applications depending upon their size and capabilities. Drones can vary from nano, micro, and mini version to a larger one, from radio, and tele-operated to satellite guided. They can be used for the purpose of surveying and monitoring, performing actions in the reported areas, and as a medium of assistance during the times of disaster. Imaging systems employed in the drones are capable of acquiring high quality images and videos to run analysis and generate insights for decision making.

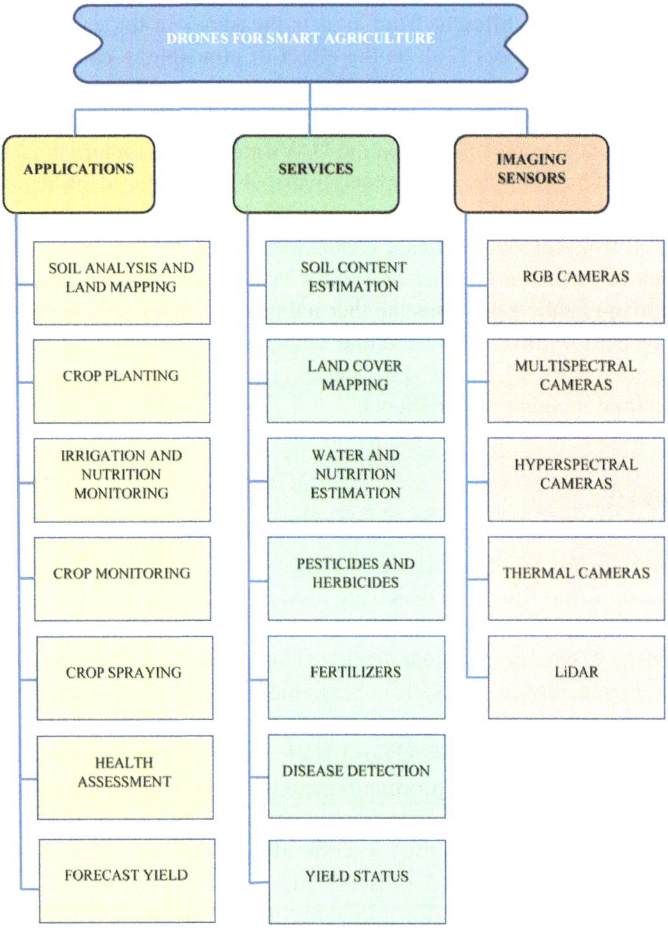

Fig. 4 General hierarchy of applications, services and imaging sensors using drones

UAV systems have applications in the field of agriculture for the different farming activities involved, which are still performed in inefficient and inaccurate traditional ways [4].

Drone technology and the imaging systems have the potential to transform the agricultural system into a smart, profit making, market oriented, and analysis driven business fulfilling the gaps of productions between present and the future [28].

Figure 5 shows the hierarchy of IoD for smart farming. Agricultural drones are not only equipped with sensors but also equipped with short range communication and mobile devices to gather ground level information for smart farming. Wi-Fi has the highest demand of usage in agriculture and farming industry, followed by mobile technology due to long range of communication. Other technologies as Raspberry pi, ZigBee, RFID, Bluetooth, WSN, LoRa and GPRS have less demand in the agriculture

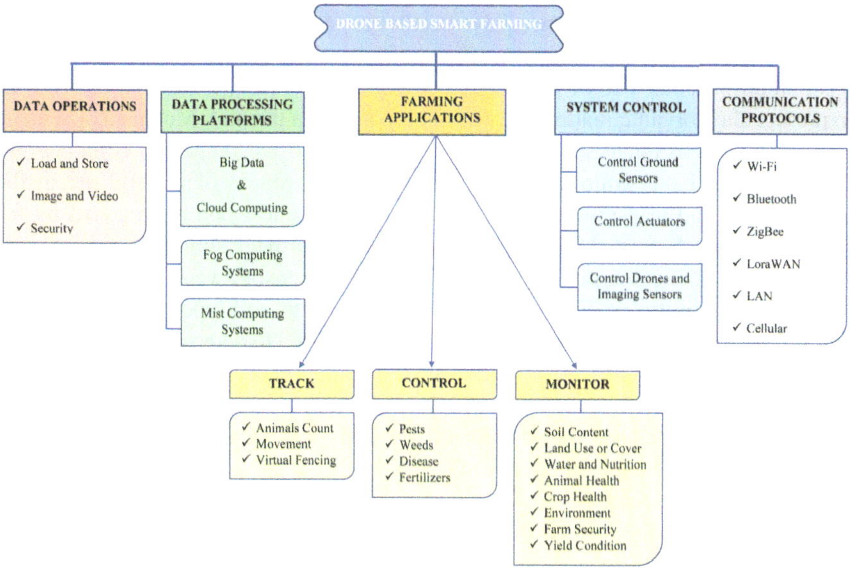

Fig. 5 Hierarchy of IoD for smart farming

and farming sectors due to its limited communication range. Various applications of drone technology in the agriculture are discussed as follows.

4.1 Soil Properties Analysis

Drones can play a major role in agriculture by being instrumental before the start of the agricultural cycle to collect the samples of the soils from the different parts of the field. Rainfall and rainwater runoff, agricultural activities, vegetative cover, wind, and slope of the land are the main causes of the erosion. Frequent soil sampling is required due to soil erosion which leads to degradation and loss of soil nutrients which are spatiotemporal changes [29, 30].

Soil samples are used to analyze and to understand the content of the soil in terms of the soil pH, texture, moisture, mineral content present, nitrogen levels, salinity etc. Besides this, drones can also collect information about the farming areas that have similar soil characteristics and produce precise maps for planning of seed planting strategy. Use of drones to perform soil and field analysis can give the fruitful insights to the farmers for planning the agricultural activities at the start and avoid any losses due to errors in the process. Various sensors like Coupled Plasma Mass Spectrometry, X-Ray diffraction, electromagnetic, optical, mechanical, electrochemical, airflow and acoustic are attached to drone with Internet connectivity for examination of soil mineralogy as well forecasting.

Electronic magic sensors are used to get the properties such as soil texture, salinity, moisture content and organic matter. Some of other soil properties such as soil pH or residual nitrates can also be predicted using these sensors. Optical sensors are used to predict clay, organic matter, and moisture content in the soil. Mechanical sensors can be used to estimate soil mechanical resistance (often related to compaction). These sensors have strain gauges to measure the hardness of the soil are at the ground level and transmit the data to the receiver drone when approached by for further analysis. Electrochemical sensors are used to provide soil nutrient levels and pH level of soil for the precision farming. Airflow sensors were used to measure soil air permeability for distinguishing between various soil types, moisture levels, and soil structure/compaction. Acoustic sensors are used to determine soil texture by measuring the change in noise level due to the interaction of a tool with soil particles. Due to low signal-to-noise ratio, this technology did not develop further. However, acoustic sensors are used for pest detection during night time. In addition to these sensors, crop health and soil condition can be analyzed using infrared, multispectral and hyperspectral sensors.

Some of the studies discuss the usage of drones for soil sampling. Huuskonen, J. and Oksanen, T. [31], uses drone for soil sampling with AR, where AR is used to locate the sample location. In this discussion exact location for physical sampling are identified using drones and AR but not on-go analysis of soil. To improve this solution, we need IoD where drones are equipped with receiver and at ground level, real-time sensor which transmit the data to the drone receivers. This method will be able to do the real-time soil analysis and recommend variable-rate fertilization.

4.2 *Planting and Weed Control*

Seed-drones designed specifically to plant the seeds from the sky [32]. These drone systems are loaded with the pods containing the germinated seeds and all the nutrients needed to sustain life, which are fired towards the ground. Firstly, a 3D map of the area is created using drones and analyzed to figure out the most efficient automated planting pattern with the help of computer programs. Such an application has the potential to grow the agricultural fields and regrow the forests suffered in disasters. When scaled correctly, a network of drones can team up to handle the activities efficiently at a faster pace to bring the planet back into a balanced and green state.

DJI drone's technology company has obtained around 80% rate of successful germination using sky seeding. Sky seeding is the biggest projects helped Alto Tajo park to start the population of new trees to over 200,000 m^2 which was burned in the major forest fire. DJI Agras MG-1 is used specifically for the herbicides, where drones identify the weed in the crop and destroy it using chemical spraying on them.

In the United States, small startup company Drone-Seed is also working on sky seeding. It is the first company operating with granted permission by US regulators to fly swarms of drones. These swarms drone generates map of the area with LIDAR and multispectral cameras before next set of swarm's drones drops seeds and spraying for weeds.

4.3 Crop Spraying

Drone technology equipped with processing components can precisely analyze the plants requirements for water, nutrients, or other chemicals and can spray to targeted area in a uniform manner. Vegetation indices calculated in the various spectral regions can help understand the kind of stress the crop plants are suffering with. An estimate about the deficiencies present in the plants can be made precisely and suitable treatment can be applied in a more appropriate and controlled manner. Experts believe that such an application can help us save water, and reduce the inclusion of chemicals into the soil and percolation in the ground. UAVs designed these days are equipped with distance measuring units so that according to the terrain, the drones can fly safely avoiding any collisions and accidents.

The drones [33] can be customized and integrated with hyperspectral camera for the detection of Ganoderma boninense, which is a serious threat to oil palm plantations in Malaysia and has caused great losses to healthy palms. Arthropod pest outbreaks are unpredictable and not uniformly distributed within fields. Early outbreak detection and treatment application are inherent to effective pest management, allowing management decisions to be implemented before pests are well-established and crop loss occurs. Pest monitoring is time-consuming and may be hampered by lack of reliable or cost-effective sampling techniques. Remote sensing technologies have been used in precision agriculture for the last few decades, with various applications, such as yield predictions and evaluation of crop phenology [34].

Due to disturbances in the wind, swarm of UAV are used for spraying to overcome non-uniform spraying of pesticide and fertilizer [35]. Swarm of UAVs were used in a control loop of algorithm for the spraying pesticides. The process of spraying the pesticides on the crop is organized by the feedback coming from the WSNs deployed in the field [36]. The algorithm to adjust the UAV route works as follows: periodically, the UAV broadcasts messages to the sensors on the fields to check the amount of pesticides being used. If the sensor receives the message, it responds with a message reporting the amount of measured pesticide and its position. Using communication link with WSN, the UAV is able to obtain wind speed and direction and concentration levels of pesticides sprayed on the crops. If imbalance is detected then drones changes the direction to enable the uniform distribution of pesticides. To coordinate among swarm, swarm head is required and selection of this is done by bio-inspired optimized leader election algorithm and is used by Ganesan et al., [37].

In December 2019, when locust invaded the parts of Gujarat, that they had destroyed crops spread over 25,000 ha of land. This kind of the problem can be easily addressed by IoT enabled drones by taking preventive measurements before damages are done to the crop. FAO is working with HEMAV to detect risk areas, inspection of pre-swarm locust development; and precise spraying of locusts at pre-swarm stages. Drones are equipped with high-resolution sensors and thermal camera can map outbreak area and rotary wing drones fitted with micro-sprayers can access these areas to inhibit the locust nest [38].

4.4 Crop Monitoring

In past, means were available to monitor large fields using airborne or spaceborne technologies. These technologies had the disadvantage in terms of the cost, timing, flexibility, etc. The photos captured using such means were of low quality, frequency was of days and mostly suitable for the large fields. But with the large-scale advancements done today, not only large but even small fields can be monitored with the help of drone technology. Fields are monitored on regular basis for changing weather conditions using drones which can take up the photos in high resolution for decision making on detected problems and effects on yields. Drone technology helps in minimizing any risks associated in farming and reduces the monitoring maintenance cost to the low. NDVI or NIR sensors are used to detect the height, density and stress in the crop and actions are then taken accordingly [39]. Multirotor UAV are used in Malaysia to monitor rice crop [40]. SenseFly Ag 360 drone technology are commercially available for the crop monitoring purpose. IAEG C35 [41] is another example for crop monitoring drone which create 3D model of area of interest for analysis.

4.5 Irrigation

Drones systems can acquire the information using multispectral, hyperspectral, and thermal cameras about the parts of the fields having adequate amount of water or which parts are dry. As soon as the crop starts to grow, the relative density and health of the crop can be described by calculating the vegetation indices. Heat signature detected in the fields by the cameras can show the amount of energy being emitted by the crops. Drones in the situations can be used to spray water in the targeted dried up areas, saving extra water going into fields and the chances of fertilizers being run off can also be reduced. Careful planning using the technology for irrigation can help the farmers save water, chemicals, and the cost associated with the activity. Drones are also used to detect possible pooling or leaks in irrigation to increase watering efficiency [42]. Zenmuse XT Thermal Imagery drones are available for this purpose. Satellite-based, proximal and remote sensing technologies are used to determine crop water requirements in order to enhance irrigation efficiency and monitoring crop water status [43].

4.6 Health Assessment

UAV assisted routine health check-ups can be done for the crop plants in order to get assessment reports. Drones can capture the images in visible and near infrared spectral ranges very closely at high resolution with the ability to fly on low altitudes and calculate the vegetation indices to discover any infection or diseases in the crops and report the issues to the farmers. Farmers can use these images to apply the remedial action in the field at detected locations and save the crop from failure or spreading the infection. Farmers can use drones to maximize their returns on crops and minimize on losses by accepting the beneficial inclusion of the technologies in agricultural. In case of the damages or losses happened to the crops, farmers can also use these images to make claims for insurance [28]. Tomato spot are detected using drones to access the health of the crop [44]. Forest health monitoring using drones are also designed [45, 46] to check on density and fire incidence.

In an IoT-based plants disease and pest prediction system was presented to minimize the excessive use of fungicides and insecticides [47]. Weather condition monitoring sensors, i.e., temperature, dew, humidity and wind speed, are used to monitor weather parameters to find a correlation between pest growths with weather. The sensors have been deployed in orchards, and data collected from these sensors are sent to the cloud. The farmer is informed about the alarming condition of the pest attack on the crops.

Generally hyper-spectral images are used to analyze crop's health and pest attack using manned or unmanned vehicles on which spectral cameras are mounted. Machine Learning (ML) techniques are used to identify the disease in the plants. Advanced Artificial Neural Networks (ANNs) are used due to their ability to learn complex structures and patterns. Using hyper-spectral images, a system was presented by Golhani et al. [48] to identify disease or pest attack in crops.

The UAV designed [49] to detect the pests and diseases periodically by capturing the images of crops with a spectral camera. Subsequently, these images will be sent to the cloud data center and stored for further data analysis. With the aid of image processing technology, these images can be analyzed to determine whether there are pests and diseases or not. Currently, a UAV usually is equipped with a smaller fuel tank, which causes the flight time to be short, so that the flight time of a UAV is generally about 30 min in China. The level of nitrogen nutrition of corn is detected by the digital camera [50]. Different growing stages of plant have different nitrogen nutrition and this help accessing the health of the crop.

5 Case Study

In the context of the technology drones are widely known as Unmanned Aerial Vehicles (UAVs), which have the capabilities to be operated remotely from the distance and fly on their own. These systems house different sensors and units to perform a variety of operations in the agriculture sector. Drones are capable to capture the high-quality images of the agricultural land from the sky, in order to help the farmers for understanding the specific issues with the crops and take some remedial actions in time. Two case studies are discussed one is drone-based challenge and services for agricultural upliftment [4] and another one of agricultural insurance claim assessment using drones and GIS [51].

Case Study 1: Drone-Based Challenges and Services for Agricultural Upliftment

India is a country with regions having diversity in terms of climate, soil, and terrain. Issues arise from the small sized fields of the farmers, where they do crops and livestock management round the year. With such current farming practices and potential to perform better, the agricultural industry is contributing a large share to country's annual growth. However, there are certain challenges to the implementation of this revolutionary technology such as quality, features, and support of the software involved, regulatory laws related to the use of UAVs, acceptance of the new technology by the farmers, performance of the drone in terms of endurance and coverage, drone purchasing cost to the farmers, conflict with manned aircraft airspace, connectivity between farmlands, and dependency on weather conditions.

The technology is promising as it offers fruitful results to all the stakeholders involved, from crop production to food processing. These stakeholders are farmers as they get real time information about the changing conditions, the people involved for bringing input products and taking output from the lands, institutions that can support with credit and insurance benefits, industry involved to provide mechanization and optimization support, people managing supply chain for adequate storage and transport facility, and the food processing industry to plan the outputs early.

Considering the challenges mentioned above, Tata Consultancy Services (TCS) has designed an in-house fully autonomous multirotor drone system [4], which is customizable in terms of the payloads with different cameras, range, and radio frequencies. The drone is suitable to carry out applications such as precision agriculture, forest plantation, wildlife conservation, etc. The TCS drones design challenges, stakeholder roles and services provided is as shown in Fig. 6.

Specifications of the typical UAV designed by TCS:

- Quad-rotor.
- Battery powered.
- Size as 2 ft × 2 ft × 1.5 ft.
- Weight as 4.5 kg.
- Payload as camera for capturing images.
- Equipped with safety features such as geo-fencing, automatic return to home.

Services provided are as follows:

(i) Precision Agriculture: In order to take actions in time, TCS uses multispectral cameras mounted on drones to take images of the lands under investigation for understanding the conditions on soil, water, nutrients level, pest incidence, to assess the overall health of the crop and estimate the crop production. The data

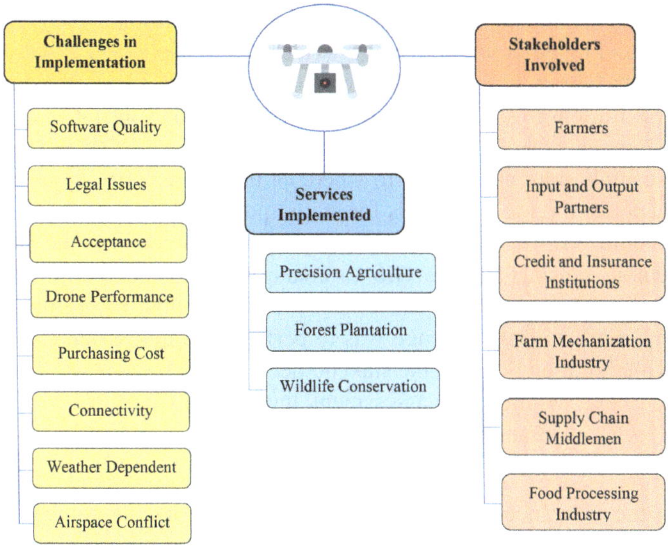

Fig. 6 Challenges, stakeholders and services for agriculture sector

captured is analyzed using the cloud-based infrastructure to get the insights as early as possible for farmer.

(ii) Forest Plantation: TCS has applied the drone technology to estimate the characteristic information for planting the forests. Drones help in capturing the images related to elevation maps, tree counts and height measurements, estimation of canopy diameters, recognition and proximity assessment of tree species. The data is analyzed using deep learning algorithms for automatic identification.

(iii) Wildlife Conservation: Drones are used in the initiatives to conserve wildlife to support healthy ecosystem. TCS employed the technology to perform the activities such as training, routine surveillance, anti-poaching operations, and monitoring of wildlife.

With so much potential to change every aspect of the farming industry, the farmers are needed to be given proper training and awareness about the technology for the deeper involvement and use. The rate at which the innovations are going on, the technology will become better accessible to the farmers in the coming years. New advancements will keep our faith in the technology, as what more can be done using drones.

Case Study 2: Drone-Enabled Cloud-Based Agricultural Insurance Claim Assessment

It is estimated according to the various agencies in the world that the food production worldwide is indicating a decline in the pattern due to the fluctuating climate conditions and natural disasters, raising the alarms for the food security to feed the population. So, it is becoming necessary for farmers to apply the risk management tools and strategies in the agriculture to avoid losses and sufferings.

The modern technology such as GIS (Geographical Information Systems) along with drones and cloud computing can provide a great support to assess the losses happened to the crops in the fields. These technologies are currently used by the insurance institutions to assess the insurance claims raised by the farmers and inform the government agencies [51].

The case study presents a scenario where an insurance company assess the hail damages happened to the corn fields for the purpose of the insurance claim assessment. The main goal of the insurance company was to combine the technologies such as remote sensing and the GIS to create a solution that can help them in keeping the assessment cost at low and make settlements in the minimum time without any conflict. The challenge was to perform the field inspection, share the data with collaboration between the stakeholders involved.

In order to execute the claim assessment, the company used the MaVinci Sirius Pro drone to the assessment. The process followed is as shown in Fig. 7 and its description as follows:

1. Firstly, the insurance claim application is received by the company from the farmer.
2. Company then sends the drone to assess the damaged fields.
3. A GPS-enabled drone captures the images of the affected farm and marks the location. The NDVI images are captured using the multispectral high-resolution camera.
4. The data is received by the company for analysis purpose.
5. Analysts combine the data with GIS technology to map the scenario in a proper manner.

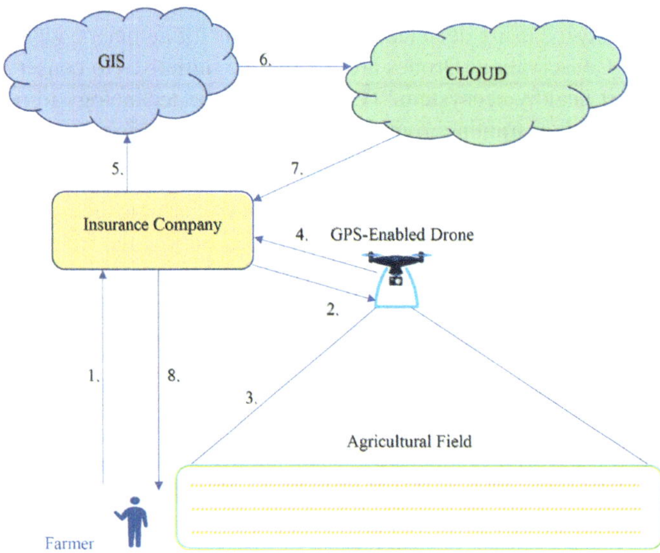

Fig. 7 Drone-Enabled Agriculture Insurance Claim Assessment Process

6. The data is sent to the cloud for large scale processing and quick analysis reports.
7. Cloud system sends the report of the insurance case back to the company.
8. After careful look on the report, the company can decide the damage and the insurance amount to be given to the farmer and settle the claim, along with information sent to government agencies.

The following are the conclusions drawn from the above case studies are that

(i) Multispectral imagery can very well serve the purpose of detecting the affected areas in agriculture sector.
(ii) A very high spatial resolution images have been captured with the drone systems to better understand the condition in the field.
(iii) Assessment using the aerial remote sensing is fast and more accurate in comparison to the current means of performing insurance claim settlements.
(iv) Combination of drone images, GIS and the cloud technology can be successful risk management solution in agriculture.
(v) All the stakeholder can be involved in the claim assessment and settlement process using the cloud and GIS based solution.
(vi) UAV with real-time applications has serious security threats at communication protocols that need to be addressed in the near future.

6 Challenges in IoD

First most challenge of using Drones/UAV is the restriction for flying that has been imposed by various countries which hinders the development of IoD. Flying of larger UAV above ground level of 21,000 ft requires clearance from air traffic management systems [4]. Drones with short flight ranges has limited field coverage. Agricultural drones are used in the open field which may have to face harsh weather environment. During monsoon and windy weather drones need more complex algorithm to maintain balance and sometimes needs to work against the wind speed. Some research project in the flight plan, utilize wind speed to their flight benefit.

IoD are subjected to additional challenges such as stabilization, payload, flight time, external disturbances and communication with ground sensors to receive data for analysis. To resolve the non-uniform spraying of pesticides due to wind speed, swarm of drones are utilized. Due to air pollution and quality of air some of the image sensor may not work properly in such cases thermal and near-infrared sensors of drones resolve these issues. Due to site specific variation in location setting, fixed set of parameters for earth observation processing are not reliable for monitoring purpose.

Data sent to cloud for analysis by IoD may have lag in the response and this lag is the major concern of IoD. In addition, data strategy and data management have its own challenges due to size of datasets generated up to 140 GB for one square kilometer with ground sampling distance of 1 cm. These voluminous data are generated by advanced image data analysis and processing.

Due to communication failure or any technical glitches data interpretation results will be different from the expected results. Mobile phone is used to upload drone image data to cloud. With limited mobile phone coverage, real-time analysis is not possible. Integration of multiple data sources from different protocol is another challenge that need to be addressed by IoD.

7 Conclusion

UAV/Drones are not only used in the military purpose but also in agricultural purpose and expected to exceed military use by 2025 to fulfil the growing population food demand. IoD are equipped with smart sensors, communication module, smart imaging system, cloud computing with big data analysis is going to increase the span of agricultural services provided by them. IoD has the potential to perform various agricultural management activities such as water management, crop management, smart farming, livestock management, and irrigation management with ease and accuracy.

Since agricultural drones needs to operate in harsh environment the selection of drone controller board, communication module, smart sensors, propeller, motor and power supply selection is crucial and it should work on low power to increase the flight duration. For the auto navigation drone, GPS are enabled to coordinate the pesticides spray over infected areas by detection and identification using NDVI. This technique reduces the wasting of water and chemicals and hence achieves the goal of precision agriculture.

Drones collect the environmental information along with space-borne, air-borne and ground-based remote sensing technologies. These techniques collect ultra-clear images, geometric shape and multi-sensor information from different altitude over agriculture field for analysis. Selection of imaging camera is also important for remote sensing which may belong to any of the category such as RGB, multispectral, hyperspectral, thermal and LiDAR.

IoD has gained popularity with wide variety of application such as soil properties analysis, planting and weed control, crop spraying, crop monitoring, remote irrigation control and health assessment. Swarm of connected drones operate in cooperative and coordinated manner to provide optimized, reliable and safe cum secure results. These wide applications will benefit millions of farmers by acquiring real-time farming information especially disaster warning and weather information.

8 Future Work

IoD technology is still not well established for large-scale application. Full-fledged IoD research and development is still required, including the development of other applications of farming such as fishing, poultry and farming enterprises. IoD technology, still needs improvement in terms of lower costs, flying times, batteries, image processing techniques, new camera designs, low and uniform volume sprayers, and different nozzle types for different crop types.

The current communication protocols still lack security mechanism to encrypt messages and this may result in serious consequences. Therefore, there is a need for new secure communication protocol which can overcome this problem. However, applying the encryption/security in turn increases the overhead and complexity which affects the overall performance and efficiency and this needs to be handled separately by designing new protocol.

Another challenge of real-time IoD is to balance between the UAV cost and the performance. High performance of the IoD with long flight time, stability, with limited interference will be costly and prevent farmers from adopting the application especially in developing countries. Moreover, farmers need time to accept this new technology and convincing them that profits are guaranteed against investment is another challenge.

References

1. Yao, H., Qin, R., Chen, X.: Unmanned aerial vehicle for remote sensing applications—a review. Remote Sens. **11**(12), 1443 (2019)
2. Puri, V., Nayyar, A., Raja, L.: Agriculture drones: a modern breakthrough in precision agriculture. J. Stat. Manage. Syst. **20**(4), 507–518 (2017)
3. Eriksson, S., Lundin, M.: The drone market in Japan. EU-Japan Centre for Industrial Cooperation (2016)
4. Sylvester, G. (ed.): E-agriculture in Action: Drones for Agriculture. Food and Agriculture Organization of the United Nations and International Telecommunication Union (2018)
5. Nayyar, A., Nguyen, B.L., Nguyen, N.G.: The internet of drone things (IoDT): future envision of smart drones. In First International Conference on Sustainable Technologies for Computational Intelligence, pp. 563–580. Springer, Singapore (2020)
6. Eastman, J.R.: Guide to GIS and Image Processing Volume. Clark University, USA (2001)
7. Wójtowicz, M., Wójtowicz, A., Piekarczyk, J.: Application of remote sensing methods in agriculture. Commun. Biometry Crop Sci. **11**(1), 31–50 (2016)
8. Walz, U.: Remote sensing and digital image processing, In: Bastian, O., Steinhardt, U. (eds.) Development and Perspectives of Landscape Ecology, pp. 282–294. Kluwer Academic (2002)
9. Khan, N.A., Jhanjhi, N.Z., Brohi, S.N., Nayyar, A.: Emerging use of UAV's: secure communication protocol issues and challenges. Drones in Smart-Cities, pp. 37–55. Elsevier, Amsterdam (2020)
10. Tsouros, D.C., Bibi, S., Sarigiannidis, P.G.: A review on UAV-based applications for precision agriculture. Information **10**(11), 349 (2019)
11. Andrew, N.: Best Drones for Agriculture 2017: The Ultimate Buyer's Guide (2017)
12. Mogili, U.R., Deepak, B.B.V.L.: Intelligent drone for agriculture applications with the aid of the MAV link protocol. In: Innovative Product Design and Intelligent Manufacturing Systems, pp. 195–205. Springer, Singapore (2020)
13. Mogili, U.R., Deepak, B.B.V.L.: Review on application of drone systems in precision agriculture. Procedia Comput. Sci. **133**, 502–509 (2018)
14. Saheb, S.H., Babu, G.S.: Design and analysis of light weight agriculture robot. Glob. J. Res. Eng. (2017)

15. UAV & Drone Propellers [Online]. Available: https://www.unmannedsystemstechnology.com/category/supplier-directory/propulsion-power/propellers/. Accessed 7 June 2020
16. Multirotor Motor Guide [Online]. Available: https://www.rotordronepro.com/guide-multirotor-motors/#visitor_pref_pop. Accessed 10 July 2020
17. Selecting a Drone Flight Controller [Online]. Available: https://dojofordrones.com/drone-flight-controller/. Accessed 15 July 2020
18. Nayyar, A.: Flying adhoc network (FANETs): simulation based performance comparison of routing protocols: AODV, DSDV, DSR, OLSR, AOMDV and HWMP. In 2018 International Conference on Advances in Big Data, Computing and Data Communication Systems (icABCD), pp. 1–9. IEEE (2018)
19. Radoglou-Grammatikis, P., Sarigiannidis, P., Lagkas, T., Moscholios, I.: A compilation of UAV applications for precision agriculture. Comput. Netw. **172**, 107148 (2020)
20. Allaw, K., Al-Shami, L.: Geographic information system-based map for agricultural management in South-Lebanon. In: 2018 International Arab Conference on Information Technology (ACIT), pp. 1–11. IEEE (2018)
21. Tripicchio, P., Satler, M., Dabisias, G., Ruffaldi, E., Avizzano, C.A.: Towards smart farming and sustainable agriculture with drones. In: 2015 International Conference on Intelligent Environments, pp. 140–143. IEEE
22. Lum, C., Mackenzie, M., Shaw-Feather, C., Luker, E., Dunbabin, M.: Multispectral imaging and elevation mapping from an unmanned aerial system for precision agriculture applications. In: Proceedings of the 13th International Conference on Precision Agriculture (2016)
23. King, A.: The future of agriculture. Nature **544**(7651), S21–S23 (2017)
24. Honkavaara, E., Kaivosoja, J., Mäkynen, J., Pellikka, I., Pesonen, L., Saari, H., Salo, H., Hakala, T., Marklelin, L., Rosnell, T.: Hyperspectral reflectance signatures and point clouds for precision agriculture by light weight UAV imaging system. ISPRS Ann. Photogramm. Remote Sens. Spat. Inf. Sci. **7**, 353–358 (2012)
25. Sagan, V., Maimaitijiang, M., Sidike, P., Eblimit, K., Peterson, K.T., Hartling, S., Esposito, F., Khanal, K., Newcomb, M., Pauli, D., Ward, R.: UAV-based high resolution thermal imaging for vegetation monitoring, and plant phenotyping using ICI 8640 P, FLIR Vue Pro R 640, and thermomap cameras. Remote Sens. **11**(3), 330 (2019)
26. Ribeiro-Gomes, K., Hernández-López, D., Ortega, J.F., Ballesteros, R., Poblete, T., Moreno, M.A.: Uncooled thermal camera calibration and optimization of the photogrammetry process for UAV applications in agriculture. Sensors **17**(10), 2173 (2017)
27. Narváez, F.J.Y., del Pedregal, J.S., Prieto, P.A., Torres-Torriti, M., Cheein, F.A.A.: LiDAR and thermal images fusion for ground-based 3D characterisation of fruit trees. Biosyst. Eng. **151**, 479–494 (2016)
28. Gupta, S.K., Kumar, S., Thombare, P.B.: Drones for future agriculture. In: Sharma, S. (ed.) Advances & Challenges in Agricultural Extension & Rural Development, pp. 41–62 New India Publishing (2009)
29. Gholami, M., Sharifi, Z., Karami, Z., Haghighi, S., Minouei, S.F., Zema, D.A., Lucas-Borja, M.E.: The potential impacts of soil sampling on erosion. Int. J. Environ. Sci. Technol. (2020)
30. Issaka, S., Ashraf, M.A.: Impact of soil erosion and degradation on water quality: a review. Geol. Ecol. Landscapes **1**(1), 1–11 (2017)
31. Huuskonen, J., Oksanen, T.: Soil sampling with drones and augmented reality in precision agriculture. Comput. Electron. Agric. **154**, 25–35 (2018)
32. Stone, E.: Drones spray tree seeds from the sky to fight deforestation. National Geographic (2017)
33. Shamshiri, R.R., Hameed, I.A., Balasundram, S.K., Ahmad, D., Weltzien, C., Yamin, M.: Fundamental research on unmanned aerial vehicles to support precision agriculture in oil palm plantations. Agricultural Robots-Fundamentals and Application (2018)
34. Mulla, D.J.: Twenty-five years of remote sensing in precision agriculture: key advances and remaining knowledge gaps. Biosyst. Eng. **114**(4), 358–371 (2013)
35. Yao, L., Jiang, Y., Zhiyao, Z., Shuaishuai, Y., Quan, Q.: A pesticide spraying mission assignment performed by multi-quadcopters and its simulation platform establishment. In: 2016

IEEE Chinese Guidance, Navigation and Control Conference (CGNCC), pp. 1980–1985. IEEE (2016)
36. Faiçal, B.S., Costa, F.G., Pessin, G., Ueyama, J., Freitas, H., Colombo, A., Fini, P.H., Villas, L., Osório, F.S., Vargas, P.A., Braun, T.: The use of unmanned aerial vehicles and wireless sensor networks for spraying pesticides. J. Syst. Architect. **60**(4), 393–404 (2014)
37. Ganesan, R., Raajini, X.M., Nayyar, A., Sanjeevikumar, P., Hossain, E., Ertas, A.H.: BOLD: bio-inspired optimized leader election for multiple drones. Sensors **20**(11), 3134 (2020)
38. Monitoring Locust Swarms Using Drones [Online]. Available: https://www.thegeospatial.in/monitoring-locust-swarms-using-drones. Accessed 7 June 2020
39. Veroustraete, F.: The rise of the drones in agriculture. EC Agri. **2**(2), 325–327 (2015)
40. Norasma, C.Y.N., Sari, M.A., Fadzilah, M.A., Ismail, M.R., Omar, M.H., Zulkarami, B., Hassim, Y.M.M., Tarmidi, Z.: Rice crop monitoring using multirotor UAV and RGB digital camera at early stage of growth. In: IOP Conference Series: Earth and Environmental Science, Vol. 169, no. 1 (2018)
41. Giordan, D., Adams, M.S., Aicardi, I., Alicandro, M., Allasia, P., Baldo, M., De Berardinis, P., et al.: The use of unmanned aerial vehicles (UAVs) for engineering geology applications. Bull. Eng. Geol. Environ., pp. 1–45 (2020)
42. DeBell, L., Anderson, K., Brazier, R.E., King, N., Jones, L.: Water resource management at catchment scales using lightweight UAVs: current capabilities and future perspectives. J. Unmanned Veh. Syst. **4**(1), 7–30 (2015)
43. Cancela, J.J., González, X.P., Vilanova, M., Mirás-Avalos, J.M.: Water Management Using Drones and Satellites in Agriculture 2019
44. Patrick, A., Pelham, S., Culbreath, A., Holbrook, C.C., De Godoy, I.J., Li, C.: High throughput phenotyping of tomato spot wilt disease in peanuts using unmanned aerial systems and multispectral imaging. IEEE Instrum. Meas. Mag. **20**(3), 4–12 (2017)
45. Dash, J.P., Watt, M.S., Pearse, G.D., Heaphy, M., Dungey, H.S.: Assessing very high-resolution UAV imagery for monitoring forest health during a simulated disease outbreak. ISPRS J. Photogrammetry Remote Sens. **131**, 1–14 (2017)
46. Smigaj, M., Gaulton, R., Barr, S.L., Suárez, J.C.: UAV-borne thermal imaging for forest health monitoring: detection of disease-induced canopy temperature increase. Int. Arch. Photogrammetry Remote Sens. Spat. Inf. Sci. **40**(3), 349 (2015)
47. Shafi, U., Mumtaz, R., García-Nieto, J., Hassan, S.A., Zaidi, S.A.R., Iqbal, N.: Precision agriculture techniques and practices: From considerations to applications. Sensors **19**(17), 3796 (2019)
48. Golhani, K., Balasundram, S.K., Vadamalai, G., Pradhan, B.: A review of neural networks in plant disease detection using hyperspectral data. Inf. Process. Agric. **5**(3), 354–371 (2018)
49. Gao, Demin, Sun, Quan, Bin, Hu, Zhang, Shuo: A framework for agricultural pest and disease monitoring based on internet-of-things and unmanned aerial vehicles. Sensors **20**(5), 1487 (2020)
50. Korobiichuk, I., Lysenko, V., Opryshko, O., Komarchyk, D., Pasichnyk, N., Juś, A.: Crop monitoring for nitrogen nutrition level by digital camera. In: Conference on Automation, pp. 595–603. Springer, Cham (2018)
51. Case Study: Combining Remote Sensing & Cloud technology is the future of farming [Online]. Available: https://www.geoawesomeness.com/case-study-combining-remote-sensing-cloud-technology-future-farming/. Accessed 15 July 2020

Towards a Smarter Surveillance Solution: The Convergence of Smart City and Energy Efficient Unmanned Aerial Vehicle Technologies

Rachna Jain, Preeti Nagrath, Narina Thakur, Dharmender Saini, Nitika Sharma, and D. Jude Hemanth

Abstract Recent research illustrates that surveillance is in a development phase, massively increasing overall security solutions with the usage of advanced computer vision and Unmanned Aerial Vehicles (UAVs). Automation technologies are making significant progress upon its global Smart City Surveillance sector, coupled with high-growth sub-segments such as Video Surveillance, Robotic Aerial security, Drones surveillance systems, etc. Drone fleets augment traditional security technology solutions for human security guards or fixed cameras by powerful legitimate approaches to fully autonomous action strategies. Obtaining an electronic eye-in-the-sky is as similar as it appears at quite a magic solution for security detection, recognition, and monitoring. Drone surveillance systems have made significant strides in recent years, such as leveraging advanced computer vision and cameras to provide 24/7 operational monitoring capabilities and autonomous deployment. However, there are still many challenges to the drone to include data obtained during surveillance activities with cognitive capacity and machine learning computing power, drone recharging under adverse weather conditions, and data security and safety. Another research area identified is to improve the energy conservation in UAVs to enhance the flight time and

R. Jain (✉) · P. Nagrath · N. Thakur · D. Saini · N. Sharma
Department of Computer Science and Engineering, Bharati Vidyapeeth's College of Engineering, New Delhi, India
e-mail: rachna.jain@bharatividyapeeth.edu

P. Nagrath
e-mail: preetinagrath1@gmail.com

N. Thakur
e-mail: narinat@gmail.com

D. Saini
e-mail: dsaini77@gmail.com

N. Sharma
e-mail: nitika.sharma.2792@gmail.com

D. J. Hemanth
Department of Electronics and Communication Engineering, Karunya University, Coimbatore, Tamil Nadu, India
e-mail: jude_hemanth@rediffmail.com

R. Krishnamurthi et al. (eds.), *Development and Future of Internet of Drones (IoD): Insights, Trends and Road Ahead*, Studies in Systems, Decision and Control 332,
https://doi.org/10.1007/978-3-030-63339-4_4

save its power during flying or hovering. This chapter proposes an energy-efficient framework for autonomous intelligent drones for Smart City surveillance. The underlying motivation for this chapter is to identify methods that allow the development of new strategies to the Internet of Drones frameworks and the effective utilization of drones in surveillance for smart cities highlighting the different areas of research. Furthermore, this chapter concludes with the recommendations and suggestions for future research in Smart City Surveillance.

Keywords Smart cities · Unmanned aerial vehicles · Surveillance · Energy-efficient drones · Tilt-rotor drones

1 Introduction

The fundamental inception of a smart city primarily aims to augment the lifestyle of its residents by providing efficacious infrastructure and enhanced security. The ideation of "*The use of discrete new technology applications such as RFID and Internet of things through a more holistic conception of intelligent, integrated working that is closely linked to the concept of living and user-generated services*" is accepted as the outline of the smart city by the European Network of Living Labs and the European Platform for Intelligent Cities. In general, the smart city aims to achieve the solutions for the goals of the future city derived from the trends of Information and Communication Technology (ICT) [1]. Depending on the degree of growth, resources, and desires of the city dwellers, and the capacity to change and reform, the idea of Smart City differs from region to region. For sustainable development, the urban planners preferably aim at creating the entire urban ecosystem, which consists of four pillars for holistic, social, physical, institutional, and economic growth, to meet the ambition and the needs of the people [2]. This will be a long-term goal, and communities should operate progressively to build this robust network and steadily connect to 'smart' levels [3]. Figure 1 shows the adoption of the idea of smart cities across the world.

The cities have been ranked based on scores obtained in ten categories namely: budget, talent readiness, smart policies, vision, financial incentives, people centricity, track records, leadership, support programs, and innovation ecosystem. These ten elements offer local officials a wide set of criteria when formulating their smart city strategies [4].

Recent studies have elaborated on how integrated monitoring solutions will enhance the establishment of law and order and improve the security in smart cities. The modern monitoring solutions intelligently interconnect the entire infrastructure of the city from public places to homes [2]. The whole data from various monitoring devices is gathered at a centralized control room that can be traced and prompt steps to counter any security threats. The authorities are notified, via pop-ups, to any unusual activities or actions, such as unexplained luggage, reckless driving,

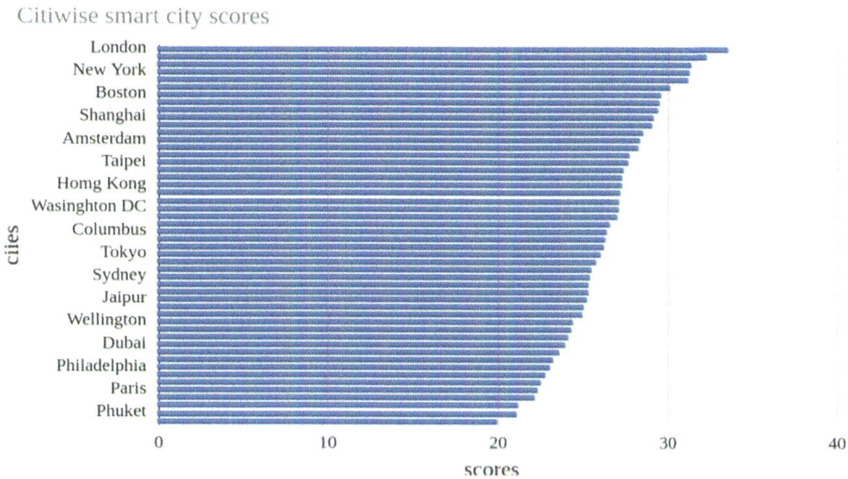

Fig. 1 City-wise smart city score across the world

etc. The data obtained in real-time allows officials to stop the violence or apprehend the perpetrators as they escape. Although vigilance is a substantial benefit of video surveillance solutions, it might also play a significant role in delivering truly sophisticated public infrastructure in the smart cities [5]. Taking the example of a traffic monitoring system, where a smart video surveillance system would inspect and generate real-time traffic reports, capable of detecting any traffic congestion and would immediately advise drivers suggesting alternate routes. Many linked transport networks (rail, subway, air) within a city or area, also at the national level, may even adapt and boost their productivity appropriately and benefit instantaneously from the organized network.

Some of the examples that cite the challenges faced by city management that triggered the idea of building a smart surveillance system are listed below:

- **A city hall region challenged with a crime**: City leaders in Hayward, California, discovered while working on improving their downtown district that the region surrounding the city hall required extra protective precautions to combat drug trafficking, robbery, and other violent crimes taking place [6].
- **Unlawful dumping**: California's Bay Area Region encountered a major issue: around 200 illicit locations where citizens dumped discarded appliances, contaminants, and trash. The dumping cost the community nearly 1 million dollars a year, with 50 calls a day received from people [7].

Monitoring through cameras emerged as an efficient solution to the problems faced by the city. The infallible electronic eye is almost as sensitive as it pertains to detection, identification, and mitigation in the scope of the defense. In addition to real-time monitoring, defense and law enforcement agencies utilize video camera images to assist facilitate post-event operations, offer visual documentation to prosecute murders and analyze multiple-source data information to establish lawsuits. Installing security devices culminated in nil service calls in the areas because the suspects realize they are indeed being looked at. The cessation of illicit dumping has also greatly improved the quality of life for residents.

- **Strengthening public security**: Mankato County, Minnesota, has adhered to the latest public safety and property security protocols and managed to reduce the prevalence of public security violations, and also episodes of the destruction of property [8]. A major widely reported incident of George Floyd killing that took place in Minnesota was caught by security cameras throughout the area, presenting evidence to law enforcement that helped in the constitutional proceeding [9].
- **Small Police Force on wide college premises**: Police Department at San Jose State University (SJSU) is fully accountable for a 19-block campus in central San Jose, California, which houses around 33,000 faculty members, staff, and students [10]. The university installed portable HD video monitoring systems with gunshot detection because of increased complexity to enable the Police department to track the entire premises efficiently and to be able to take rapid action in an active shooter scenario.

Video is constantly emerging as a primary Protection and Security sensor. Through advanced analytics it boosts the efficiency of conventional surveillance and provides strategies to most of the primary problems in safety management, organizations can optimize the current monitoring systems. Video analytics enables reliable, constructive, and proactive monitoring, allowing somewhat more secure and accurate video analysis.

Smart Cities are intended to improve operational efficiency and effectiveness, it may potentially pose dangerous threats to citizens and authorities when surveillance is overlooked [11]. Even for most adroit officers surveillance and security constitute a substantial challenge. It's quite probable that an operator will eventually lose attention or get overwhelmed while watching continuous video streams, thereby losing a piece of critical information or event on display. Often the human eye can't recognize the crucial evidence caught on camera, no matter how vigilantly the investigating officer is monitoring the live video feed. Another daunting obstacle is the amount of manpower and time that will be expended to view the footage [12]. The video proofs from multiple locations, cameras, and days can be gathered and then checked particularly in the post-event survey. Such a vast volume of visual data turns out into manual inspection hours. The officers must limit the amount of video evidence they consider, to manage time and resources effectively. Nevertheless, running the risk of losing and ignoring critical pieces of evidence and still bound to man-hours to viewing video footage. The officers must limit the amount of video evidence they consider, to manage time and resources effectively. Nevertheless, running the risk of losing and ignoring critical pieces of evidence and still bound to man-hours to view video footage. With an overarching aim to minimize violence, ensuring protection and enhancing the standard of life for people in rapidly developing cities, there are several promising aspects to community surveillance approaches, specifically as it relates to particular problems.

Driven by artificial intelligence, technology is storing all of the goals and objectives provided by law enforcement around the globe [13]. The protection team may also leverage AI-driven video analytics technologies in situations where there is no identified perpetrator or there is no guidance to lead the video search. An AI-powered video analytics algorithm will identify, retrieve, and recognize video instances to construct a standardized database archive from unstructured video evidence. This retrievable database lets observers grasp all of the video activities easily and comprehensively. Computers are trained in video surveillance systems to identify and segregate video objects and evaluate them based on their patterns and features to classify data so that it may assist in strategic decisions making and security preparedness.

Current surveillance is done through static infrastructure-based sensing applications demanding an enormous amount of energy contributing to various expenses and network failures. These challenges can be addressed with the use of on-demand nodes, including drones, etc. Conventional monitoring and surveillance methods, such as fixed security cameras can't compete with UAVs monitoring, particularly when it comes to distressful situations and tough weather conditions [12]. Static cameras demand an on-site investigation by the security personnel. This can be time-consuming and place the responding individuals in dangerous situations. Drones can significantly minimize the risk factors involved in security and surveillance exercises. Helicopters however can be seen as an alternative to static surveillance method but this can be exorbitantly expensive. UAVs majorly were used for military applications. Lately, these are deployed in various civil applications such as environmental protection, traffic monitoring, and public safety, etc.

UAVs are finding its wide application in smart cities, particularly for surveillance because of the added advantages that it offers to the security methods [14]. Some of the areas of deployment are listed below:

- Perimeter control
- Risk estimation
- Traffic Surveillance and administration
- Event Security
- Maritime Surveillance
- Railway inspection
- Anti-poaching Operations
- Remote Area Inspection
- Inspections

UAVs are highly responsive than on-foot responders. These help investigate threats through aerial monitoring by offering assistance to evaluating teams to make the best decisions during critical circumstances. Unmanned Aerial Vehicles unified to artificial intelligence is soaring growth in drone surveillance swiftly framing its mark in the global security sector. UAVs are capable of providing instant alerts to the monitoring squad on the detection of any threats and anomalies [15, 16]. Surveillance through drones leaves no blind spots providing an all-encompassing coverage in contrast to fixed CCTVs. Longer flight time of a drone in air promises sweep surveillance of a large area.

Smart cities are the way forward and technology is expected to play a greater role in driving the government's plan to make their country future-ready from intelligent management of traffic to using Artificial Intelligence (AI) to monitor crowd density. Further drones and UAVs will be deployed for surveillance for cities and managing security.

Besides the overarching aim of minimizing violence, ensuring protection and enhancing the standard of life for people in rapidly developing cities, there are several pragmatic ramifications of urban surveillance systems when it comes to a certain set of problems.

Potential Benefits of drones equipped with AI in smart cities

- Drones with AI will assist local planners to identify the pattern of movement of people in the city that will allow them to bring more focused developments [13].
- Before getting on the scene, fire department services can discern the certitude of the fire emergency.
- Police forces have the option to deploy a drone to validate a condition and collect license plate video before calling out personnel that would save them time.
- A huge volume of data collection will be possible allowing the compilation, research, and review to be carried out in useful order to facilitate the public.
- The need for public protection tasks in smart cities is expanded with Drones.

Community infrastructure will become informed, more responsive, creative, and accessible to everyone with a growing population, and that is the Smart Cities' key goal. It mostly comprises the security factor which renders information monitoring openly highly relevant. The network of video monitoring enhances the protection of the city and a safe method of traffic management. The video monitoring network will reduce street crime by up to 30% [18].

It is undeniable to presume that surveillance systems will be deployed at every node of the network and considering the enormity of smart cities the coming years will see an upsurge in need of video surveillance technology [17]. Some of the applications of artificial intelligence in the smart city are shown in Fig. 2.

Smarter Security for Smart Cities requires intelligent surveillance that includes autonomous cameras equipped with built-in algorithms, which can itself identify and infer many possibilities without the requirement for a back-end server to contribute to results. Currently deployed surveillance equipment constitutes sensors that can detect a car, track crowd, and can interpret the license plate of a car in the traffic.

This chapter is structured in the following order. The work is essentially concentrated on the role of drones in smart cities. This also describes various developments that are used in developing an energy-efficient smart city framework. Section 2 presents a comprehensive literature review discussing the techniques and technologies used in UAVs. Section 3 proposes the energy management framework for UAVs. With the aid of case studies in Sect. 4, the application of UAVs in smart cities is further illustrated. There are, however some challenges in surveillance addressed in Sect. 5. The conclusion is given in Sect. 6.

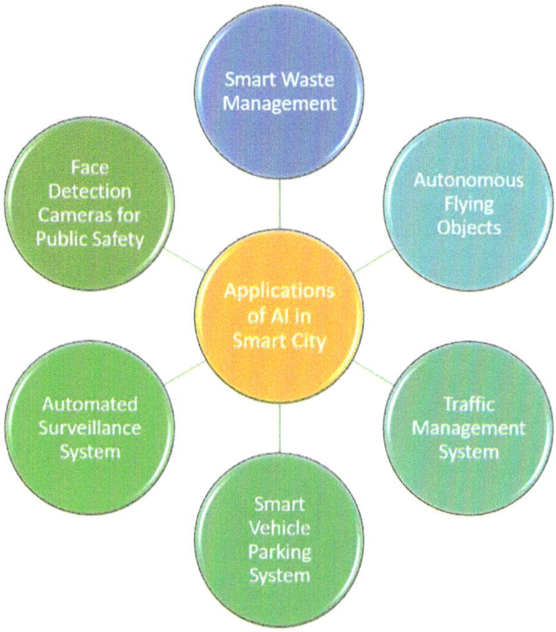

Fig. 2 Applications of AI in smart city

2 Related Work

2.1 Smart City and Challenges in Surveillance Use Case

The Smart City idea was developed in the 1980s as a new solution to urban history management and has evolved rapidly in subsequent times [19, 20]. The concept of smart cities derives from a modern vision of a city where city officials, companies and citizens utilize technology to improve as well as strengthen their place throughout the emerging service-based economy, reinforce their engagement and connectivity, build new opportunities for the local population and increase living standards [21]. The concept is also focused on the idea of the usage of smart technologies to enhance living standards in cities, developing digital technology and infrastructures through the usage of computers in general in metropolitan environments.

The essential elements of development are social, institutional, physical, and economic infrastructure. These are the four pillars that form a complete Smart City infrastructure includes home automation, a healthy environment with good climatic conditions, enhanced disaster recovery management system, manufacturing, better transportation facilities, best healthcare support, more security, and an energy-efficient infrastructure as shown in Fig. 3.

A smart city is a metropolitan region that utilizes various forms of electronic sensors to acquire and analyze data to efficiently monitor and maintain properties, capital, and services in return for optimizing community-wide operations. It covers

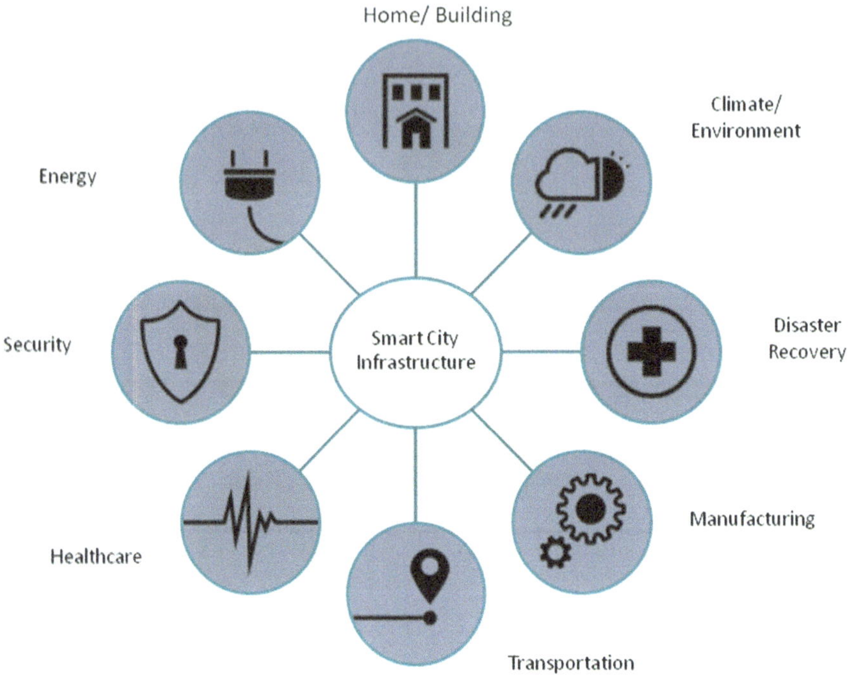

Fig. 3 Smart city infrastructure representation

data obtained from people, digital facilities such as smartphones, houses and property being stored and evaluated for tracking and maintaining clinics, libraries, classrooms, traffic and transport structures, waste collection systems, water delivery networks, power plants, the information technology industry and other municipal services [22, 23]. IBM describes Smart City as the usage of information and communication technologies to evaluate and implement key core knowledge structures in operating communities. At the same time, smart cities will adapt intelligently to various types of needs, including everyday life, environmental security, public safety, and community infrastructure, industrial and commercial activities [24]. In other terms, Smart City can be described as a policy that applies to various regions and achieves insightful and effective urban governance. This may also be seen to be an efficient blend of strategic design principles, smart building models, smart management strategies, and smart development approaches. Through managing the data grid of urban geography, land, landscape, political, environmental, social and other systems, as well as data and knowledge technology and public infrastructure usage and the basic infrastructure system, smart public management and services can be accomplished, thereby facilitating the more effective, comfortable and peaceful functioning of modern cities. The city becomes smart, according to A. Caragliu et al. [25], "when investments in human and social capital and traditional and modern information and communication technology (ICT) fuel sustainable economic growth and high quality of life through participatory governance, with a wise management of natural resources". S. P. Mohanty describes Smart City as an area where conventional networks and services are simply made more versatile, more effective, and more sustainable by using information, digital, and telecommunications technology to enhance their operations for the benefit of their citizens. They are greener, cleaner, quicker, and more welcoming [26].

Through managing the data grid of urban geography, land, landscape, political, environmental, social and other systems, as well as data and knowledge technology and public infrastructure usage and the basic infrastructure system, smart public management and services can be accomplished, thereby facilitating the more effective, comfortable and peaceful functioning of modern cities [27, 28]. It helps to track the activities taking place in the city and corresponding situations that will take place in the city. ICT is used to increase the productivity, interactivity, and effectiveness of public facilities, reduce costs and usage of energy, and promote engagement between citizens and government [29]. Smart city networks are built to control public movements and promote real-time responses [30]. A smart community will also be best prepared to respond to issues and threats than a mere 'transaction-based' relationship of its citizens [31, 32]. Nevertheless, the concept itself remains vague to its details and hence open to a lot more definitions. The Web plays an important role in knowledge exchange, collaboration, and the collection, data storage, and study of Smart Cities and digital networking to tackle multi-faceted and cross-domain problems impacting sustainable cities. The advent of the Internet of Things and the rapid acceptance of Web technology in metropolitan areas have demonstrated that Internet-based innovations can address social challenges effectively. However, systemic solutions to open issues will tackle the intrinsic complexity of urban environments [33]. Smart city networks contain levels of awareness, network layers, and device layers, all of which will render the urban environment progressively appreciable and observable, increasingly integrated and interoperable and increasingly smart [34].

In recent years, Surveillance Systems have evolved from being manned to automated and self-capable. For instance, automated video control applications in video monitoring technologies carry over traditional closed-circuit television (CCTV) cameras. Video surveillance technology enables remote control by utilizing video cameras in a specified circumscribed region. The usage of video monitoring devices would also allow the police to further solve certain societal problems and crimes [35]. The current video monitoring approach is typically based on the standard CCTV device. A typical CCTV system consists of a surveillance camera network covering the cameras' whole field of view. Such devices detect the existence of a scene automatically and video streams are transmitted to the central location for processing in one or more computer displays and analyzed by security officers or captured for future surveillance or review. [36]. Thanks to further surveillance and safety concerns, more and more devices have been mounted, creating large quantities of video data resulting in their storing problems on the one side and, on the other side, a substantial amount of processing time is required when the police or security forces decide to investigate it. Besides, this rise in the number of cameras used to monitor the incurring costs of the supervision staff would usually enable one person to display four cameras with fair precision at a time when detecting events [37]. Such issues created fundamental questions about the impact of the current CCTV-centered video surveillance systems in security, protection, and public operations. The implementation of automatic video monitoring systems was important to address these challenges, as specialists all over the world worked diligently to build surveillance systems that are simple to deploy, reliable, usable, easy to manage and ready to operate with real-time video-scan processing to produce automatic real-time and accurate recovery warnings. Automatic video surveillance is a quickly evolving field and has been particularly relevant in recent years for the academic environment because of their ability to track smart video processing algorithms more effectively and reliably. A remote video surveillance system benefits in contrast to a traditional CCTV-based network by having a self-connected computer that can be used to evaluate the scene in real-time [38] and can intelligently determine when to produce an alarm for suspicious behavior, what to store for further review, what to communicate and what to monitor. Such technologies minimize the expense of human capital to track camera performance and remove the risk of human error in analyzing multiple video streams and making objective decisions. In contrast, the automated video-monitoring device provides features produced from pictures taken or a high-level summary of the scene that flows through the automatic decision-making panel while the traditional CCTV video monitoring offers recorded images or video feeds that involve a high output spectrum. Many research projects for the design of automated video surveillance systems have been funded worldwide over the last few years. There are several commercially available surveillance devices on the market [39]. Many of them are costly and have been installed as a separate software framework that is connected to many cameras—for example, the new S3 IBM surveillance system [40]. Many automatic video monitoring devices are equipped to pick and monitor motion-based objectives on a video stream. Therefore, two main components apply to most automatic video surveillance systems: an object detector and an object-tracking device. Architecturally, they may be further separated into various categories, such as the development of a scenario, area of detection, object tracking, camera input, warning decision generation, and the results display [35]. Automated video monitoring devices are part of the Intelligent Video Monitoring system.

Intelligent video monitoring upgrades conventional passive security technologies by automated incident identification and recording, scenario interpretation, and visual occurrence indexing/retrieval. Visual inspection methods have implemented a broad variety of technologies in the fields of access management, human-specific diagnosis, anomaly detection in the scientific sector, and business as well as the government [41, 42]. Wide video monitoring technology programs have driven the creation of practical visual control systems. Successful video monitoring technologies such as the application of machine vision, device architecture, and connectivity methods have been incorporated in combination with the Annotated Intelligent Surveillance and Optimized Recovery System [43] and IBM's Smart Surveillance System [44]. As a control device of the third generation surveillance system (3GSS) [45] concept, recent work in Intelligent Visual Surveillance (IVS) in a wide field is the subject of visual surveillance. The work findings in the IVS can be separated mainly into picture processing and large field tracking techniques. The purpose of the picture analysis is to derive high-level details from a complex visual event scenario. Object identification often includes motion tracking, shape awareness, control, and human interpretation. Current imaging experiments concentrate on accurate computer vision methods such as target recognition in occlusion and non-rigid deformation conditions, human motion interpretation behavioral understanding, and motion identification in circumstances with shifts in light and temperature. Wide area management approaches to expand the field of detection to a wider region [46]. Several sensor monitoring and co-operative camera systems are currently the focus of work to extend the area of surveillance. Specifically, camera configuration and camera integration approaches have been established utilizing numerous sensor controls to reduce redundant camera installations. Data processing strategies are critical challenges in the network of mutual cameras to resolve occlusion concerns and expand the area of surveillance.

2.2 Unmanned Aerial Vehicles

UAVs widely identified for a drone is an aircraft without an onboard professional operator and a form of an autonomous, and an unmanned aerial vehicle. UAVs are a feature of the UAS, which comprises a UAV, a land-based controller, and a communication network. UAV flight may be performed at different autonomous stages, i.e. either remotely controlled by a human pilot or onboard independently utilizing computers. [47]. The major advantage and benefit are that there is a line-of-sight (LoS) connectivity provided to users which can consequently contribute to rate enhancement and coverage. Also, they are easily implementable and deployed quickly as well as efficiently, because of their endurance and mobility to improve their quality of service (QoS) and facilitate cellular networks. In temporary hotspots or events like outdoor events, sports grounds and stadiums, aerial base stations based at UAV can be implemented to boost wireless network connectivity and capacity. Furthermore, UAVs can be designed to use in public safety situations to assist emergency relief efforts, as well as communications in case damage, is done to traditional terrestrial networks [48]. The important use of UAVs is in the Internet of Things (IoT), where the systems have little transmission power therefore communication between the systems might not be possible for larger distances. To resolve this issue, a UAV provides a means of collecting and transmitting the IoT data from one computer to the intended receiver [49, 50]. Last but not least, the deployment of UAVs is highly beneficial in countries where the construction of complete cellular infrastructure is not affordable and expensive, since it eliminates the necessity of cables and towers. A range of technological issues must be tackled to properly utilize the benefits of UAV deployments which consist of channel modeling, energy efficiency, performance analysis, and optimum deployment [51, 52]. The air-to-ground mapping of a system is a very significant recent research on UAV connectivity. In [51, 52], the possibility of a Line Of Sight (LOS) for air-to-ground communication was calculated by the elevation angle and the average height of the clustered urban buildings. The air-to-ground path failure model has been further investigated [53, 54]. The air-to-ground channel characteristics are strongly contingent upon the height of the base stations, because of path loss and shadowing. A. Hourani et al. [55] derived the optimal altitude to address the UAV deployment challenge, enabling optimum coverage radius of a single and static UAV. Yet when a route failure with a certain level is measured, the authors have clearly described a deterministic coverage and did not consider the probability of coverage. The study of M. Mozaffari et al. expands the findings to the situation of two UAVs while considering UAV interference [56]. J. Kosmerl and A. Vilhar researched the best positioning of UAVs for public safety communications to enhance the efficiency of the range and coverage [57]. Although, the findings are based on simulations, and there is no substantial review. Furthermore, K. Daniel and C. Wiefeld addressed the application of UAVs to complement already existing cellular infrastructure that gives a general overview of practical considerations to incorporate UAVs into cellular networks [58]. S. Rohde along with C. Wietfeld proposed another utilization of UAVs to combat cell congestion and cellular network outages [59]. The earlier study, however, does not include any research on UAV coverage efficiency and their optimum deployment methodologies. Z. Han et al. discussed how UAVs can be optimally relocated to improve ad hoc network connectivity [60]. However, they focused only on an ad hoc network and suggested that the UAV had full node location knowledge. F. Jiang and A. L. Swindlehurst extracted the UAVs with several antennas that provide the optimal direction and moving into consideration of static ground users [61]. E. Yaacoub. and O. Kubbar recommended that cell downlink data communications for various UAVs be implemented and integrated effectively [62].

2.2.1 Use Cases of UAVs

The primary work of UAVs is feature extraction. The key aim of the extraction functionality is to obtain descriptive characteristics drawn from raw data obtained via UAV cameras evaluation. Within the field of robotics, feature extraction systems that were CNN based models have primarily used in object recognition [63–65] and scene classification [47, 48]. Recent developments concerning object identification have combined object detection techniques inside the same CNN paradigm, by the regression of the bounding box and the classification of objects [66, 67]. Unsupervised feature learning for object recognition has been implemented in [68], with fewer criteria for manually labeled training results, which can be an incredibly time-consuming and expensive method to access. O. A. B. Penatti has demonstrated that the trained characteristics of pre-trained CNN models have been able to accurately make generalizations to those in which they have been practiced, even in significantly different domains, For e.g., aerial image recognition. The Parrot AR quadrotor drone was also used for scene classification in [69] where a CNN 10 layer was used to classify a three-class feedback image of the Forest Trail, both of which reflected actions to be taken to maintain the aerial robot traceable. The topic of automated detection and recognition of plywood objects has been discussed in [70]. This solution consists of a cascade of CNN-centered classifiers based on the Nvidia Titan X, utilizing 24 M pixel RGB images from the Nvidia Jetson TK1 mounted on a UAV equipped arm. By embedding the detection step on processor cores, the ADLC algorithm has been modified to enable the GPU to concentrate on classification tasks.

Images taken from UAVs utilizing deep learning strategies have attained significant importance in tracking and search as well as rescue operations including UAV path monitoring [71], terrorist identification [72] and supporting search operations with UAV imagery in critical environmental conditions such as avalanche and floods [73]. The famous Inception model was used in both situations [74]. M. Bejiga et al. used the Inception model to identify potential survivors by making use of the Support Vector Machine (SVM) classifier, while A. Sawarkar et al. used a transfer-learning strategy to fine-tune the Inception Network to recognize possible perpetrators.

Planning and Situational Awareness is yet another field that includes tasks such as navigation and path manipulation. In these tasks, the programming is not manually coded as the problems are complex so the robot or model is required to learn and eventually train from its environment. Situational Awareness tasks require the robots to conFig. the state in which they are present corresponding to the situation they are in. These tasks can include mapping, robot state estimation, and self-localization. One of the examples can be traffic detection. Traffic detection has been implemented either with stationary ground cameras or on dynamic UAVs. It plans the path as the route for the UAVs to observe the traffic as cooperation between the UAVs is necessary if the sensors are installed [75–77]. The DARPA Grand Challenge featured environment-observing sensors on mobile unattended ground vehicles (UGVs) [78]. Mobile sensing has been used for orientation, simultaneous position, and mapping (SLAM) tracking of the environment or other targets [79–81]. Present work consists of Mobile-Ad Hoc Networks (MANET) as well as controllable bots, from which several efforts are made, like approaches to theoretical entropy in information [82]. Another example of path planning is rescue missions. J. Delmerico et al. used deep learning to explore the traversable path [83]. The research reveals that the UAV investigates and tracks the world aiming to locate a route that can be reached by a land robot, focusing on reducing the maximum launch period for both direction and exploration. A CNN for terrain classification is proposed for mapping the terrain and finding a reachable path. Training is performed on the spot instead of using a pre-trained CNN, allowing the classifier to be trained on the spot with the terrain at the disaster site. Another study by F. Aznar used Convolutional Networks for visual navigation of the UAV with map references [84]. It proposes a controlled global motion plan that suggests the measures to be taken using a potential function, given a position on the map. CNN's objective is to study to map from pictures to position-dependent behavior and then learn from them. The procedure will be similar to doing image registration and then generating control behavior given by the global motion system, but the operation is taught to be conveniently interpreted by CNN, providing improved outcomes for traditional image registration approaches.

2.2.2 Applications and Challenges

While UAVs originated mainly in military applications, its utilization is increasingly growing to include industrial, research [85–88] recreational, agrarian, and other platforms [89] like police and monitoring, product delivery, aerial photography, infrastructure inspections, piracy and recreational competitions like drone racing. These days, the most effective agricultural research is based on the installation of electronic sensors and tools for the remote measurement of seed and soil products in near-real-time. While recent changes have been made to the spatial accuracy of other satellite detectors, like Quickbird and Ikonos, certain major issues still rely on the accuracy of repetitive measurements during the crop cycle [90]. Over the last ten years, the progress of the UAV systems, which are known to be limited in scale, have provided modern agricultural practices and tracking system worthy of being used. providing high-resolution images on time, particularly where tiny productive regions need to be monitored [91]. Unmanned Aerial Vehicles benefit from a thriving global drone industry. UAV applied in a variety of ways, for instance, security, monitoring and surveillance, search and rescue, and studying sports. Unlike conventional security cameras, UAV with a moving camera naturally has many benefits, such as ease of deployment, high mobility, huge view as well as uniform scale. A UAV vision system typically improves a suite of sensors that could consist of Inertial Navigation Sensors (INS), optical compass and sonar, laser range finders, and Global Positioning System (GPS) [92]. Due to its organized nature which recently became a major research subject the Autonomous Landing function is well suited for vision dependent state estimation and regulation [93].

Vision-based Robot Control was an important research topic [94]. For unmanned helicopter flight that employs sight throughout the circuit to control the helicopter, a vision-enhanced navigation device is addressed in [95]. The visual odometer [96], that delivers precise directional knowledge (position and velocity) coupled with inertial measurements, is a groundbreaking vision-based tool used in autonomous helicopter flight. The early solution to autonomous landing [97] eliminated the problem of landing from vision-based control.

2.3 Internet of Drones

The Internet of Drones is a network built to provide UAVs with structured access to secure airspace [98]. With the compactness of the sensors and the advances in deep learning, the Internet of Drones is affecting the society in remarkable fields and applications. One of the main uses of drones is Surveillance. There is a need to shield people from security threats in busy areas such as sporting stadiums or during festivities. In subsequent years, crime levels have increased, particularly in urban communities, such as street robberies, robbery and vandalism, and acts of terrorism. Detection of criminality by the detection and acknowledgment of offenders in groups of citizens is also an essential activity. For modern surveillance networks, numerous security officers and an immense amount of human capital are required to provide the public with the appropriate defense. UAVs may be used in this way to enable security forces to detect people remotely in areas of interest. UAVs are not only able to defend from danger and help but also map, identify, and recognize criminals using face-recognition models [99]. UAVs set up with suitable IoT equipment, such as CCTV cameras, will include an efficient crowd monitoring network that can track any suspicious behavior and illegal conduct and recognize suspect faces. As a consequence, the health and protection of the audience will be greatly increased. It would be necessary to reduce the number of guards deployed on the field. Others can deal with UAV's high versatility and gain great face accuracy. Several faces may even be identified concurrently. The video capture can be decoded on local and remote servers for facial recognition, enabling MEC to unload a face recognition operation. OpenCV offers noticeable face recognition strategies. It applies to the learning algorithm to look for scanned faces present in the database in the monitored footages. For its associated sets and libraries, OpenCV uses explicitly LBPH. The LBPH technique is to equate the pixel with the surrounding pixels and summarize the local sense of the color.

I. Maza et al. [100] proposed a synchronization model for deployment of multiple drones under the project named AWARE with the context of a distributed decision-making model appropriate for a coordinated multi-UAV system. A collection of activities is established and delegated to carry out a particular project securely and successfully as per the preparation strategies. These objectives are semi-goals which are important to the accomplishment of the ultimate purpose of the program and which can be completed independently of certain semi-goals. The primary consideration is the commitment of the UAV node such that every mission is carried out to ensure good efficiency and shared cooperation with the other UAVs. A. Wada et al. [101] addressed the usage of tiny UAVs for various monitoring scenarios, such as post-disaster assessment. This type of surveillance system involves a computer that incorporates a broad variety of technology including the following techniques, i.e. networking, connectivity, sensing, image processing, and control.

Throughout the development of IoT infrastructure, N. H. Motlagh et al. addressed the future use of UAVs. A high-level description is presented and a test scenario is added to illustrate how UAVs can be used to track the crowd based on face recognition. In addition to the local processing of UAV-based video data, they have researched the retrieval of the video data transferred through the Mobile Edge Computing (MEC). The findings demonstrate the efficiency of the energy usage, identification period, and fast recognition of offenders for a MEC-based discharge device [99]. R. L. Finn, D. Wright identified the effect of UAV-based surveillance on privacy and civil liberties in civil applications. Furthermore, the Commission states that current regulatory systems, e.g. US Fourth Amendment, European and court rulings, and British law, do not adequately address privacy and civil liberties problems. This is primarily attributed to the fact that UAVs are complex, multimodal surveillance systems, integrating technologies and resources. Second, the inadequacy of existing policy strategies leads to excessive impacts on civil rights for already vulnerable groups [102]. Multi-layer active hierarchical frameworks that combine regulatory safeguards with realistic privacy and ethical testing framework provide the easiest approach to effectively address the intricacies and stratification of unmanned aerial vehicles and their planned deployments. D. Kingston et al. also presented a shared perimeter monitoring issue and are proposing a cooperative approach that allows for perimeter development and the addition or elimination of teammates [103]. The challenge with organized landscape surveillance is to obtain data on the perimeter's status and send data to the base center with minimal interruption instantaneously.

The Internet of Drones certainly provides an opportunity to enhance the surveillance methods but the process involves multiple flights of drones and synchronization between these aerial vehicles. Due to the repetitive nature of the surveillance process, energy management becomes the primary objective to ensure longer flight times and avoidance of any power failure that could cause hindrance in the surveillance mission.

2.4 Energy Conservation in Drones

In recent years, an excessive need for flying drones has been observed. The production of light plastics, low power consumption devices, and highly efficient processing systems contributed to the development of mobile flying robots. It may be used in many ways, including vehicle tracking, fire detection, and traffic management. The drones can fly autonomously at various altitudes and they are normally fitted with sensors to track the atmosphere and communication systems to exchange information with other drones or central stations. The overuse of utilities, mainly energy, increases the network's running costs and reduces the reliability of the system. Such a situation poses concerns regarding uncertainty and health and node development and reliability during continuing network operations. The usage of software-orientated systems and networks can help to reduce energy use in intelligent decision-making [104, 105].

Wu et al. [106] have introduced an industrial drone delivery platform. Their standardized solution utilizes structured training approaches to manage the risk of transmission and lack of information in connection with each other. They are effective in the method which can be utilized to boost the network's life.

The program currently relies on the usage of onboard sensors to gather data on urban areas, although efficiency and energy consumption issues have not yet been evaluated. Zeng et al. [107] focused on energy through the trajectory optimization of drones. By enforcing multiple architecture restrictions the authors introduced a plan to save energy for drones. This technically efficient solution can be used as a positioning device in LoRaWAN-like networks for the usage of drones. Nevertheless, durability and power should not just be ensured depending on their preferred approach, in terms of bit rate and toughness. Sharma et al. [108] have developed an energy-efficient data collection method utilizing drones. They used a firefly optimization algorithm to address data transmission problems while maintaining network services. Naqvi et al. [109] proposed a drone-enabled public safety network touch. They addressed various concerns and problems relating to drones in the future. The study reveals that the power management and performance control of drones depends on the requirements on the interference and data volume.

Throughout previous years, the usage of drones has largely been confined to combat observation and industrial uses. Big businesses, such as Amazon, are looking at the use of aircraft for consumer shipping. To deliver products weighing below 5 lb to the customers directly within 30 min of the arrival date Amazon Prime Air is preparing to use multi-rotor UAVs [110]. DHL, a logistics company, has revealed a practical template by leveraging shipping drones [111]. Nevertheless, the implementation of sales is hampered by several technical and legal issues. The big challenge is the construction of a drone or a UAV. A drone is a type of quadcopter featuring, as the name implies, four rotors. This is the main function of a drone or UAV system.

The Multicopter design comprises mainly the motor, battery, propeller, electronic speed controller (ESC), frame, monitor, and sensors. Controllers and sensors are affecting the flight efficiency of the vehicle, while others explicitly influence the load-carrying capacity of the vehicle. Many components can be integrated, rather than certain essential components, depending on the planned usage and the applicable specifications [112]. In general, the design parameters, such as the size of the rotor, the duration of the arm, and the total load ability of the machine, are determined by taking into consideration the operating conditions and performance areas of the unit. Although four-rotor drones typically are ideal for pure photography, in cases where the load size is significant, such as technical imagery, agricultural spraying, and freight, the number of rotors must be increased. Nevertheless, it is somewhat rare to carry forth a roadmap for the introduction of a standardized and autonomous program that could be subject to local working conditions. Nevertheless, experiments have been performed about the usage of more than four rotors, such as hexarotor and octorotor [113–117].

Niemiec et al. [118] suggested a configuration between 4 and 10 rotors with interchangeable rotor arrangement. In the proposed framework the rotors added are in the same plane. Bringing more than one rotor into a single-arm reduces the rotor count. In the study, in terms of energy use, it is worse for the hover process in the 10-rotor model within contrast to others, yet positive outcomes are seen for forward motion. The design of twelve rotors was studied by S. Zabunov and G. Mardirossian. [119]. In the first model below the frame, the 12 rotors used were mounted in the reverse role. The rotors are positioned in the same direction in the second case. They were linked to the power ratio from Model 1 to Model 2. The scientists observed that the mass-to-power ratio was higher in the first case. The concept of drones which could be modified modularly between 3 and 8 was proposed by Brischetto et al. [120]. They claimed that a 3D printer may be used to build structural components of this system. Through adjusting the positions of the arms mounted on the rotors, Xiu et al. changed the distance between the rotors and the center of the vehicle. [121]. Survival and restoration have been linked. The scientists found that the longest-armed configuration produces greater outcomes, according to their efficiencies. In the case of a single rotor failure, McKay et al. [122] checked the efficiency of the hexacopter by adjusting the configuration of the drone. The adjustment suggested enhancing the hexacopter's efficiency in compliance with the findings. Many control systems are used in the UAV study and analysis. Kuric et al. suggested a modern approach focused on Recursive Least Squares (RLS) to identify and locate faults in octorotor propulsion systems [123]. They also witnessed noteworthy monitoring efficiency according to the reports.

3 Energy Management Proposed Framework for UAVs

Currently, our city's surveillance is majorly dependent on human cops and static nodes stationed and installed around the city. This is both a costly and comparatively inefficient practice as compared to an autonomous intelligent aerial drone that would eliminate the need for multiple nodes or human eyes [124].

3.1 Tilt-Rotor Drone-Based Energy Conservation Framework for UAVs in Smart City Surveillance

Tilt-rotor UAV is a design developed by combining the features of a multi-copter and a plane that possesses the capability to switch between the flight modes instantaneously as and when required. The lift generated in these aerial vehicles is dependent on the rotors and wings. Tilt-rotor vehicles have three modes of flight which are helicopter mode, transition mode, and airplane mode. These three modes of flight give this aircraft exceptional speeds of an airplane, make a smooth transition with the variable tilt angle of the rotor, and an ability to hover like a helicopter in a place. The maneuverability of aircraft with these flight modes makes it dynamic and capable to perform a range of tasks.

The general approach to surveillance using UAVs usually employs a multi-rotor aircraft as it has the ability to hover over a point, unlike a fixed-wing aircraft. However, this results in very poor flight times, as these drones also need to move between locations (otherwise a pole-mounted camera could be used to serve the purpose of surveillance) and multi-rotor drones generate both the forward-thrust and lift using their propellers. In a fixed-wing aircraft, the lift is generated due to the flow of air over the wings; the aircraft can continue to glide in the air for a while with the thrust motors turned off (a multi-rotor, on the other hand, will fall like a rock), making them very efficient for covering distances. Fixed-wing VTOLs are a combination of both of these designs; they are capable of hovering at the same place using their hover motors and cruise like a plane using the thrust motors. There are many variations to fixed-wing VTOL designs; Skyprowler from Krossblade Aerospace is one such example.

The framework proposes a tilt-rotor drone with fixed wings to extend the overall flight time of the drone instead of a standard quad-rotor or hexa-rotor. The drone

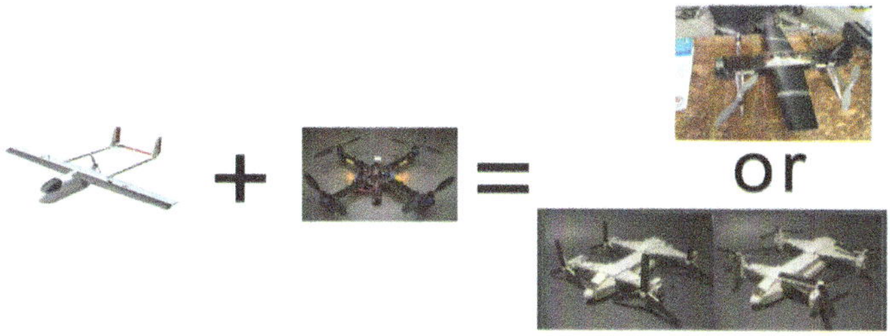

Fig. 4 Tiltrotor drone with extended wings

with wings as shown in Fig. 4, can take off like a quadcopter and fly efficiently in a straight line like a fixed-wing, albeit at high speeds (>15 m/s).

The higher level of kinematic control of VTOL has two parts: Fixed-wing mode and stationary mode.

Hover mode

In hover mode, the hover motors are used for generating the lift to keep the aircraft airborne. The translational and rotational movements are controlled by controlling the pitch, roll, and heading angle of the aircraft, while the altitude is controlled by the net thrust of all the hover motors. Moving the nose down, or pitching down, results in forward-acceleration, while "rolling" the aircraft to the right, results in the aircraft accelerating to the right side. These movements are executed by varying the rotational speeds of the hover motors. The control signals are generated by employing a control loop that could be PID, PI-PID, MPC, LQR etc. to control the orientation. Additionally, an extra control loop is used to control the position by controlling the orientation. These control algorithms are executed by the flight controller on-board the aircraft, which produces control signals by estimating the state of the drone using Inertial navigation systems (INS) and GPS.

Fixed-wing mode

In fixed-wing mode, the thrust motor is used to push the plane forward against the aerodynamic drag. In this mode, there exist non-holonomic constraints, i.e., the aircraft cannot be made to slide to the left or the right (similar to how a car cannot move sideways under normal conditions). Thus, the aircraft can only move in the direction in which it points. Turning right or left requires turning the nose of the aircraft right or left. This is typically done by rolling in the direction of the turn and pitching into the turn, often known as a "bank and yank" approach, as it would require the pilot to first bank the plane into the turn and then "yank" (pull) the control stick towards themselves. The altitude is controlled by controlling the pitch angle; pitching the nose up results in the aircraft going up and vis-a-vis. These movements are executed by "control" surfaces such as ailerons, elevons, rudders, etc.; which are essentially flaps that redirect the air to generate the desired movement. The control signals for these movements are generated by an on-board flight controller, which uses some control loop (P-PI, PID, etc.) to control the orientation. Additionally, another control loop is used is to control the velocity, heading, and altitude by controlling the roll, pitch, and yaw motion. The flight controller generates these control signals by estimating the state using the INS, GPS, and in some cases, an airspeed sensor.

Energy Saving Protocol

The approach described here, takes advantage of this configuration to increase the surveillance flight time. By using a fleet of drones instead of just one, multiple points can be covered simultaneously. The drones move in a chain-like fashion; if there are n drones, n-1 of them will continue to hover while the last drone will move to a location ahead of the first drone in fixed-wing mode as shown in Fig. 5. The movement is such that each drone stays in hover mode and fixed-wing mode for the same amount of time. This can theoretically allow each drone to have a 50% increase in overall flight time, thus increasing surveillance time and coverage.

The energy-saving protocol has been explained in the flowchart given in Fig. 6. According to the approach, the drones are capable of making decisions for the locations to maneuver to and the mode of flight based on its battery conditions. The drone itself returns to the base to get itself charge if the battery is low. In case of sufficient charge in the battery, it maneuvers to the next waypoint and sends the video feed

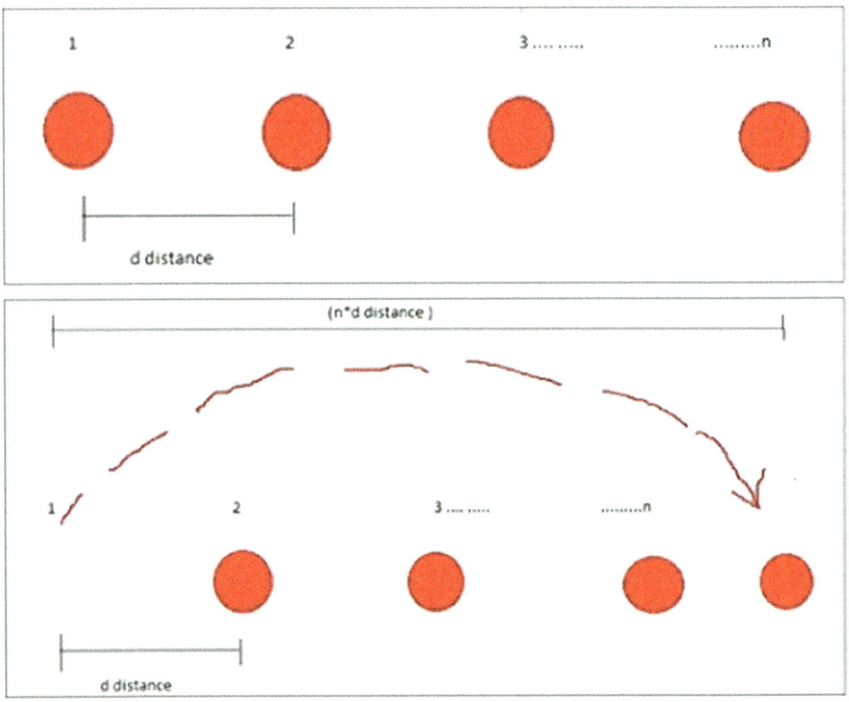

Fig. 5 Protocol for the fleet of more than two drones to increase the flight time

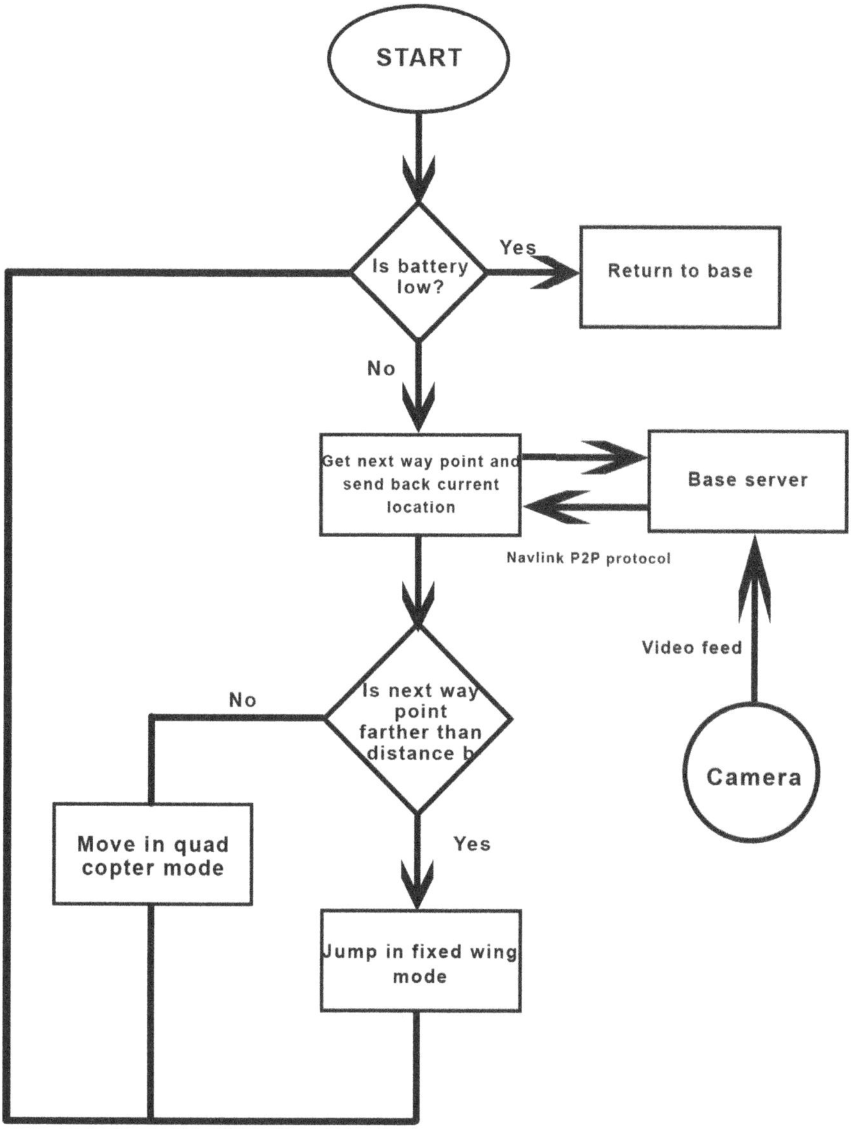

Fig. 6 Flowchart of the energy-saving UAV protocol

to the base server communicated through point to point protocol (P2P). In case the distance to next waypoint is greater than distance 'd', the flight mode is changed to the fixed-wing mode and the drone returns to the base else it remains in the stationary mode/quadcopter mode.

4 Case Study on IoD Based Surveillance

The previous year, Comparitech published a ranking of "the most surveilled cities in the world" in 150 cities focused on the number of public CCTV cameras installed [125]. Out of the top ten rankings of the cities, eight are from China. The only cities outside China to make the top 10 were London and Atlanta. By 2022 it is estimated that China will have one public CCTV camera for every two citizens. Video Surveillance is more successful in preventing violence in car parks and suburban neighborhoods, even when paired with various measures, in a new report, focused on 40 years of studies. Active surveillance through installed camera solutions has been more successful than passive methods. Throughout the case of the Boston bombing, investigators utilized federal, media, and private images and videos to identify the perpetrators. During the coordinated bomb blasts in Sri Lanka during April 2019, CCTV footage was used to recognize the suspects. However, the emergence of facial recognition technology in surveillance systems has given rise to privacy issues among people. Two case studies have been presented further discussing surveillance in Singapore and India.

4.1 Case Study on Singapore: An Energy-Efficient Smart City

In 2014, the Singapore Government introduced the Smart Nation project by announcing that a range of smart city policies will be adopted as a solution to environmental problems, including population growth and energy sustainability. Transportation and social connectivity are the most sophisticated cognitive utilities in Singapore. Local councils have focused during the past decade on the introduction of a smart transit network [126].

Singapore also established a program to boost traffic management and to ensure the smooth operation of road transport. One Monitoring, a comprehensive database servicing all drivers and vehicle owners in the region, is one of the main initiatives in the automotive industry. The Portal helps people to view the traffic data obtained by GPS-installed security cameras. In addition, the program tells about the parts of roadwork, travel time calculator, traffic maps of roads, traffic reports, road directions, and parking details.

A surveillance program is often provided by the Land Transportation Agency that hires control systems for traffic incidents. LTA triggers the automobile recovery facility to move the car into the closest marked parking garage on the road when an incident is identified by the device. The Parking Guidance System is another intelligent city initiative implemented by the authorities [127]. Initially launched in 2008, this smart service provides drivers with real-time parking availability detail. The program is built to reduce the amount of traffic traveling in pursuit of accessible spaces and aims to allow the usage of current parking facilities more effectively. A variety of smart city programs in the area of infrastructure and public welfare have already been initiated by the City. The Singapore Police Force operates a web-based automated policing center for people to collect details, submit online police reports, and manage administrative matters such as filing for a verified copy of police reports and criminal records.

The Intelligent Energy System is being extended in the arena of surveillance by the Singapore Police Force by deploying drones. The Singapore Police Force (SPF) is developing a self-operating system that lets a drone take off and land and also refuel the battery on its own. The drone rests in a 2.2 m-tall shell, which can be relocated to anywhere it wants to be installed. The autonomous control feature of the kit guarantees that the 10 batteries are completely charged and ready to be mounted and even endows the UAV with video surveillance equipment such as cameras before it is deployed. Industrial fields are fairly abandoned throughout the circuit-breaker time, yet the police will remain alert. These environments are vulnerable to offenses such as burglary and robbery. Precautionary plans are formulated and implemented to deter offenders from having the chance to hack into these abandoned warehouses or office buildings. The aerial vehicles, which are 1.79 m wide and weigh around 10 kg, track the industrial region, and minimize the need for police officers. Plans for expanding the program to housing developments are kept in the pipeline for later trials. For cameras that have high-zoom capabilities, the drones will operate within a range of a few hundred meters. Based on the weight of gear the drone holds, and environmental factors including strong wind or fog, flights may long for at least 30 min. The police used traditional drones to support land activities such as crowd management and surveillance at the Marina Bay countdown events in 2020.

4.2 Case Study on India: Smart City Surveillance in India

The video surveillance market of India is projected to expand by more than 22.7% in 2019 and 2025, according to research and consultancy firm 6Wresearch. By 2020, the demand is expected to hit 2.4 billion dollars [128]. Major video monitoring players are positioning themselves for their creative approaches to exploit this massive market potential. The image of a smart city includes a wish list of infrastructures and facilities such as smart traffic monitoring [129], agrarian [130], real estate monitoring, etc. representing their degree of ambition in the mind of every citizen in India.

There has been limited usage of commercial drones in the Indian airspace in various fields. Some of the examples of usage of drones in India are: for 3D mapping of Raebareli-Allahabad Highway by an autonomous firm of the government of India. A proprietary positioning system used drones developed by DeTect in association with GAIL, Hindustan Petroleum, and Indian Oil for monitoring structures like boilers in the plant. Drones were always used in Nagpur city by Indian Forest Department for monitoring the 13 crore plantation. AI was successfully used to manage crowds during the recently held ‘Kumbh Mela’ in Prayagraj (earlier Allahabad). With this, it became easier to map the crowd. About 1100 cameras were installed at the Kumbh Village, Prayagraj Smart City adding the scientific study of traffic and roads through the new AI technology in the future.

To support India's 100 Smart City mission, a centralized infrastructure configured for multilevel large-scale administration and inter-organizational channels with high-definition video surveillance has been built by extensive research and evaluation to provide outstanding benefits. Honeywell, operating primarily in aerospace, is working with 14 Indian cities, including Bhubaneshwar and 11 Madhya Pradesh cities, to develop a smart city infrastructure. It provides Odisha Police with a city-wide monitoring system that helps boost security, deter crime, enforce the rule of law, and manage traffic in the capital city of the state. In the form of the Smart Cities program, it is now being discussed explicitly in India, taking account of the number of opportunities that are created. The integrated and comprehensive system combines Surveillance, facial detection, access management, intrusion warning, camera monitoring, adaptive mobility, and management center under one roof. The technology allows for quick emergency response and cross-system or cross-departmental knowledge exchange as an automated network with a single command and centralized data processing.

UAVs have also been used in the past in emergency circumstances. During the Kargil War in 1999 for the first time, India used the military-grade drones. When grappling with the Pakistan-sponsored terror attacks in Mumbai (2008), the northern Siachen avalanche (2016), the forest fires in Bandipur (2019), the Bihar floods (2019) and the Pulwama attack (2019)—to mention a few—drones were deployed. Drones had a significant part to play in achieving results. But the spectrum of its implementation has been extended perhaps for the first time during the COVID-19 pandemic.

The unprecedented situations such as the COVID pandemic showed an amazing usage of drones for various tasks. Photographs and news accounts from across the globe show officials placing drones in several applications when handling the pandemic. These were deployed during the lock-down periods for surveillance and tracking, to relay critical information, to hunt down restriction violators, to decontaminate hotspots, to distribute medications and other vital products. Chhattisgarh government has hired Chennai based venture to disinfect certain regions in the state. Drones equipped with loudspeakers and sirens were used by Indian Police in some states to alert people to stay indoors.

Owing to their capabilities for Vertical Takeoff and Landing Operation (VTOL), and Beyond Visible Line of Sight (BVLOS), drones have been suitable in these circumstances. These are cost-effective, compact, and support the first take in extreme emergencies, providing imagery, analysis, and evaluation and severity of the damage.

5 Potential Implications and Suggestion for Future Research in Smart City Surveillance

No one can claim that the benefits of smart city technologies do not occur in pursuit of more stable communities. It is necessary to bring technology into an appropriate application which should be morally the right thing to do. The technology should keep people safe. With security as a concern, it is the social obligation to take into account the risks involved, the severity of these risks, and to estimate the impact it has on the society [16] [131]. It is of utmost importance to understand the sensitivity of data.

Some of the issues addressed in drone surveillance in smart city applications include:

- *Many privacy issues:* The degree of willingness to be tracked by others is specific and through data usage and control can contribute to more lawsuits. Consequently, in the immediate future, there are several big legislative problems in terms of personal protection.
- *Reputational risk*: When a company's cameras catch a crime, what are the consequences that cause the local authorities to spy into their cameras? This might actually subject individuals to lawsuits. And would challenge the willingness to share the details, too.
- *Network definition*: Defining new methods in which everything from smart city sensors to homes, would be tied together to connect in a network is going to be challenging
- *A larger number of data nodes*: There are several more access points and gateways accessible for drones in smart cities, which lead to further risk to security breaches. The general understanding of the usage of data and data security would be the problem.
- *Unused data*: Privacy experts discussed the problem of unused data which could have a significant effect if evaluated and implemented. Officials have to work with the way their data should be interpreted and prioritized.
- *Technical challenges*: City workers would need to be educated on how to utilize and be comfortable with this new technology. Criminals may find their new methods to breach drone security. Another challenging part would be safeguarding these portable safety devices itself.
- *Unanticipated nuances:* As you bring up and raise the stakes with sensors that track criminality, someone has to control it, react to it, manage it and repair it, have to try and ensure the license contract and distribution for the goods is upheld. People will have to be aware round the clock because there is no anonymity, so they have to be comfortable about that.

A Cyber-Physical System (CPS) compactly takes onboard internet, methods and its users through a process managed and tracked by computer-based algorithms [132]. It is an engineered structure that develops and relies on, the seamless assimilation of physical components and computational algorithms. Broadly, Cyber means discrete, logical computing that is conceivably capable of regulating and establishing effective communication. Physical constitutes the management of man-made and natural structures that operate over continuous-time and are regulated by the laws of physics. Thus, Cyber-Physical Systems refer to exquisite coherence of physical infrastructure with the communication and computing systems. CPS can provide a safe and secure network to manage privacy of data. Thus, as of now CPS is capable of solving most of the data related issues encountered in surveillance systems. Besides, CPS also provides network management system to efficiently manage multiple drones and set-up an effective communication between the aerial devices as well as with the base station [133, 134]. Smart Cities are only in really initial phases and addressing the nuances which could propagate the expected swell of urban security and smart cities might prove valuable in the coming years.

6 Conclusion

Smart cities utilize the data as well as information technology to deliver operational excellence, promote efficiency, establish sustainable growth, and increase the quality of lifestyle aspects of the native population and the workforce. This also establishes that the city has an intelligent infrastructure. Smart cities are clearly the future living pattern which is being speedily evolved and expanding across the world. Surveillance is an integral part of the smart city. Static surveillance methods have its own limitation that can be resolved using dynamic aerial vehicles UAVs. This chapter highlights the evolving technology of Internet of Drones which is a magnificent combination of UAV technology combined with artificial intelligence that makes the surveillance mission even more impactful widening the area of application in smart cities such as traffic monitoring, remote area inspection, facial recognition surveillance, maritime monitoring etc. This chapter presents an energy-efficient protocol of drones integrated with artificial intelligence as the smarter solution towards surveillance. The framework presented tilt-rotor UAVs with an energy-efficient algorithm for the trajectory of drones to maximize the time of flight of the drones theoretically to 50%. The work also discusses the case study on the use of drones in the smart cities of Singapore and India. Subsequently, the potential challenges that are foreseen in the deployment of drones followed by cyber physical system as the solution in the near future are also discussed.

Acknowledgements This work is supported by the grant from Department of Science and Technology, Government of India, against CFP launched under Interdisciplinary Cyber Physical Systems (ICPS) Programme, DST/ICPS/CPS-Individual/ 2018/181(G).

References

1. Komninos, N.: Intelligent cities: variable geometries of spatial intelligence. Intell. Build. Int. **3**(3), 172–188 (2011)
2. Nam, T., Pardo, T.A.: Conceptualizing smart city with dimensions of technology, people, and institutions. In: Proceedings of the 12th annual international digital government research conference: digital government innovation in challenging times, pp. 282–291 (2011)
3. Vanolo, A.: Smartmentality: The smart city as disciplinary strategy. Urban Stud. **51**(5), 883–898 (2014)
4. Eden Strategy Institute and ONG&ONG Pte Ltd.: TOP 50 Smart City Governments. Eden Strategy Institute (2018). [online] Singapore: Eden Strategy Institute and ONG&ONG Pte Ltd. Available at: https://www.smartcitygovt.com/. Accessed 30 July 2020
5. Calavia, L., Baladrón, C., Aguiar, J.M., Carro, B., Sánchez-Esguevillas, A.: A semantic autonomous video surveillance system for dense camera networks in smart cities. Sensors **12**(8), 10407–10429 (2012)
6. Brandt, A.A.: Illegal Dumping as an Indicator for Community Social Disorganization and Crime (2017)
7. Boissevain, C.: Smart city lighting. In: Smart Cities, pp. 181–195. Springer, Cham (2018)
8. Chace, R.W.: An overview on the guidelines for closed circuit television (CCTV) for public safety and community policing. Retrieved 23 Sept, p. 2005 (2001)
9. Barbot, O.: George Floyd and Our Collective Moral Injury (2020)
10. Shepard, S.: San Jose State Installs New Gunshot Detection System, Surveillance—Security Today. [online] Security Today (2017). Available at: https://securitytoday.com/articles/2017/10/25/san-jose-state-installs-new-gunshot-detection-system-surveillance.aspx. Accessed 25 July 2020
11. Sookhak, M., Tang, H., He, Y., Yu, F.R.: Security and privacy of smart cities: a survey, research issues and challenges. IEEE Commun. Surv. Tutor. **21**(2), 1718–1743 (2018)
12. Witwicki, S., Castillo, J.C., Messias, J., Capitan, J., Melo, F.S., Lima, P.U., Veloso, M.: Autonomous surveillance robots: A decision-making framework for networked muiltiagent systems. IEEE Robot. Autom. Mag. **24**(3), 52–64 (2017)
13. Srivastava, S., Bisht, A., Narayan, N.: Safety and security in smart cities using artificial intelligence—A review. In: 2017 7th International Conference on Cloud Computing, Data Science & Engineering-Confluence, pp. 130–133. IEEE (2017)
14. Menouar, H., Guvenc, I., Akkaya, K., Uluagac, A.S., Kadri, A., Tuncer, A.: UAV-enabled intelligent transportation systems for the smart city: applications and challenges. IEEE Commun. Mag. **55**(3), 22–28 (2017)
15. Alsamhi, S.H., Ma, O., Ansari, M.S., Almalki, F.A.: Survey on collaborative smart drones and internet of things for improving smartness of smart cities. IEEE Access **7**, 128125–128152 (2019). https://doi.org/10.1109/ACCESS.2019.2934998
16. Mohammed, F., Idries, A., Mohamed, N., Al-Jaroodi, J., Jawhar, I.: UAVs for smart cities: opportunities and challenges. In: 2014 International Conference on Unmanned Aircraft Systems (ICUAS), pp. 267–273. IEEE (2014)
17. Gharibi, M., Boutaba, R., Waslander, S.L.: Internet of drones. IEEE Access **4**, 1148–1162 (2016). https://doi.org/10.1109/ACCESS.2016.2537208
18. Bang, J., Lee, Y., Lee, Y.T., Park, W.: AR/VR based smart policing for fast response to crimes in safe city. In: 2019 IEEE International Symposium on Mixed and Augmented Reality Adjunct (ISMAR-Adjunct), pp. 470–475. IEEE (2019)
19. Samuelsson, M.: Advanced Intelligent Network Products Bring New Services Faster (1991)
20. Batty, M.: The computable city. Int. Plann. Stud. **2**(2), 155–173 (1997)
21. Anttiroiko, A.V., Valkama, P., Bailey, S.J.: Smart cities in the new service economy: building platforms for smart services. AI Soc. **29**(3), 323–334 (2014)
22. McLaren, D., Agyeman, J.: Sharing Cities: A Case for Truly Smart and Sustainable Cities. Mit Press (2015)

23. Sam, M.: Smart City Roadmap (2016)
24. Qin, H., Li, H., Zhao, X.: Development status of domestic and foreign smart city. Glob. Presence **9**, 50–52 (2010)
25. Caragliu, A., Ch. Del Bo, P. Nijkamp, Smart cities in Europe, (2011): J. Coelho, N. Cacho, F. Lopes, E. Loiola, T. Tayrony, T. Andrade, M. Mendonca, M. Oliveira, D. Estaregue, B. Moura, ROTA: A Smart City Platform to Improve Public Safety
26. Mohanty, S.P., Choppali, U., Kougianos, E.: Everything you wanted to know about smart cities: the internet of things is the backbone. IEEE Consum. Electron. Mag. **5**(3), 60–70 (2016)
27. Cohen, B.: The 3 Generations of Smart Cities. Inside the Development of the Technology Driven City (2015)
28. Peris-Ortiz, M., Bennett, D.R., Yábar, D.P.B.: Sustainable Smart Cities. Innovation, Technology, and Knowledge Management. Springer International Publishing Switzerland, Cham (2017)
29. Forward, N.Y.C.: Building a Smart City, Equitable City (2016)
30. Komninos, N.: What makes cities intelligent, pp. 77–87. Governing, modelling and analysing the transition, Smart Cities (2013)
31. UK, G.: Smart cities: Background Paper (2013)
32. Chan, K.: What Is A 'Smart City'?. Expatriate Lifestyle (2017)
33. Celino, I., Kotoulas, S.: Smart cities [Guest editors' introduction]. IEEE Internet Comput. **17**(6), 8–11 (2013)
34. Rongxu, L.Y.H.: About the sensing layer in internet of things. Comput. Study **5** (2010)
35. Singh, S., Saurav, S., Shekhar, C., Vohra, A.: Prototyping an automated video surveillance system using FPGAs. Int. J. ImageGraph. Signal Process. **8**(8), 37 (2016)
36. Shi, Y., Lichman, S.: Smart cameras: a review. In: Proceedings of, pp. 95–100 (2005)
37. Delp, E.J.: Smart Video Surveillance: Mission, Solution, and Impact, PURVAC
38. Jiang, H., Ardo, H., Owall, V.: A hardware architecture for real-time video segmentation utilizing memory reduction techniques. IEEE Trans. Circ. Syst. Video Technol. **19**(2), 226–236 (2008)
39. Video Surveillance Homepage: (http://www.videosurveillance.com/manufacturers/)
40. IBM Official Homepage: http://researcher.watson.ibm.com/researcher/view_group.php?id=1394 (2007)
41. Hu, W., Tan, T., Wang, L., Maybank, S.: A survey on visual surveillance of object motion and behaviors. IEEE Trans. Syst. Man Cybern. Part C **34**(3), 334–352 (2004)
42. Collins, R.T., Lipton, A.J., Kanade, T., Fujiyoshi, H., Duggins, D., Tsin, Y., Tolliver, D., Enomoto, N., Hasegawa, O., Burt, P., Wixson, L.: A System for Video Surveillance and Monitoring. Carnegie Mellon University Technical Report, CMU-RI-TR00-12 (2000)
43. Siebel, N.T., Maybank, S.: The advisor visual surveillance system. In: Proceedings of the ECCV Workshop on Applications of Computer Vision, pp. 103–111 (2004)
44. Shu, C.F., Hampapur, A., Lu, M., Brown, L., Connell, J., Senior, A., Tian, Y.: IBM smart surveillance system (S3): an open and extensible framework for event based surveillance. In: Proceedings of IEEE Conference on Advanced Video and Signal Based Surveillance, pp. 318–323 (2005)
45. Regazzoni, C., Ramesh, V., Foresti, G.L.: Special issue on video communications, processing, and understanding for third generation surveillance systems. Proc. IEEE **89**(10), 1355–1367 (2001)
46. Kim, I.S., Choi, H.S., Yi, K.M., et al.: Intelligent visual surveillance—A survey. Int. J. Control Autom. Syst. **8**, 926–939 (2010). https://doi.org/10.1007/s12555-010-0501-4
47. ICAO's circular 328 AN/190: Unmanned Aircraft Systems (PDF). ICAO. Retrieved 3 Feb 2016
48. Bucaille, I., Hethuin, S., Munari, A., Hermenier, R., Rasheed, T., Allsopp, S.: Rapidly deployable network for tactical applications: Aerial base station with opportunistic links for unattended and temporary events absolute example. In: Proceedings of IEEE Military Communications Conference (MILCOM), San Diego, CA, USA (2013)

49. Lien, S.Y., Chen, K.C., Lin, Y.: Toward ubiquitous massive accesses in 3GPP machine-to-machine communications. IEEE Commun. Mag. **49**(4), 66–74 (2011)
50. Dhillon, H.S., Huang, H., Viswanathan, H.: Wide-Area Wireless Communication Challenges for the Internet of Things. Available online: arxiv.org/abs/1504.03242., 2015
51. Hourani, A., Kandeepan, S., Jamalipour, A.: Modeling air-to-ground path loss for low altitude platforms in urban environments. In: Proceedings of IEEE Global Telecommunications Conference (GLOBECOM), Austin, TX, USA (2014)
52. Feng, A., Tameh, E.K., Nix, A.R., McGeehan, J.: Modelling the likelihood of line-of-sight for air-to-ground radio propagation in urban environments. In: Proceedings of IEEE Global Telecommunications Conference (GLOBECOM), San Diego, CA, USA (2006)
53. Feng, Q., McGeehan, J., Tameh, E.K., Nix, A.R.: Path loss models for air-to-ground radio channels in urban environments. In: Proceedings of IEEE Vehicular Technology Conference (VTC), Melbourne, Vic, Australia (2006)
54. Holis, J., Pechac, P.: Elevation dependent shadowing model for mobile communications via high altitude platforms in built-up areas. IEEE Trans. Antennas Propag. **56**(4), 1078–1084 (2008)
55. Hourani, A., Sithamparanathan, K., Lardner, S.: Optimal LAP altitude for maximum coverage. IEEE Wirel. Commun. Lett. **3**(6), 569–572 (2014)
56. Mozaffari, M., Saad, W., Bennis, M., Debbah, M.: Drone small cells in the clouds: design, deployment and performance analysis. In: Proceedings of IEEE Global Communications Conference (GLOBECOM), San Diego, CA, USA (2015)
57. Kosmerl, J., Vilhar, A.: Base stations placement optimization in wireless networks for emergency communications. In: Proceedings of IEEE International Conference on Communications (ICC), Sydney, Australia (2014)
58. Daniel, K., Wietfeld, C.: Using Public Network Infrastructures for UAV Remote Sensing in Civilian Security Operations. DTIC Document, Tech. Rep. (2011)
59. Rohde, S., Wietfeld, C.: Interference aware positioning of aerial relays for cell overload and outage compensation. In: Proceedings of IEEE Vehicular Technology Conference (VTC), Quebec, QC, Canada (2012)
60. Han, Z., Swindlehurst, A.L., Liu, K.: Optimization of MANET connectivity via smart deployment/movement of unmanned air vehicles. IEEE Trans. Veh. Technol. **58**(7), 3533–3546 (2009)
61. Jiang, F., Swindlehurst, A.L.: Optimization of UAV heading for the ground-to-air uplink. IEEE J. Sel. Areas Commun. **30**(5), 993–1005 (2012)
62. Mozaffari, M., Saad, W., Bennis, M., Debbah, M.: Optimal transport theory for power-efficient deployment of unmanned aerial vehicles. In: Proceedings of IEEE International Conference on Communications (ICC), Kuala Lumpur, Malaysia (2016)
63. Girshick, R., Donahue, J., Darrell, T., Malik, J.: Rich feature hierarchies for accurate object detection and semantic segmentation. In: Proceedings of the 27th IEEE Conference on Computer Vision and Pattern Recognition (CVPR'14), pp. 580–587, Columbus, Ohio, USA (2014)
64. Ren, S., He, K., Girshick, R., Sun, J.: Faster R-CNN: towards real-time object detection with region proposal networks. Adv. Neural. Inf. Process. Syst. **28**, 91–99 (2015)
65. Lee, J., Wang, J., Crandall, D., Sabanovic, S., Fox, G.: Real-time, cloud-based object detection for unmanned aerial vehicles. In: Proceedings of the 1st IEEE International Conference on Robotic Computing (IRC), pp. 36–43, Taichung, Taiwan (2017)
66. Zhou, B., Lapedriza, A., Xiao, J., Torralba, A., Oliva, A.: Learning deep features for scene recognition using places database. In: Proceedings of the 28th Annual Conference on Neural Information Processing Systems 2014, NIPS 2014, pp. 487–495 (2014)
67. Hu, F., Xia, G.-S., Hu, J., Zhang, L.: Transferring deep convolutional neural networks for the scene classification of high-resolution remote sensing imagery. Remote Sens. **7**(11), 14680–14707 (2015)
68. Ghaderi, A., Athitsos, V.: Selective unsupervised feature learning with convolutional neural network (S-CNN). In: Proceedings of the 2016 23rd International Conference on Pattern Recognition (ICPR), pp. 2486–2490 (2016)

69. Giusti, A., Guzzi, J., Ciresan, D.C., et al.: A machine learning approach to visual perception of forest trails for mobile robots. IEEE Robot. Autom. Lett. **1**(2), 661–667 (2016)
70. The Technion—Israel Institute of Technology. Technion aerial systems 2016. In Journal Paper for AUVSI Student UAS Competition (2016)
71. Kim, N.V., Chervonenkis, M.A.: Situation control of unmanned aerial vehicles for road traffic monitoring. Modern Appl. Sci. **9**(5), 1–13 (2015)
72. Sawarkar, A., Chaudhari, V., Chavan, R., Zope, V., Budale, A., Kazi, F.: HMD Vision-Based Teleoperating UGV and UAV for Hostile Environment Using Deep Learning. CoRR abs/1609.04147. URL http://arxiv.org/abs/1609.04147
73. Bejiga, M., Zeggada, A., Nouffidj, A., Melgani, F.: A convolutional neural network approach for assisting avalanche search and rescue operations with UAV imagery. Remote Sens. **9**(2), 100 (2017)
74. Szegedy, C., Liu, W., Jia, Y. et al.: Going deeper with convolutions. In: Proceedings of the IEEE Conference on Computer Vision and Pattern Recognition (CVPR'15), pp. 1–9, Boston, Mass, USA (2015)
75. Blasch, E., Kondor, S., Gordon, M., Hsu, R.: Georgia tech aerial robotics team competition entry. J. Aerial Unmanned Veh. Syst. (1994)
76. Blasch, E.P.: Flexible vision-based navigation system for unmanned aerial vehicles. In: Proc. SPIE. (1994)
77. Seetharaman, G., Lakhotia, A., Blasch, E.: Unmanned vehicles come of age: the DARPA grand challenge. IEEE Comput. Soc. Mag. (2006)
78. Shen, D., Chen, G., Cruz, J., Blasch, E.: A game theoretic data fusion aided path planning approach for cooperative UAV control. In: IEEE Aerospace Conference, Big Sky, MT (2008)
79. Durrant-Whyte, H., Bailey, T.: Simultaneous localization and mapping (SLAM): Part I the essential algorithms. Robot. Autom. Mag. **13**, 99–110 (2006)
80. Lee, K.M., Zhi, Z., Blenis, R., Blasch, E.P.: Realtime vision-based tracking control of an unmanned vehicle. IEEE J. Mech. Intell. Motion Control (1995)
81. Balch, T., Boone, G., Collins, T., Forbes, H., MacKenzie, D., Santamaria, J.C.: Io, Ganymede, and Callisto—a multiagent robot trash-collecting team. AI Magazine (1995)
82. Kim, S.C., Shin, K.H., Woo, C.W., Eom, Y.S., Lee, J.M.: Performance analysis of entropy-based multi-robot cooperative systems in a MANET. Int. J. Control Autom. Syst. **6**(5), 722–730 (2008)
83. Delmerico, J., Mueggler, E., Nitsch, J., Scaramuzza, D.: Active autonomous aerial exploration for ground robot path planning. IEEE Robot. Autom. Lett. **2**(2), 664–671 (2017)
84. Aznar, F., Pujol, M., Rizo, R.: Visual navigation for UAV with map references using ConvNets. In: Advances in artificial intelligence, vol. 9868 of Lecture Notes in Computer Science, pp. 13–22, Springer (2016)
85. Koparan, C., Koc, A.B., Privette, C.V., Sawyer, C.B.: Adaptive water sampling device for aerial robots. Drones **4**(1), 5 (2020). https://doi.org/10.3390/drones4010005
86. Koparan, C., Koc, A.B., Privette, C.V., Sawyer, C.B., Sharp, J.L.: Evaluation of a UAV-assisted autonomous water sampling. Water **10**(5), 655 (2018). https://doi.org/10.3390/w10050655
87. Koparan, C., Koc, A.B., Privette, C.V., Sawyer, C.B.: In situ water quality measurements using an unmanned aerial vehicle (UAV) system. Water **10**(3), 264 (2018). https://doi.org/10.3390/w10030264
88. Koparan, C., Koc, A.B., Privette, C.V., Sawyer, C.B.: Autonomous in situ measurements of noncontaminant water quality indicators and sample collection with a UAV. Water **11**(3), 604 (2019). https://doi.org/10.3390/w11030604
89. Franke, U.E.: Civilian Drones: Fixing an Image Problem?. ISN Blog. International Relations and Security Network. Retrieved 5 March 2015
90. Moran, M.S., Inoue, Y., Barnes, Edward: Opportunities and limitations for image-based remote sensing in precision crop management. Remote Sens. Environ. **61**, 319–346 (1997). https://doi.org/10.1016/S0034-4257(97)00045-X
91. Lelong, Camille, Burger, Philippe, Jubelin, Guillaume, Bruno, Roux, Sylvain, Labbe, Frederic, Baret: Assessment of unmanned aerial vehicles imagery for quantitative monitoring of wheat crop in small plots. Sensors **8**, 3557–3585 (2008). https://doi.org/10.3390/s8053557

92. Bosse, M., Karl, W.C., Castanon, D., DeBitetto, P.: A vision augmented navigation system. In: IEEE Conference on Intelligent Transportation Systems, pp. 1028–1033 (1997)
93. Kaminer, I., Pascoal, A.M., Kang, W., Yakimenko, O.: Integrated vision/inertial navigation system design using nonlinear filtering. To Appear IEEE Trans. Aerosp. Electron.
94. Ma, Y., Kosecka, J., Sastry, S.S.: Vision-guided navigation for a nonholonomic mobile robot. IEEE Trans. Robot. Automat. **15**, 521–537 (1999)
95. Bosse, M.: A Vision-Augmented Navigation System for an Autonomous Helicopter. Master's thesis, Boston Univ., Boston, MA (1997)
96. Amidi, O., Kanade, T., Fujita, K.: A visual odometer for autonomous helicopter flight. Robot. Autom. Syst. **28**, 185–193 (1999)
97. Miller, R., Mettler, B., Amidi, O.: Carnegie Mellon University's 1997 international aerial robotics competition entry. In: Proceedings of International Aerial Robotics Competition (1997)
98. D'Andrea, R.: Guest editorial can drones deliver? IEEE Trans. Autom. Sci. Eng. **11**(3), 647–648 (2014)
99. Motlagh, N.H., Bagaa, M., Taleb, T.: UAV-based IoT platform: a crowd surveillance use case. IEEE Commun. Mag. **55**(2), 128–134 (2017). https://doi.org/10.1109/MCOM.2017.1600587CM
100. Maza, I., Caballero, F., Capitán, J., Martínez-de-Dios, J.R., Ollero, A.: Experimental results in multi-UAV coordination for disaster management and civil security applications. J. Intell. Robot. Syst. **61**(1), 563–585 (2011)
101. Wada, A., Yamashita, T., Maruyama, M., Arai, T., Adachi, H., Tsuji, H.: A surveillance system using small unmanned aerial vehicle (UAV) related technologies. NEC Tech. J. **8**(1), 68–72 (2015)
102. Finn, R.L., Wright, D.: Unmanned aircraft systems: Surveillance, ethics and privacy in civil applications. Comput. Law Secur. Rev. **28**(2), 184–194 (2012)
103. Kingston, D., Beard, R.W., Holt, R.S.: Decentralized perimeter surveillance using a team of UAVs. IEEE Trans. Robot. **24**(6), 1394–1404 (2008)
104. Hakiri, A., Berthou, P., Gokhale, A., Abdellatif, S.: Publish/subscribe-enabled software defined networking for efficient and scalable IoT communications. IEEE Commun. Mag. **53**, 48–54 (2015)
105. Shin, D., Sharma, V., Kim, J., Kwon, S., You, I.: Secure and efficient protocol for route optimization in PMIPv6-based smart home IoT networks. IEEE Access **5**, 11100–11117 (2017)
106. Wu, D., Arkhipov, D.I., Kim, M., Talcott, C.L., Regan, A.C., McCann, J.A., Venkatasubramanian, N.: ADDSEN: adaptive data processing and dissemination for drone swarms in urban sensing. IEEE Trans. Comput. **66**, 183–198 (2017)
107. Zeng, Y., Zhang, R.: Energy-efficient UAV communication with trajectory optimization. IEEE Trans. Wirel. Commun. **16**, 3747–3760 (2017)
108. Sharma, V., You, I., Kumar, R.: Energy efficient data dissemination in multi-UAV coordinated wireless sensor networks. Mob. Inf. Syst. **2016**, 8475820 (2016)
109. Naqvi, S.A.R., Hassan, S.A., Pervaiz, H., Ni, Q.: Drone-aided communication as a key enabler for 5G and resilient public safety networks. IEEE Commun. Mag. **56**, 36–42 (2018)
110. Amazon. Amazon prime air. http://www.amazon.com
111. Bristeau, P-J, Callou, F., Vissiere, D., Petit, N., et al.: The navigation and control technology inside the ar. drone micro uav. In: 18th IFAC World Congress, vol. 18, pp. 1477–1484 (2011)
112. Yıldırım, Ş., Çabuk, N., Bakırcıoğlu, V.: Design and trajectory control of universal drone system. Measurement **147**, 106834 (2019)
113. Moussid, M., Sayouti, A., Medromi, H.: Dynamic modeling and control of a hexarotor using linear and nonlinear methods. Int. J. Appl. Inf. Syst. **9**(5), 9–17 (2015). https://doi.org/10.5120/ijais2015451411
114. Li, S., Wang, Y., Tan, J., Zheng, Y.: Adaptive RBFNNs/integral sliding mode control for a quadrotor aircraft. Neurocomputing **216**, 126–134 (2016). https://doi.org/10.1016/j.neucom.2016.07.033

115. Kotarski, D., Benic, Z., Krznar, M.: Control design for unmanned aerial vehicles with four rotors. Interdiscip. Descr. Complex Syst. **14**, 236–245 (2016). https://doi.org/10.7906/indecs.14.2.12
116. Hadi, N., Ramz, A.: Tuning of PID controllers for quadcopter system using hybrid memory based gravitational search algorithm—particle swarm optimization. Int. J. Comput. Appl. **172**, 9–18 (2017). https://doi.org/10.5120/ijca2017915125
117. Alwi, H., Edwards, C.: Fault tolerant control of an octorotor using LPV based sliding mode control allocation. In: 2013 American Control Conference, IEEE, pp. 6505–6510 (2013). https://doi.org/10.1109/ACC.2013.6580859
118. Niemiec, R., Gandhi, F., Singh, R.: Control and performance of a reconfigurable multicopter. J. Aircr. 1–12 (2018). https://doi.org/10.2514/1.C034731
119. Zabunov, S., Mardirossian, G.: Innovative dodecacopter design–Bulgarian knight. Int. J. Aviat. Aeronaut. Aerosp. **5**(4), 9 (2018). https://doi.org/10.15394/ijaaa.2018.1293
120. Brischetto, S., Ciano, A., Ferro, C.G.: A multipurpose modular drone with adjustable arms produced via the FDM additive manufacturing process. Curved Layer. Struct. **3**, 202–213 (2016). https://doi.org/10.1515/cls-2016-0016
121. Xiu, H., Xu, T., Jones, A.H., Wei, G., Ren, L.: A reconfigurable quadcopter with foldable rotor arms and a deployable carrier. In: 2017 IEEE International Conference Robotics Biomimetics, IEEE, pp. 1412–1417. https://doi.org/10.1109/ROBIO.2017.8324615
122. McKay, M., Niemiec, R., Gandhi, F.: An analysis of classical and alternate hexacopter configurations with single rotor failure. In: 73rd AHS International Annual Forum, pp. 1–11 (2017). https://doi.org/10.2514/1.C035005
123. Kuric, M., Lacevic, B., Osmic, N., Tahirovic, A.: RLS-based fault-tolerant tracking control of multirotor aerial vehicles. In: IEEE/ASME International Conference on Advanced Intelligent Mechatronics, AIM, IEEE, pp. 1148–1153 (2017). https://doi.org/10.1109/AIM.2017.8014173
124. Wu, J., Ma, J., Rou, Y., Zhao, L., Ahmad, R.: An energy-aware transmission target selection mechanism for UAV networking. IEEE Access **7**, 67367–67379 (2019)
125. Bischoff, P.: Surveillance Camera Statistics: Which Cities Have The Most CCTV Cameras? (2020). [online] United Kingdom: Comparitech. Available at: https://www.comparitech.com/vpn-privacy/the-worlds-most-surveilled-cities/. Accessed 23 July 2020
126. Hoe, S.L.: Defining a smart nation: the case of Singapore. J. Inf. Commun. Ethics Soc. (2016)
127. Rahman, M.F.B.A.: Protecting the vertical space of cities: perspectives for Singapore. In: Lee Kuan Yew School of Public Policy Research Paper, (17–19) (2017)
128. India Video Surveillance Market (2019–2025). [online] New Delhi, India: 6Wresearch, p. 140. Available at: https://www.6wresearch.com/industry-report/india-video-surveillance-market-2019-2025. Accessed 26 July 2020
129. Khan, N.A., Jhanjhi, N.Z., Brohi, S.N., Usmani, R.S.A., Nayyar, A.: Smart traffic monitoring system using Unmanned Aerial Vehicles (UAVs). In: Computer Communications (2020)
130. Puri, V., Nayyar, A., Raja, L.: Agriculture drones: a modern breakthrough in precision agriculture. J. Stat. Manage. Syst. **20**(4), 507–518 (2017)
131. Khan, N.A., Jhanjhi, N.Z., Brohi, S.N., Nayyar, A.: Emerging use of UAV's: secure communication protocol issues and challenges. In: Drones in Smart-Cities, pp. 37–55. Elsevier (2020)
132. Baheti, R., Gill, H.: Cyber-physical systems. Impact Control Technol. **12**(1), 161–166 (2011)
133. Banerjee, A., Venkatasubramanian, K.K., Mukherjee, T., Gupta, S.K.S.: Ensuring safety, security, and sustainability of mission-critical cyber–physical systems. Proc. IEEE **100**(1), 283–299 (2011)
134. Nayyar, A., Nguyen, B.L., Nguyen, N.G.: The internet of drone things (IoDT): future envision of smart drones. In: First International Conference on Sustainable Technologies for Computational Intelligence, pp. 563–580. Springer, Singapore (2020)

Efficient Design of Airfoil Employing XFLR for Smooth Aerodynamics of Drone

Shubhanshu Verma, Sachin Kumar, Pooja Khanna, and Pragya

Abstract With limitations in access to places with extreme environmental conditions and unfavorable places, role of unmanned aerial vehicle have found a vital and potential role in reaching out to such places. Earlier the only use unmanned aerial vehicle UAV's was found only in the fields of military but with the advancement in time these flying sets of technology are ready to serve in almost every fields of work whether it be agriculture, transportation, door to door delivery, 3D mapping, photography, military needs, providing urgent medical help etc. UAV's can be quite vast in terms of their designs ranging from quadcopters to octocopters, fixed winger drones to a combination of both the multi-rotors and fixed winger, the V-to-l's. Technological innovations creating diversified applications areas for UAV's, generating more complex additions to the design to fulfil applications specific requirements. Additions involve different electronic components and accessories to the UAV's architecture i.e. like camera with anti-noise, night vision and flight stabilization supports, sensors for smooth flight, installation of relays, clamps or automatic guns etc., however every new addition to the design, poses new issues and challenges in the design architecture and functioning of the drones, thus the process involves complex permutations & combinations and skilled hard work. The developments in the field of manufacturing materials and electronic equipment's like long life batteries has boosted this industry to open its doors for the public use. This paper is an effort to give a brief about the various variety of drones available, designing aspect of these machines, the grassroot manufacturing practices, the basic electronic equipment's required for its flight etc.

S. Verma · S. Kumar (✉) · P. Khanna
Amity University, Lucknow Campus, Lucknow, India
e-mail: skumar3@lko.amity.edu

S. Verma
e-mail: shubhanshuv49@gmail.com

P. Khanna
e-mail: pkhanna@lko.amity.edu

Pragya
MVPG College, Lucknow University, Lucknow, India
e-mail: dr.pragya2011@gmail.com

R. Krishnamurthi et al. (eds.), *Development and Future of Internet of Drones (IoD): Insights, Trends and Road Ahead*, Studies in Systems, Decision and Control 332,
https://doi.org/10.1007/978-3-030-63339-4_5

The paper also simulates a design for air foil and aircraft for efficient and smooth flying operations, simulation was performed using XFLR simulator software.

Keywords Airfoil · XFLR · NACA · Coefficient of lift · Angle of attack · Lift

1 Introduction

With limitations in access to places with extreme environment and unfavorable places, role of unmanned aerial vehicle has found a vital and potential role. Earlier the only use unmanned aerial vehicle UAV's found was in the fields of the military but with the advancement in time these flying sets of technology are ready to serve in almost every fields of work whether it be agriculture, transportation, door to door delivery, 3D mapping, photography, military needs, providing urgent medical help etc.

At first sight these UAV's will only seem out to be just small planes attached with some motor and propellers flying in the sky, with a person standing on the ground controlling it. But it's not just this, it involves a great knowledge of the aerodynamics of the plane, the designing of it, understanding of electronics involved in it. Initially fixed wing drones were employed, depicted in Fig. 1 and various terminologies are:

- Fuselage—it is the main body part of every plane. Comparing it to the real plane it serves as the location for the crew, passengers, cockpit and all the cargo that is carried. However when it comes to drones the fuselage serves as the base where the various electronics are fitted such as the battery, esc, motor, the servo motors for the tails etc. The front of the fuselage i.e. the nose of the plane is used to mount the motor.

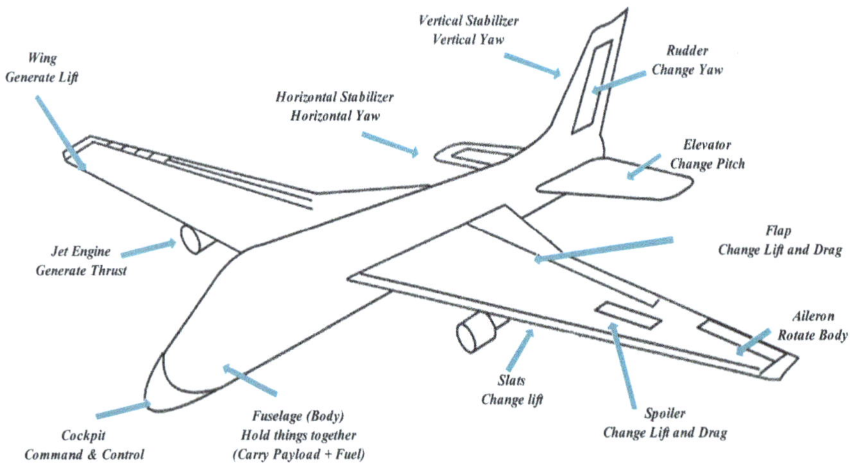

Fig. 1 Fixed wing drone

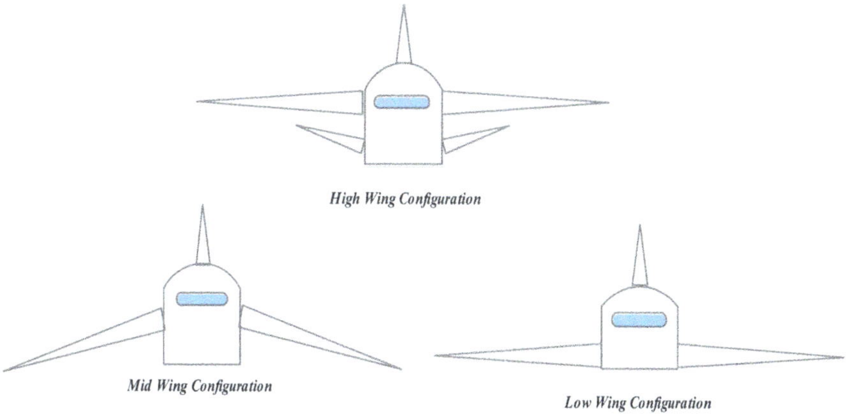

Fig. 2 Different configuration of wings

- Main wing—it is the most Important part of any aircraft whether it be a real plane or a fixed wing drone. The function of wing is to generate lift for the aircraft so that it can fly in the sky. The science involved behind designing a wing is very complicated as it involves the analysis of the various airfoils of which the main wing is made of. These wings can be mounted at different locations on the fuselage. Figure 2 depicts different configuration of wings.
- Tails—the traditional fixed wingers consist of 2 tails that are the horizontal and the vertical tails. The main function of the tail is to balance/stabilize the drone
  - Horizontal tail/stabilizer—stabilizes the drone along the pitching axis
  - Vertical tail/stabilizer—stabilizes the drone along the yawing axis.
- Propeller—the propeller is used along with motor and provides the drone the thrust or the speed for its forward motion.
- Control surface—the control surfaces refers to those surfaces of the drone which can be controlled manually using the remote control.

– Each of the traditional fixed wingers consists of 3 control surfaces.

  a. Ailerons—the ailerons refers to the control surfaces present on the main wing. They help the aircraft for its rolling motion.
  b. Elevators—it is the control surfaces present on the horizontal tail. It controls the aircraft's pitch motion.
  c. Rudder—it is the control surface present on the vertical tail. It controls the aircrafts yaw motion.

An aircraft basically supports three axis motion, which are -

- The pitching movement, in layman language the up and down movement of the plane is performed using the Elevator.

- The rolling motion, i.e. the sideways turning motion of it is performed using the Ailerons.
- The yaw motion that is turning about while staying at same position is performed using Rudder.

  All these motions of the aircraft are performed along its center of gravity [1–4].

1.1 Airfoil: Heart of Aircraft

An airfoil can also be defined as the best possible shape of the wing which is suitable for creating low pressure over its upper surface so that it could generate the required lift for the flight, Fig. 3 depicts the generalized structure and different components of airfoil

- Leading edge—the front portion of any airfoil is termed as the leading edge.
- Trailing edge—the ending point of any airfoil is termed as the trailing edge.
- Chord or chord length—the total length of any airfoil/wing is termed as its chord
- Chord line—it refers to a straight line which joins the leading edge and the trailing edge and thus divides the airfoil into 2 parts called as the upper camber and the lower camber.
  If the chord line divides the airfoil into 2 equal parts that is if the upper camber and the lower camber are equal then the chord line is called the "mean aerodynamic chord".

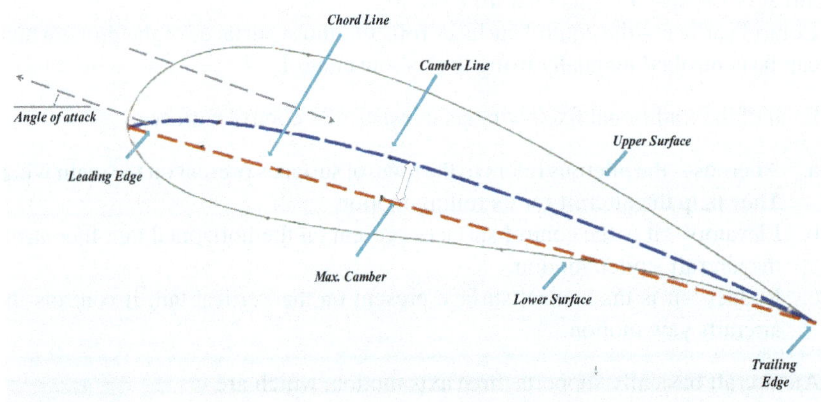

Fig. 3 Airfoil

Airfoil can be categorized into two types.

1. Symmetric airfoil—airfoil in which the upper camber area and the lower camber area are equals.
2. Un-symmetric or cambered—airfoil in which the upper camber and lower camber area is unequal.

There is also a special series of airfoils designed by the "national advisory committee of aeronautics" at NASA. These airfoils were termed as the NACA airfoils. Across the globe these airfoils are highly considered due to their excellent characteristics [2–7]. The chapter is organized as follows; Sect. 2 presents the motivation for taking up the work, Sect. 3 provides the science behind drone flight, Sect. 4 explores in brief; basics of airfoil—the heart of aircraft, Sect. 5 presents the result obtained, Sect. 6 discusses about future scope of the work conducted and finally paper concludes with optimum choices for design in Sect. 7.

2 Motivation

There were various experiments going on with various methods to reduce drag and to improve efficiency of wind turbine, airfoil of flight vehicle etc. A detailed study was done on airfoil to reduce drag on the trailing side. Since one of the main reasons of drag or pressure drag is the formation of turbulent on the trailing side. Roughness was created in turbulent region of the smooth airfoil. This result in decrease in drag and gives better aerodynamic efficiency. Experimental work was done using a wind tunnel and it was found that lower roughness gives the better result than upper roughness compared to smooth airfoil [1]. Airfoil four-digit nomenclatures tell camber and thickness of airfoil. From this, we can find airfoil series. This helps to change the camber of airfoil and location of maximum thickness [2]. At various Reynolds no aerodynamic property of smooth and rough airfoil was compared. It was observed that aerodynamic losses were increased at high Reynolds no and losses were decreased at low Reynolds no. in rough airfoil. Separation bubble becomes weak due to roughness of surface at low Reynolds no [3]. Two row and eight rows dimples on flat surface were investigated and observed that dimples were effective to convert laminar into turbulent at low Reynolds no. Multiple rows increases strength of mixing flows. The analysis was also done on sphere where it was found that drag was decreased in dimpled sphere compared to smooth sphere [4]. Sphere is a bluff body and dimple pattern on the spherical body was studied by understanding the flow of hydrodynamics. Experimental and numerical investigation was performed. By using various turbulent models, K-w turbulent model gives good results and k-ε is not useful for curved body. Geometry of dimple also affects significantly aerodynamic characteristics. Formation of small separation bubble leads to delay of the flow separation [5]. In this paper active flow control system was used. The vortex generator jet added to the main wing to suppress the separation. Slatless high lift airfoil was used with a flap. VG location was on the pressure side close to leading edge. Two test were performed with constant mach no. Reynolds no varies during study. It was found that lift is inversely proportional to Reynolds no. Dynamic blowing results higher lift [6]. NACA 2412 dimpled and smooth 3d printed airfoil was tested at various angles of attack. Velocity varies 5, 10 m/s with different angle of attack that varies from 60 to 180. The complete evaluation was at low angle of attack. Printed 3D model of airfoil was cured with acetone vapour bath for smoothness of surface. 3-D printer was used to save time and money. After experimenting it was observed that dimpled airfoil give increase in stall then smoothed airfoil. Stall angle of attack is increased by 2.10 [7]. A car model was used and dimple ratio (depth to diameter) also designed on the car body. In this paper 5 different dimple ratio were used. Simulation was done on ANSYS simulation tool [8]. Computational fluid dynamics was used to study of airfoil. Other researcher's drawbacks were the motivation to do study of passive device to improve the lift and decrease in drag of airfoil. Most efficient position of the dimple and cylinder was found out. The dimple model shows good result than cylinder. Modeling was done in CATIA V5 R20, preprocessing was done on ANSYS ICEM CFD 14.0 and post processing done on ANSYS FLUENT 14.0. If the location of dimple or cylinder was at the optimum position, then it gives better outcomes. NACA 4412 studied and model prepared in CATIA V5 R20. Both models were analyzed but it was found that dimple model gives better result. Fine meshing gives more accurate result, but computer memory and time are the limitations. Result showed that dimple airfoil gives the better result compared to cylinder [9]. Turbulence was created by using dimples due to generation of vortices. This result in decreased pressure drags and improves maneuverability. Different shapes of dimples were used. A comparative study was done between inward and outward dimple. A combination of two (semi spherical followed by square) dimple with constant height and depth ratio studied. Inward dimpled shape gives better lift coefficient in both single and compound dimples [10]. NACA00012 airfoil was tested under different turbulent models i.e. [Spalart—Allmaras, Realizable k-ε and k-ω shear stress transport (SST)]. These turbulent models were compared and validated with experimental data. It was found that k-ω shear stress transport (SST) gives the best result for given airfoil. 80,000 cells were taken for simulation. Air velocity was taken constant. Before solving it the main important work was to find out the transition point. Transition point should be modeled to get more accurate result. Here commercial CFD software was used. According to this if much amount of nodes are used, result will be more accurate, but huge amount of nodes take much time in computation. Here C type grid topology was used, and 80,000 quadrilateral cells were taken. Transition point was determined by hit and trial method. If the value of simulated CD is greater than experimental value, it means that transition point chosen is wrong, turbulent region is larger. So, author have to shift it in right side and accordingly he can determine transition point. In result there was a disagreement between the data at near stall. The predicted drag coefficient was higher than experimental data. This is because the actual airfoil has laminar over the half of leading side. Turbulent models consider turbulent boundary layer throughout its length [11]. NACA4315 model was used to analyze the aerodynamic properties. Bumps were used at upper surface on trailing side. The bumps were created 80% from leading edge. Regular and bumped airfoils were compared. It shows that stall angle increased due to controlling on flow separation [12]. With the help of nose flap, slats, variable camber can increase lift coefficient but these are mechanically complex. The addition of roughness or vortex generator is the simplest technique to increase lift coefficient. Vortex generator can be taken at 10% of chord from leading edge, its height and length is about 1% of chord and 2–3% of chord respectively [13]. Vortex generator was provided on NACA 0015 airfoil to improve lift and decrease drag. Vortex generator placed on 10% from the leading edge. Experimental model developed for analysis. VG height, space between them was also considered. The vortex generator height plays an important role in analysis. The optimized geometry improves result by 14% in lift and 16% in drag [14]. Aerodynamics characteristics were measured and flow separation is determined. NACA0015 model airfoil is used and CProgramming is used. Cp and x/c graph was analyzed and found that at 120 AOA there is no flat area and at an AOA 200 it seems that value of Cp was approx smooth, showing boundary later separation. It is clear from this paper that flow separation cannot be controlled or delayed with conventional ways [15]. Design of Natural Laminar Flow on airfoil can be used to increase in the percentage of laminar flow and to decrease drag by using Boundary Layer Mixing Devices. Vortex generators were used as a boundary layer mixing device in current uses throughout aircraft industries. Vortex generators are passive flow control device and do not require any additional power to operate. In order to reduce laminar bubble several airfoil section were designed. Code was modified to design airfoil at the upper and lower surface with the help of vortex generator [16]. NACA 0012 airfoil used to study transient growth in laminar separation bubble. Laminar separation bubble formed due to separation of flow, the flow convert from transition to turbulent and reattached to surface. This LSB can affect the performance of airfoil. It was observed that flow is periodic with temporal frequency of 1.27. In two dimensional flows it was observed that monotonic energy increases with interval of time and it can reach up to around six order of larger [17]. Wing was designed with dimples on the upper surface of the wing. Analysis was done using ANSYS. It was found that skin frictions increased but drag reduced in such an amount that effect of skin friction can be neglected [18].

3 Science—Behind Drone Flight

The two basic principles involved behind the flying of any flight are the Newton's laws and the Bernoulli's principle, Newton's laws come into play while estimating various forces acting on the plane. Figure 4 depicts different forces acting on a plane, thus there is a need to balance each force having influence for a table flight. Bernoulli's principle states that pressure is inversely related to the velocity and the theorem explains the reason behind the generation of lift by the wing.

Airfoil shape ensures that the velocity of wind flowing over its surface is more than the velocity of the wind flowing underneath it, thus due to the inverse relation between the pressure and the velocity the pressure due to the greater velocity of wind

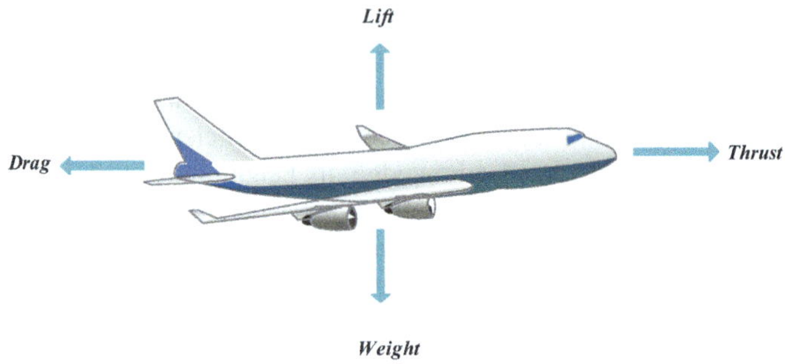

Flight Condition	*Effect*
Lift > Weight	*Plane Rises*
Weight > Lift	*Plane Falls*
Drag > Thrust	*Plane Slows*
Thrust > Drag	*Plane Accelerates*

Fig. 4 Forces acting on plane

above the surface of wings becomes less there. Following the same principle less velocity of wind under the wing the pressure there becomes more as a result of which the pressure under the wings become more and thus the wings generate the necessary lift required.

Amount of lift and drag experienced by the plane can be estimated from following mathematical expression given in Eqs. 1 and 2 respectively:

$$Lift = C_L S\left(\frac{1}{2}\rho V^2\right) \tag{1}$$

CL = Coefficient of Lift
S = Surface Area
ρ = Density of Air (Altitude)
V = velocity.

$$F_D = \frac{1}{2}\rho V^2 C_D A \tag{2}$$

FD = drag force
P = Density of the fluid
V = speed of object relative to the fluid
A = cross sectional area

CD = drag coefficient.

Above equations are employed to predict lift and drag experienced and generated by aircraft respectively, values of CL and CD here is calculated by the analysis of your airfoil on various software's. Wings have certain important terminologies associated:

Swept wing—a wing that angles backwards or occasionally forwards from the point of its root chord, depicted in Fig. 5.

Aspect ratio—it is defined as the ratio of the square of wing span to the area of the wing, depicted in Fig. 6.

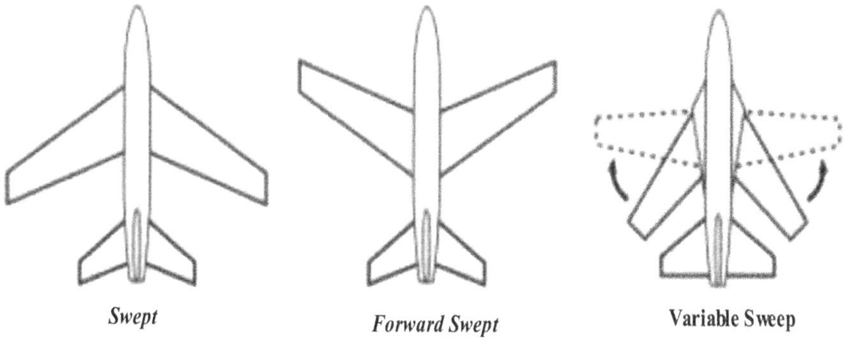

Fig. 5 Swept ring

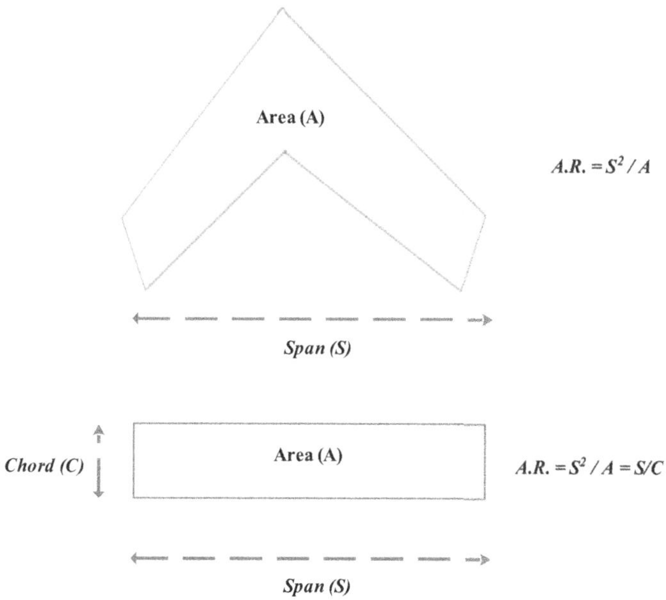

Fig. 6 Aspect ratio

Fig. 7 Taper ratio

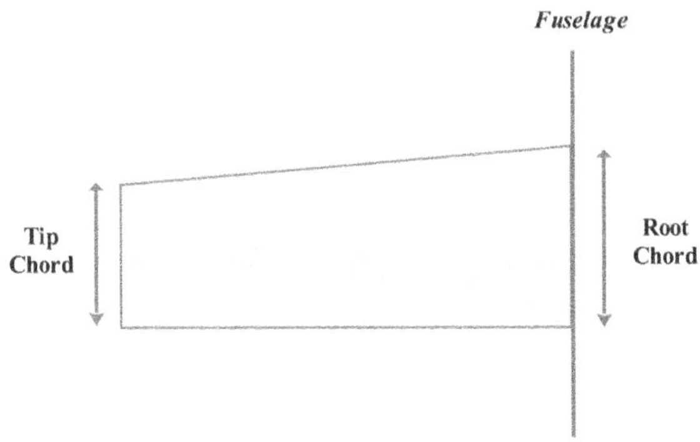

Aspect ratio plays a vital role when it comes to the turning a plane during flight navigation, an easy turn would require lower aspect ratio. Aspect ratio is directly dependent on the wingspan thus for plane to have low aspect ratio, it should have a smaller wing span.

Taper ratio—it is defined as the ratio of the length of tip chord to the length of root chord, depicted in Fig. 7.

Reynold's number—Reynold's number is a dimensionless quantity, that explains mechanics involved in the flow of liquid. However when it comes to the flying of an aircraft this dimension less quantity is used to understand the behavior or flow of air over the surface of plane, mathematical expression for the same is given in Eq. 3.

$$Re = \frac{\rho ul}{\mu} = \frac{ul}{\upsilon} \tag{3}$$

where

ρ = density of liquid
μ = dynamic viscosity of fluid
l = characteristic linear dimension
u = flow speed
υ = Kinematic viscosity

Reynold's number decide whether the flow of air over the surface of aircraft will be laminar/smooth or turbulent/disturbed at low Reynold's numbers the flow is generally laminar but at high Reynold's number the flow is generally turbulent. turbulent airflow sometime may result in formation of vortex, which is basically circular loops of air formed behind the trailing edge of the wings.

Angle of attack—angle of attack is defined as the angle between the chord line of an airfoil and the vector representing the flow of air, depicted in Fig. 4.

The lift generated by any wing of an aircraft depends directly upon the angle of attack of air on its surface. With the increase in the angle of attack of air the lift generated by the wings also increases consequently. When the angle of attack of air is increased the air flow above the airfoil's surface becomes more pronounced, that is the air begins to flow less smoothly and the point of separation of airfoil and air changes. The increase in Cl value with increase in angle of attack is only up to a certain limit and this point is known as the "critical angle of attack "or the "stalling point".

The general angle of attack for the commercial planes is between 0° and 15° however in case of drones the wings are analyzed generally between 0° and 10°. For the fighter aircraft the angle of attack increases more than 15° at some [19–23].

4 Understanding the Heart of Aircraft—Airfoil

During initial design days of aircraft power capacity of aircraft engines was increased in an effort to reduce drag, and further to increase lift and velocity of aircraft. Importance of aircraft aerodynamics came in role in the twelfth century. Modification in airfoil design plays a vital role in aerodynamics. Diversified airfoil design series were developed to address different job targets which aircraft addressed, i.e. carrier craft, payload craft, fighter craft etc. Every series had its own specific aerodynamics characteristics to smooth flight. The flow separation on airfoil increases pressure drag. During flight, increment or decrement in lift will create incremental start or stop of vortices, always with the result that a smooth parallel flow is maintained at trailing edge. Therefore, process of choosing the appropriate airfoil is a very important, as its determines the over-all flight behavior of your aircraft. Whether your aircraft will be a good glider or heavy pay lifter or a good arcobacter, all depends upon the airfoil design, thus selection of optimal airfoil plays a vital role. Before beginning the process you need to first finalize the tasks which you will perform with your aircraft and then select the airfoil design which can help you with the same.

To design a drone for photography, thus we would like that wings of aircraft should be such that they could generate sufficient lift so that aircraft with a heavy camera mounted on it, can fly smoothly. Moreover we would also like that our drone should be a good glider too so that it helps me to capture excellent shots. Since there are millions of airfoils designed, therefore it would be best to identify application and look for specific airfoil designs which can meet our requirements, lift coefficient (C_L) is a coefficient without dimension, that establishes a relation between lift generated between force on lifting a mass against the density of fluid around it the mass, fluid velocity and reference value set up. C_L is a dependent variable, it is basically a function of body angle and the fluid flow, Reynolds value associated with it and its Mach quantity. Highest value of lift coefficient that can be attained depends on the airfoil geometrical section, surface exposed whether its smooth or rough. Reynolds value and Mach No. There are two categories, first are those where the separation starts from the trailing edge and second are those where it starts just behind the leading edge. Maximum value of high lift having rear separation is decided by the geometrical design of the rear part of the section; Highest values are achieved by airfoils having big thicknesses and leading-edge radii and so have rear separations. Standard values of the order of 1.6 for early conventional models and about 2.0 for more advanced airfoils, recent ones have been particularly designed for optimum high lift is achieved.

Lift curve slope is a quantity of how quickly the wings generates lift with change in α, maximum theoretical value that can be achieved is 2π, mostly advanced airfoils deviate from it. Lift curve slope of a wings is mostly less than that of the airfoils it features.

Lift-drag ratio $(L/D)_{max.}$ is a quantity that measures the aerodynamic cruising efficiency of the airplane, a higher value is favorable L/D ratio is typically required as one of the major targets of airplane design; Amount of lift required by a specific is defined by its weight, achieving required lift with relatively lower drag amounts to improved fuel efficiency in airplane. Value of L/D is estimated by calculating the lift generated, then dividing the same by the drag at that velocity [20–23].

Aerodynamic force because of angle of attack acts primarily at quarter chord point and depends directly on at angles below the stall. This explains the reason behind small negative value of zero lift angle of attack for cambered airfoil as here lift because of camber is same as well as opposite to the lift because of angle of attack. Thus, at relatively low velocity, when the angle of attack is mostly large, large amount of the aerodynamic force is because of angle of attack and quarter chord point has center of pressure (cp) almost on it. Whereas, at high velocity, when the angle of attack is mostly small, a larger percentage of aerodynamic force is because of camber dependent component and cp is almost at the center point of the chord. Therefore cp mostly lies in between quarter chord and mid chord points. Thus, cp happens to be the function of angle of attack. Table 1 depicts few of the state-of-the-art airfoil with specifications.

With abundance of airfoil design available, to carry our analysis, we shortlisted airfoils specific for our application needs, work proposed requires to analyze each airfoil shortlisted under certain conditions; such as at various angles of attack, different Reynold's numbers etc., simulation software employed is XFLR, XFLR software is a analysis tool used for the analyzing parameters related to wings, airfoils and aircrafts flying at row Reynold's number in a simulated environment.

Table 1 Standard airfoil specifications [24]

Airfoil	Max. lift coefficient (C_{lmax})	Lift curve slope ($C_{l0\alpha}$)	Max. lift drag $(L/D)_{max}$	Max. moment about quarter chord ($C_{m1/4c\ max}$)
SD 7062	1.371	0.354	25.327	−0.101
NACA 4412	1.184	0.368	26.362	−0.118
NACA 4415	1.344	0.382	27.187	−0.128
SD 7034	1.093	0.295	28.357	−0.083
S 2027	1.045	0.23	25.678	−0.077
NACA 4418	1.515	0.397	26.032	−0.142
Eppler 68	1.096	0.389	28.589	−0.137
NACA 2412	1.008	1.084	28.217	−0.065

Analysis tools which existed prior to XFLR provided only the facility to test the airfoils and that too platform required to create airfoils by specifying the coordinates, for example XFOIL whereas XFLR overcame all these difficulties as it allows us to analyze various parameter related to airfoils, which have been pre-designed and available as tools.

XFLR software has been utilized in the work for designing wings and tails, the software gives best results for models which fly at a relatively low height (Reynolds number) with fixed wing drones [23, 25–28].

5 Results

Airfoil characteristic was simulated on XFLR, curve of Coefficient of Lift (CL) versus Alpha (angle of attack) were analyzed at different Reynolds numbers ranging from 20,000 to 200,000 with regular intervals to analyze force distribution, hence necessary lift. Simulation experiments of shortlisted airfoil models gave promising results, standard parameters with modification, evaluated to optimal theoretical values. Results of shortlisted airfoils obtained are presented below:

1. NACA-4415 Airfoil

Simulation results obtained with XFLR is depicted in Fig. 8, NACA-4415 airfoil was analyzed at different Reynolds numbers ranging from 20,000 to 200,000 with

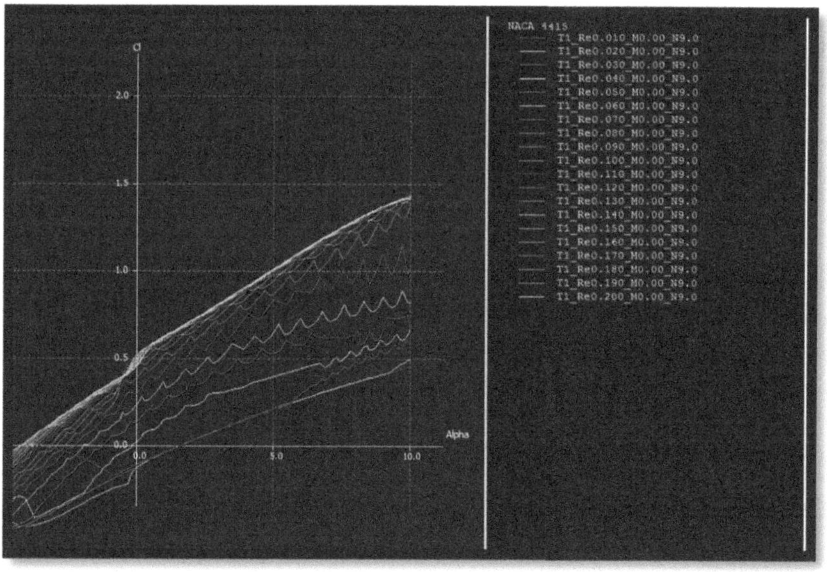

Fig. 8 C_L versus α simulation curve of NACA-4415

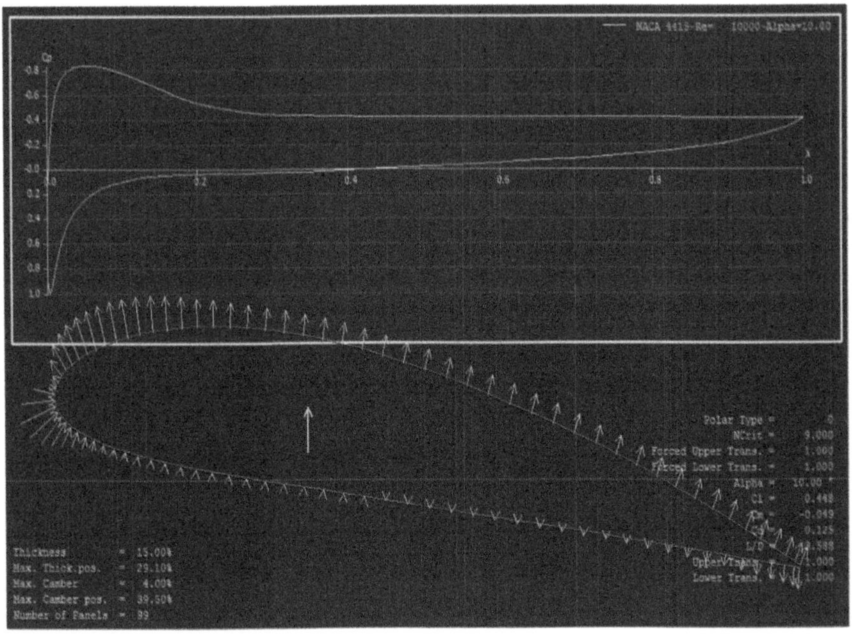

Fig. 9 Net force distribution on NACA-4415

regular intervals. Coefficient of Lift (C_L) versus Alpha (angle of attack) obtained for NACA-4415 is observed, initial slope of curves indicates normal behavior of the airfoil as the CL increases linearly with alpha values, however to attain get a high lift, high CL value at 0° of angle of attack is needed. Curves indicate C_L achieve a value of 0.5 against 0° angle of attack, thereby providing the necessary high lift.

Further, Fig. 9 depicts force distribution around the airfoil and green arrows around the airfoil indicates the net force direction which here is in upwards direction, direction indicates that airfoil will generate an upward lift necessary.

2. SD-7034

Simulation results obtained with XFLR is depicted in Fig. 10 for SD 7034, Curves indicate, that C_L at 0° angle of attack for this airfoil is slightly less than 0.5, however to attain a high lift, high C_L value at 0° of angle of attack is needed. Curves indicate C_L achieve a value slightly less than 0.5 against 0° angle of attack, but still its near to standard value thereby providing the necessary high lift.

Figure 11 Depicts force distribution along the airfoil. Curve indicates that net force in this case too is in upwards direction but the magnitude achieved is greater here as compared to NACA-4415.

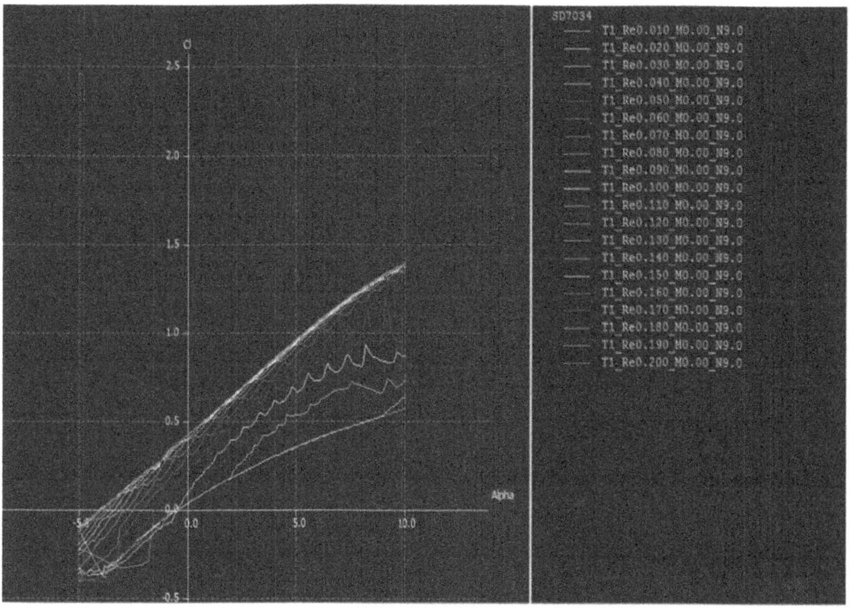

Fig. 10 C_L versus α simulation curve of SD-7034

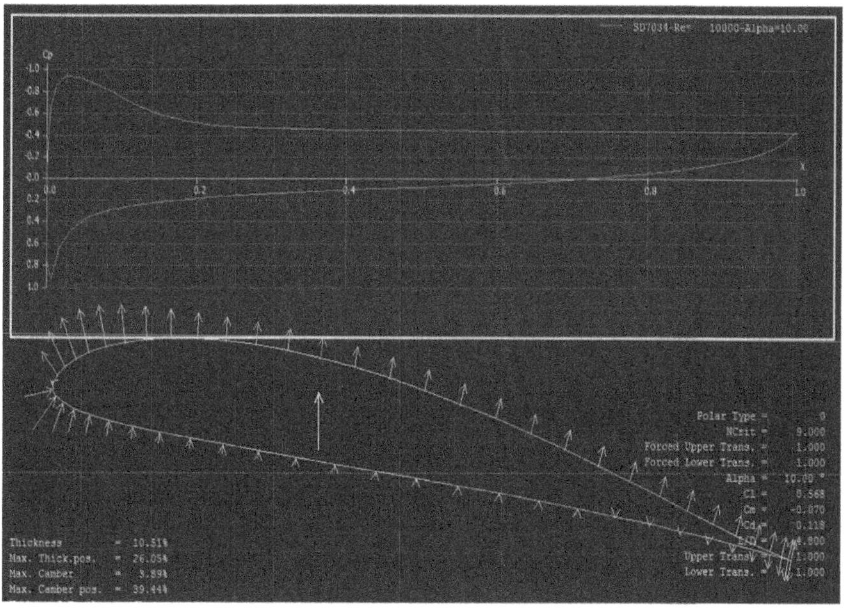

Fig. 11 Net force distribution on SD-7034

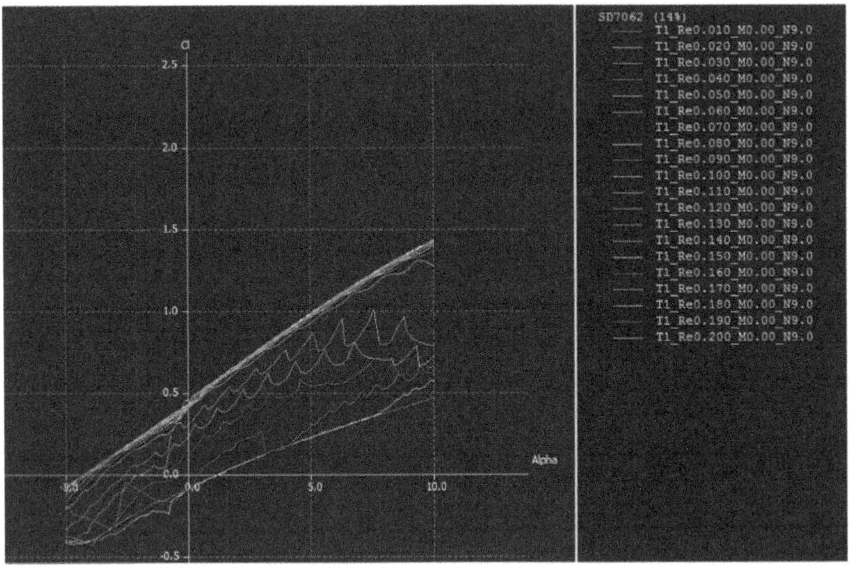

Fig. 12 C_L versus α simulation curve of SD-7062

3. SD-7062

Curve of C_L versus alpha for SD-7062 shows that the airfoil functions nearly to ideal characteristic, curve indicates airfoil works pretty well as straight lines are observed indicating a complete linear relation which is impressive further even C_L value for 0° angle of attack is almost ideal, depicted in Fig. 12.

The force distribution all over the surface of airfoil is near to ideal characteristic but the magnitude of net force is average, as visible from Fig. 13.

4. S-7075

Figure 14 depicts C_L versus alpha curves for S-7075. The curves obtained here are linear with constant velocity represented as straight line, however the value of C_L at 0° angle of attack is below average, around 0.4.

From the curve depicted in Fig. 15 it can be concluded that force distribution in this case is not good as the magnitude of upward forces is not much on the above layers whereas the magnitude of net force on overall structure is still good.

5. NACA009

Figure 16 depicts C_L versus alpha curves for NACA-009, curves show a unpredictable behavior, as can be gathered from curve values depicting a disturbed behavior further CL value at 0° angle of attack is zero which demonstrates extremely poor characteristic.

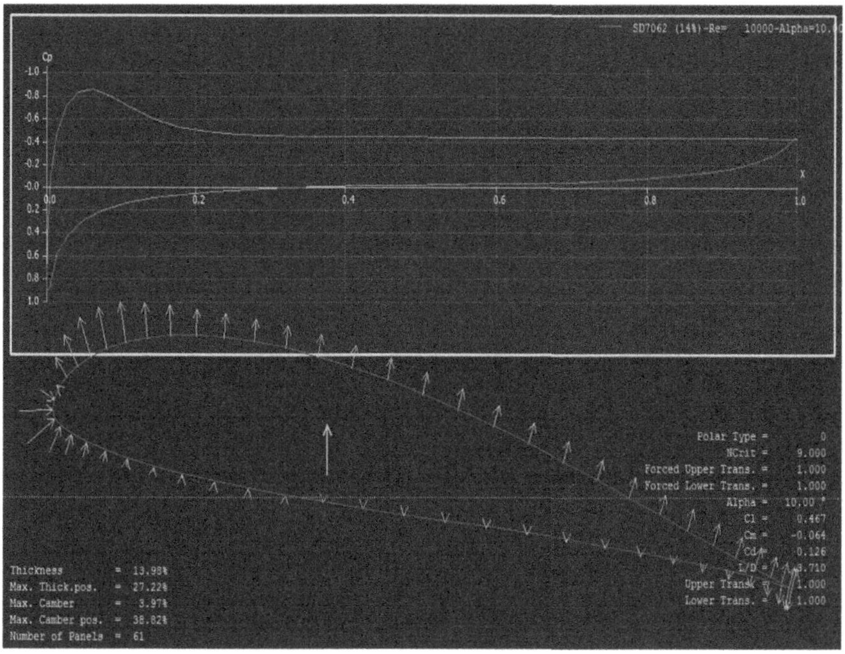

Fig. 13 Net force distribution on SD-7062

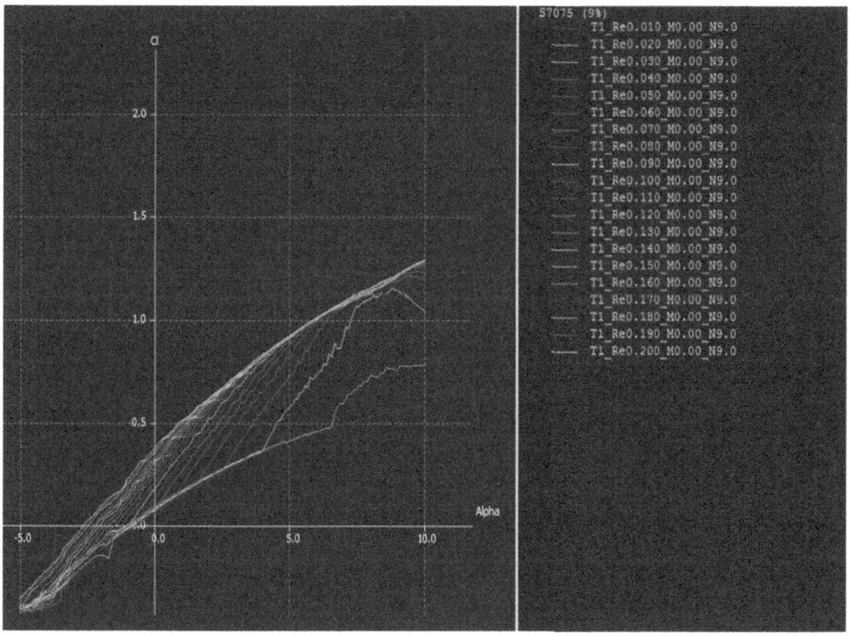

Fig. 14 C_L versus α simulation curve of S-7075

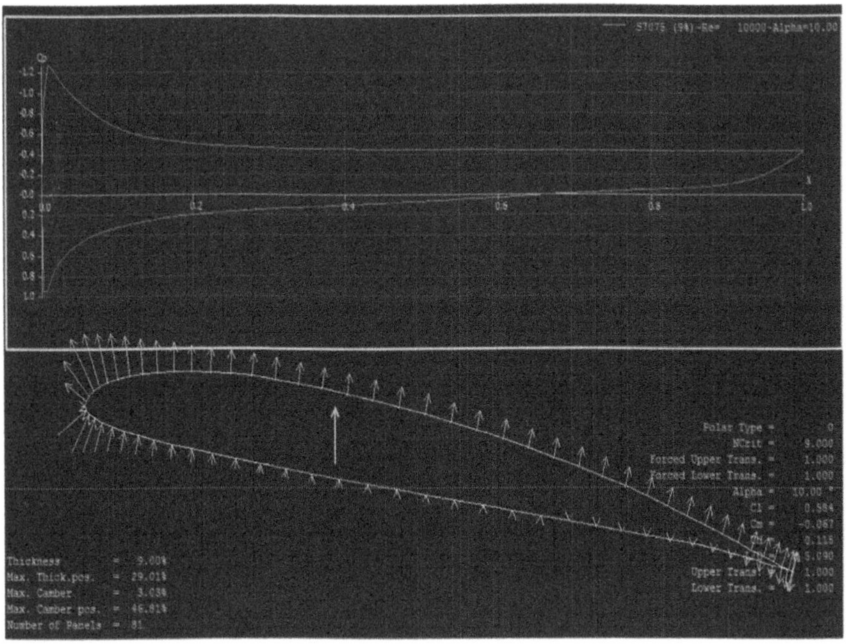

Fig. 15 Net force distribution on S-7075

From the force distribution curves depicted in Fig. 17 it can be concluded that distribution is not well structured and in addition magnitude of net force is also poor for NACA-009.

6 Future Scope

Work presented is an effort to establish and understand relationship between design of airfoils, optimum lift required and curve of Coefficient of Lift (CL) versus α (angle of attack). Results obtained can be employed as basis for designing target specific aircrafts, modification in airfoil design plays a vital role in aerodynamics, results obtained from work can be employed for developing diversified airfoil design series to address different job targets which aircraft address, i.e. carrier craft, payload craft, fighter craft etc. and parameters values obtained from work proposed can make value addition in airfoil designing. Lately, lot of applications with drones are exploring possibility of utilizing UAV to revolutionize solutions and work upon options to raise better informed decisions. Despite tremendous possibilities, the benefits that drones could potentially offer to foster effectiveness remains basically unexplored. With opportunities, there are certain open issues and challenges also present with working and design of airfoils, which are potentials research areas like determining

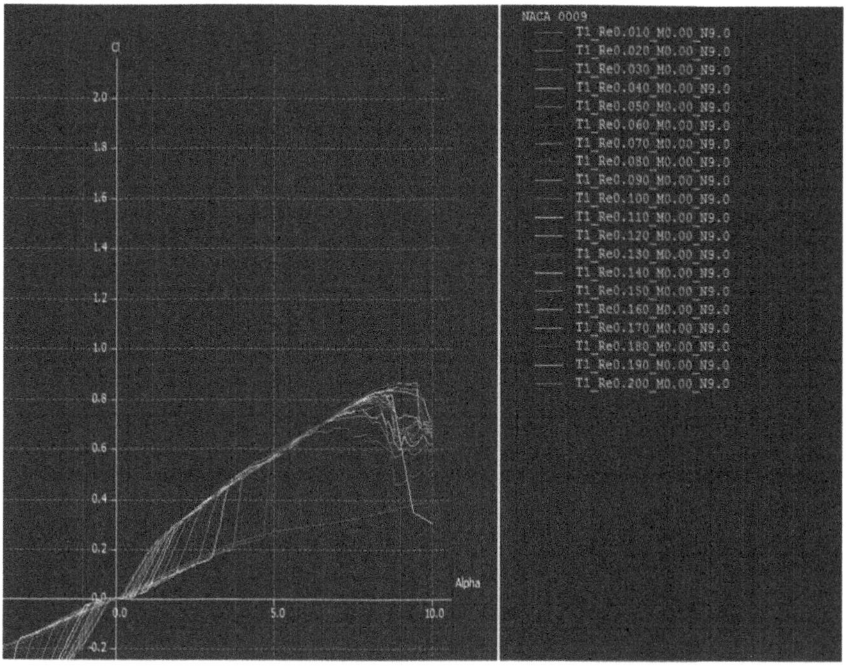

Fig. 16 C_L versus α simulation curve of NACA-009

ideal locations where launch should take place, time and infrastructure required for mapping, flight and reach to diversified remote areas with extreme environmental conditions, portability of device and design of airfoils for different wind resistance scenario. Results obtained can be extended for effective and efficient design of airfoil for different wind resistance conditions.

7 Conclusion

Airfoil characteristic was simulated on XFLR, curve of Coefficient of Lift (CL) versus α (angle of attack) were analyzed at different Reynolds numbers ranging from 20,000 to 200,000 with regular intervals to analyze force distribution, hence necessary lift. Simulation experiments of airfoil NACA-4415, SD-7034, SD-7062, S-7075 and NACA-009, models gave promising results however owing to the sensitivity, only the best are recommended. On comparing net force distribution of various airfoils it's seen that out of NACA-4415, SD-7034, SD-7062, S-7075 and NACA-009, only NACA-4415 and SD-7062 gave the maximum C_L value at 0° angle of attack, both provide the necessary lift, meeting standard values. Further force distribution curves of both are almost identical but NACA-4415 characteristic display a more distributed force pattern. Thus we can conclude that both the airfoils can be chosen for wing

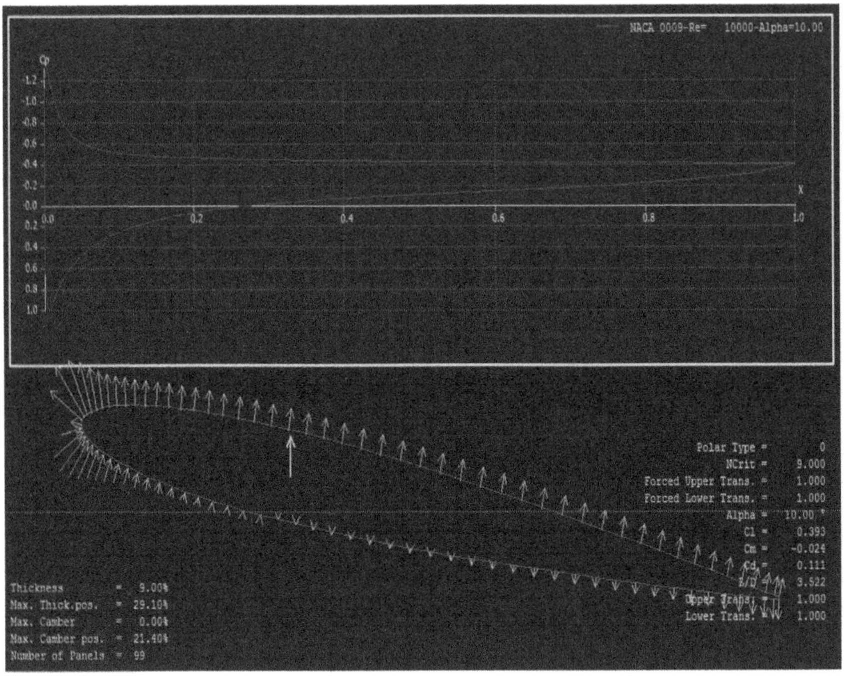

Fig. 17 Net force distribution on NACA-009

designing however NACA-4415 would be a more optimum choice for achieving optimum lift with uniform distributed pattern.

References

1. Dhiliban, P., Narasimhan, P.: Aerodynamic performance of rear roughness. In: The Eighth Asia-Pacific Conference on Wind Engineering, pp. 193–200. Research Publishing, Singapore Chennai, India (2013). ISBN: 978-981-07-8011-1
2. Anderson, J.D.: Fundamental of aerodynamics. In: Anderson, J.D. Jr. (ed.) Fundamental of Aerodynamics, 5th reprint edn., pp. 300–301. Tata McGraw-Hill, New Delhi India, New York (2012)
3. Bai, T., Liu, J., Zhang, W., Zou, Z.: Effect of surface roughness on the aerodynamic performance of turbine blade cascade. Propul. Power Res. **3**(2), 82–89 (2014)
4. Beratlis, N., Balaras, E., Squires, K.: Effects of dimples on laminar boundary layers. J. Turbul. **15**(9), 611–627 (2014)
5. Bogdanovic Jovanovic, J.B., Stamenkovic, Z.M.: Experimental and numerical investigation of flow around a sphere with dimples for various flow regimes. Therm. Sci. 1–14 (2012)
6. Casper, M., Scholz, P., Radesiel, R.: Separation control on high lift airfoil airfoil using vortex generator jet at high Reynolds numbers. In: 41st AIAA Fluid Dynamic Conference and Exhibit, pp. 1–11. American Institute of Aeronautics and Astronautics, Inc., Honolulu, Hawaii (2011)
7. Chavarin, A., Magtoto, A., Lee, D.: Experimental Comparison of Lift between a NACA. Design Laboratory Project, University of California, Irvine, pp. 1–16 (2014)

8. Chear, C., Dol, S.: Vehicle aerodynamics: drag reduction by surface dimples. Int. J. Mech. Aerosp. Ind. Mechatron. Eng. **9**(1), 202–205 (2015)
9. PrabhakaraRao, P., Sampath, S.: CFD analysis on airfoil at high angles of attack. Int. J. Eng. Res. **3**(7), 430–434 (2014)
10. Livya, E., Anitha, G., Vali, P.: Aerodynamic analysis of dimple effect on aircraft wing. Int. J. Mech. Aerosp. Ind. Mechatron. Eng. **9**(2), 350–353 (2015)
11. Eleni, D.C.: Evaluation of the turbulence models for the simulation. J. Mech. Eng. Res. **4**(3), 100–111 (2012). ISSN 2141-2383
12. Faruqui, S.H., AlBari, M.A., Emran, M.: Numerical analysis of role of bumpy surface to control the flow separation of an airfoil. In: Procedia Engineering, 10th International Conference on Mechanical Engineering, ICME 2013. 90, pp. 255–260. Elsevier, Bangladesh (2014)
13. Gyatt, G.W.: Development and testing of vortex Generators for Small Horizontal Axis. US Department of Energy, National Aeronautics and Space Administration, Lewis Research Center (1986)
14. Tebbiche, H., Boutoudj, M.: Optimized vortex generators in the flow separation control around a NACA 0015 profile. In: Proceeding of 9th International Conference on Structural Dynamics, EURODYN 2014, Porto, Portugal, pp. 3219–3226 (2014). ISSN: 23119020, ISBN: 9789727521654
15. Islam, M., Hossain, M.A., Uddin, M.N.: Experimental evaluation of aerodynamics characteristics of a baseline airfoil. Am. J. Eng. Res. **4**(1), 91–96 (2015)
16. Kerho, M., Kramer, B.: Enhanced airfoil design incorporating boundary layer mixing devices. In: 41st AIAA Aerospace Science Meeting & Exhibit, pp. 1–18. American Institute of Aeronautics and Astronautics Inc., Reno, NV (2003)
17. Loh, S., Balckburn, M.H., Sherwin, J.S.: Transient growth in an airfoil separation bubble. In: 19th Australasian Fluid Mechanics Conference, Melbourne, Australia, pp. 8–11 (2014)
18. Mohanasaravanan, P.: Flow analysis around the dimple wing on aircraft. Int. J. Eng. Res. Online **3**(2), 402–406 (2015)
19. Patacsil, M.B.: Dimples on Wing. California State Science Fair, California (2010)
20. Prince, S.A., Khodagolian, V., Singh, C.: Aerodynamic stall suppuration on airfoil section using passive air jet vortex generator. In: Coton, F. (ed.) AIAA J. **47**(9), 2232–2242 (2009)
21. Saravi, S.S.: A review of drag reduction by riblets and micro-textures in the turbulent boundary layers. Eur. Sci. J. **9**(33), 62–81 (2013)
22. Seling, M.S., Guglielmo, J.J.: High lift low Reynolds number airfoil design. J. Aircraft **34**(1), 72–79 (1997)
23. Sorensen, N.N., Zahle, F., Bak, C.: Prediction of the effect of vortex generators on airfoil performance. J. Phys.: Conf. Ser. **524**, 1–11 (2014)
24. http://airfoiltools.com/airfoil/details
25. Puri, V., Nayyar, A., Raja, L.: Agriculture drones: a modern breakthrough in precision agriculture. J. Stat. Manag. Syst. **20**(4), 507–518 (2017)
26. Nayyar, A., Nguyen, B.L., Nguyen, N.G.: The Internet of Drone Things (IoDT): future envision of smart drones. In: First International Conference on Sustainable Technologies for Computational Intelligence, pp. 563–580. Springer, Singapore (2020)
27. Khan, N.A., Jhanjhi, N.Z., Brohi, S.N., Nayyar, A.: Emerging use of UAV's: secure communication protocol issues and challenges. In: Drones in Smart-Cities, pp. 37–55. Elsevier (2020)
28. Khan, N.A., Jhanjhi, N.Z., Brohi, S.N., Usmani, R.S.A., Nayyar, A.: Smart traffic monitoring system using Unmanned Aerial Vehicles (UAVs). Comput. Commun. (2020)

Drone-Based Monitoring and Redirecting System

Adarsh Kumar and Saurabh Jain

Abstract Drones are discussed since the early 1990s. Nowadays, there is a wide range of applications. State of the artwork states mobility, positioning, propagation models, pattern movements, data collection, federated learning etc. features in the present scenario. This work presents case studies for the traffic monitoring prototype. In traffic monitoring, different traffic engineering equipment will be explored that can be integrated with drones through the Internet of Things (IoTs). Drones will be enabled through multiple means for traffic monitoring. Here, drones can collect data from sensors placed alongside the roads or they can use a video/image-based system. Opportunities for both types of systems will be explored in detail. This work also explores the possibilities to measure on-road vehicles' features, movement properties, building their profiles, constructing databases, analyzing, and visualization systems. The vehicle's feature identification starts from identifying the number plates, reading system, and sharing the collected information to the centralized database through drones. The possibilities of parallel processing in drone-based infrastructure are explored to shred the load over drones and efficient use of cloud infrastructure for experimental results that show significant accuracy, reliability, and performances. Further, the possibilities of integrating security primitives and protocols are explored. These primitives and protocols can be identified based on drones' resources. Lightweight primitives and protocols will be explored that reduces the hardware cost and give an efficient performance with high-security standards. The primitives and protocols take care of security properties like confidentiality, integrity, availability, authentication, and non-repudiation. Additionally, security measures with all primitives will be taken care of at data storage, processing, and propagation points. To control the complete infrastructure, a control room strategy can be designed. This strategy narrates the man-power requirements, operation, and duties.

Keywords Drones · Internet of Things · LiDAR · UAV · Drone Monitoring

A. Kumar (✉) · S. Jain
Department of Systemics, School of Computer Science, University of Petroleum and Energy Studies, Dehradun, India
e-mail: adarsh.kumar@ddn.upes.ac.in

S. Jain
e-mail: saurabh.jain@ddn.upes.ac.in

R. Krishnamurthi et al. (eds.), *Development and Future of Internet of Drones (IoD): Insights, Trends and Road Ahead*, Studies in Systems, Decision and Control 332,
https://doi.org/10.1007/978-3-030-63339-4_6

1 Introduction

In [1, 2], the drones, based on the wings type, are classified into two categories namely (i) Fixed (ii) Rotor (iii) Hybrid. The drones with rotor winds are further classified as (i) helicopter (ii) Multicopter. The fixed wind craft are capable of large-scale landscape coverage. Such crafts are used in the aviation industry, area surveying, and structural monitoring. These drones utilize the static wings along with forwarding air force for launching into air space. Examples of such fixed-wing systems are traditional aircraft, gliders, paper planes, and kites. The popular fixed wing-based drone is Raven. The demerits of this drone include high prices and required trained personal to lead launching, maneuvering, and landing. Next, the rotor wing crafts are capable of lifting from the ground using rotary wings. The common example of a single rotor wind craft is the helicopter. The drones have single or multiple rotor wings and are used for aerial shooting, inspection, photography, and filmography. The price of these drones is less compared to the fixed wind drones. The demerits of this rotor wind drone include that they are capable of handling less payload and limited time of flight. The Phantom drones are popular multirotor drones.

This work has discussed the recently observed challenges in the drone-based system. Thereafter, case studies are proposed and discussed for drone-based systems usage in various applications especially in traffic monitoring and pandemic handling. Various case scenarios are proposed to provide security in the system and the integration of cryptography primitives and protocols. This work discusses the various designs proposed for the drone-based system in different applications. The designs can be extended to have real-time implementation or integration with other application-based scenarios.

This work is organized as follows. Section 1 has given a brief introduction to the drone-based system. Section 2 presents related work from different dimensions. Section 3 discusses various case studies associated with drone-system based applications. Each system is having its pros and cons. This analysis can be used to implement the drone-based system for real-time applications. Finally, Sect. 4 concludes the work.

2 Related Work

Kumar et al. [3] studied various case studies in which drone-based scenarios are generated to tackle the COVID-19 situation. Here, the COVID-19 pandemic situation is monitored with indoor and outdoor operations. This model has proposed a security solution to drone operations and data collection. The security model considers the Lightbridge technology for secure channel establishment and cryptography primitives and protocols for data security. In conclusion, security aspects from different dimensions are considered to provide data privacy. In a complete security system, a secure channel is established between hospital infrastructure and data providers with proper control of the federal government. The multiple case studies provide a solution to inspect people, sanitize the indoor and outdoor areas, handle image-based data collection systems, and perform simulation-based analysis for operation control and system monitoring. Thus, the overall system is very well designed for pandemic scenarios. This work has also suggested improving the quality of work by incorporating the various existing comparative analysis and proposed the system design accordingly. The simulation-based approach is required to have experimented in real scenarios where there is a strong need to adapt in practical scenarios to handle the pandemic scenarios effectively.

Kumar et al. [1, 4–7] proposed various multi-layered approaches that can handle the dynamic network efficiently. These processes could be extended fr ad hoc connectivity based drone systems. In drone systems, the connectivity is highly dynamic. Thus, there is a strong need to handle the data with proper monitoring and management. The proposed approaches divided the functionalities based on operations. If the operation is highly dynamic then an individual layer completes its operation. This way load is distributed and the complexity of the system reduces. A similar approach is effective for a drone-based system. In a drone-based system, the operations are controlled through remote monitoring and management. Thus, approaches like distributed servers, software-defined networking (SDN) are more effective in providing unique but acceptable solutions to handle the drones' system data. This way of approaching the system makes it suitable for many real-time applications. The outlier and inlier-based system with the option to scan at different layers protect the system from various attacks as well. The attack detection process, filter the data at various points according to set parameters. If these parameters or their values vary with the application then other layers are flexible to change their operations as well. In conclusion, the dynamic ways of operating the system make the data collection, analysis, visualization, and sharing much flexible and efficient.

Suzuki et al. [2] discussed the importance of wireless sensor networks and UAV drone devices in infrastructure monitoring. This reduces the need to appoint large manpower in management and monitoring. In the drone-based approach, the major observation if found that the drone device is capable to capture large amounts of data from sensors deployed at different locations. The sensors are low-cost devices. Thus, UAV and sensor-based data collection approaches could be applied in application with different domains. Sensor devices are small devices that can be equipped with any infrastructure. This equipment gives infrastructure reading that can be read by UAVs through their movement. The major challenge in sensor-based applications is battery storage. Since resources are scarce in sensor devices. Thus, the battery requires frequent replacement that is a burden over infrastructure as there is a need to recruit manpower for this operation. In conclusion, this work has proposed a new infrastructure for monitoring and data collection with zero-standby power wireless sensor nodes. UAVs based approach is beneficial to those places/locations where signal through GPS is a major concern such as underground tunnels, building shadows, under the bridges, etc.

Colaprico et al. [8] proposed a cloud platform-based approach for patrolling large PV plants. The data is collected using video-based sources. The proposed approach does not require any human intervention and it implements an innovative diagnostic protocol. The behavior of various modules is independently collected from conditions set i.e. multiple conditions from the internal and external environment are made coexist to automatically compute the data and collect multimedia set with significant improvements. This information is useful in studying historical trends in monitoring plants. The proposed system is used to measure the aging trends in different modules. These modules are formed in a way that monitoring plant should be systematic. The concept that is under study from the long term is the massive test over plants like Green Power. This pays a large portion of the service cloud platform to the market. As a result, the overall system is considered to be efficient in-plant monitoring and observing the trends.

Cahill [9] discussed a safety alert system with one or more processors available to ensure object identification, building an object information database, and generating an alert on identification. In the proposed system mobile device identification that determines the current location is proposed. The proposed system transmits the received signal to the alert response server and this server keeps records of all devices with associated information. Thus, the overall system is automated with mobile identification. In the proposed system, multiple drones are free to move in any direction with collision avoidance using LiDAR sensor-based approach. With the drone movement, the sensor used to identify the mobile devices, their locations, and technical details are forwarded to drones. The drone system and its connectivity transmit the collected information to a long distance. Thus, the overall system is designed to be efficient and programmable to have a drone network exchange mobile device information. According to the authors, the proposed system can be used for drone-based network building, data collection, and mobile device identification. The collected data would be large to have more technical details like commuters' interest, signal requirements, environmental concerns, etc.

Koubâa et al. [10] discussed the importance of deploying drones over the cloud. The deployed drones can be used for various applications with robotics and the Internet-of-Drones. The drone deployment and construction of network help in an innovative service-oriented platform that can provide access to drones through web services. The web service accessibility and connectivity to the internet increase the range to operate the drones and collect data as and when it is required. The proposed approach has used cloud proxy server and logic between drones and their users. The automated systems help in estimating the drone's conditions and provide secure communication through MAVLink protocol. In implementation and validation of experimental results, the Dronemap Planner based approach is an efficient mechanism to access the drone over the internet and provide necessary monitoring and data collection

Kumar et al. [11–14] proposed various application-based systems that can be integrated with drone-based technology. For example, Proof of Game (PoG)-based consensus algorithm can be used to interconnect the drone's network and build a trustworthy system. The game-theory in consensus algorithm could be lightweight to have a consensus mechanism that integrated the lightweight resources with lightweight cryptography mechanisms. This improves the quality of service and efficiency of the overall system. Similarly, approaches proposed for other applications include the usage of Industry 4.0-based technology. The use of this technology is also very important to handle the data and integrated the small scale services with larger infrastructure to have global advantages. The machine learning-based approaches handle the data after a drone-based system collects, analyzes, and processes it. The time-series analysis can predict the trends in data variations and help to control and monitor the drone-based application for advantage to its large. The usage of other Industry 4.0 technology like cognitive computing, cloud computing, and AI-based approaches can give large scale implementation with fast and efficient subsystems. Various operations required for enhancing the overall system adaptability with improved security and performance is also possible. In some approach [3, 12, 14–19], computation distribution using layering is proposed. A similar multi-layered computation distribution approach is very useful in meeting the requirements of a resource-constraint secure and efficient platform. Thus, the proposed approaches can be modified and integrated with a drone-based system to have a large scale implementation [20–23].

Arnab Kumar Saha et al. [24] have introduced an IOT-based drone to monitor the agricultural sector, which provides extensive assistance in the field of agribusiness and improves crop quality. The proposed drone is designed with an RGB-D sensor, Taguchi gas sensor, Raspberry Pi model 3Band Adafruit AMG8833 IR thermal camera. These incorporated modules helped farmers accomplish simple, productive, accurate farming. A brief description of the integrated parts used in the proposed system is as follows:

- The Taguchi sensor provides an untimely ripening solution to the fruit problem that farmers need to know constantly. It is also used to detect gases such as propane, ethylene, methane.
- The RGB-D sensor captures the information of convention photos, abundance information on a per-pixel basis. An infrared sensor presents the depth information accessible with an adjusted RGB camera. This creates an RGB image with the information associated with each pixel.
- The Adafruit AMG8833 IR thermal camera can measure temperatures from 32 to 176 °F (0 to 80 °C) with an accuracy of 4.5 °F (±2.5 °C). It can identify humans up to 23 ft.
- The Raspberry Pi Model 3B is an ARM-based model created by the Raspberry Pi Foundation, it is a very low cost and single-board computer that reduces user efforts. Through this module all information can be received in digital format via the Internet, this information has been saved on cloud storage for further monitoring and analysis purposes.

Integrating the above mentions helps in achieving efficient and easy precision in agriculture. In this work, the author has proved that drone-based crop management is very easy and efficient due to proper and organized monitoring. With the upcoming innovations, production rates will increase rapidly with fewer consumption resources and energy. UAVs are not only used in soil and farming fields, but also in soil nutrient shoots and sowing seeds. The use of UAVs does not stop here when it is embedded with thermal-spectral, multispectral, and hyperspectral sensors, then drones can identify which cultivable land is dry and it helps in irrigation planning for dry land is. Additionally, drones are also able to identify crop health using visible light and near-infrared. Therefore, drones help as an impeccable aerial platform to gather the necessary information inaccuracy agriculture.

Afzal Ahmed et al. [25] has proposed a UAV based monitoring framework and object recognition technique to enhance the rescue and search operation in a hazardous situation. The spatial extent of the disaster area is normally huge in the case of the natural hazardous situation. Search and rescue operations from the ground are often difficult. Such operations are often interrupted due to localization inefficiencies due to damaged infrastructures. It is often difficult for victims to be identified, and rescued as soon as possible so that timely assistance can reach them. Generally, manned aircraft equipped to cover a wide area with extraordinary sensors are used for such purposes. From very low altitudes, manned aircraft are very difficult to operate, a telescope is commonly used to identify targets smaller than high altitude. In that case, the viewing scope becomes limited and the probability of oversight increases. Because UAVs can fly at a very low altitude and receive high-resolution images, which can be used to identify victims The primary focus of this work is to design an information-preparation framework, which will provide a logical, accurate and fast means of handling and processing a large volume of data acquired by the drone system. The information preparing framework, consequently named Data Viewer, is comprised of a few segments to carry out different responsibilities e.g., Extraction of valuable information, data integration, Flight design, information quality checking, object detection, image viewing, and so forth.

In critical analysis [8–14, 26–35], it has been observed that the major challenges in applying the drone-based system to real-time applications are: (i) data privacy and security is a major issue. The drone-based system collects data irrespective of distinguishing the criticality of information disclosure. Thus, there is a need to apply secure filters that protects data privacy, (ii) the proposed approach is found to be useful in resourceful applications. The drone-based system includes many resource-constraint devices which in-turn require lightweight security solutions. In conclusion, designing lightweight security solutions for resource-constraint drone-system is necessary for a good quality-based performance system, (iii) drone-applications are not just limited to one area of monitoring and data collection. Thus, application-specific technical requirements and challenges need to be studied for improving its usages, (iv) trajectory monitoring and the error rate is required to be computed for accurately predicting the observed data, (v) one system multiple application usage is also important to analyze the system ability in reducing the cost and increasing the domain of its applications, and (vi) quiet-drones are required to be designed for experimentation and efficiently calculating the performance of system especially in the indoor healthcare system.

3 Case Studies

This section discusses the case studies where drone-based systems could be integrated for enriching the quality of monitoring, management, control, or other application scenarios. More details are presented as follows.

Case Study 1: Drone-based Traffic Monitoring System

The drone-based system in traffic monitoring is very important because it can follow the vehicles to trace the incident rather than capturing the fixed location-based data capturing and reporting system. In this complete scenario, vehicles are supposed to be attached to sensor devices and drone-system hover over the vehicles to collect sensor-based data. Many drone-movement strategies could be implemented to improve system performance. These drones can collect data through multiple sources. Either sensor-based, image-based, electro-magnetic field-based or GPS-based data is used for information collection, analysis, and redirecting the on-road traffic to multiple directions. The movement of traffic, deployment of traffic officials, hardware devices, and environmental conditions can be considered to have an efficient system of more useful data. In this system, the major challenges are how to prepare a system with proper control under the federal government. The government officials can delegate rights to a third party which gives them proper data collector systems and provide security to the system and data t every level. Figure 1 shows various modules in the traffic monitoring system. These modules are briefly explained as follows:

- Data Collector System: This system is open to collect data through multiple processes. The collected data is fused to have refined and useful information.
- Data Analysis System: In the data analysis system, data redundancies can be removed to have data meaningful for analysis. The analysis process applies various

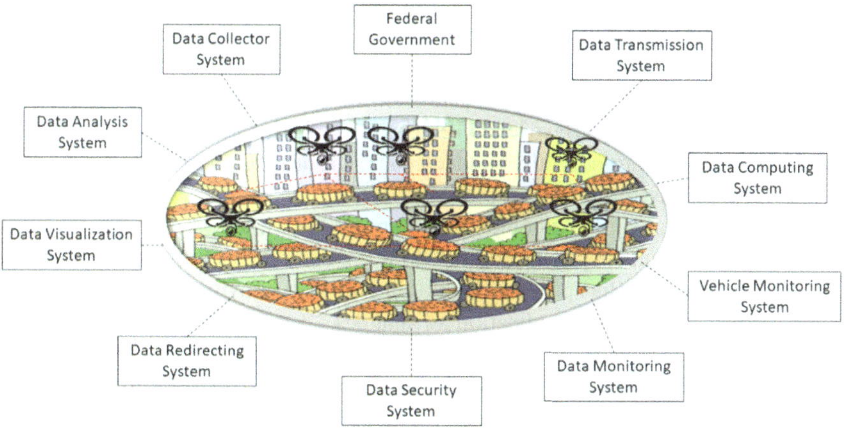

Fig. 1 Drone-based traffic monitoring system

machine learning approaches to predict or get the required information. The analysis processes can be an iterative process in nature that keep-on apply the machine learning algorithms until the required target is met.

- Data Visualization System: This system is helpful to present the monitored or collected data. This data can further be visualized through multiple ways for generating different interpretations. For example, data can be used to count the number of vehicles or it can be used to classify the type of vehicles.
- Data Redirecting System: In this system, a decision-based module redirects the received data towards computing centers. The data computing centers could be an edge, fog, or cloud computing centers that apply parallel and distributed processes to do the necessary processing.
- Data Security System: In this system, security issues are handled. Drones are capable of collecting data that can raise security or data privacy concerns. The security and data privacy concerns are handles at all three stages of data i.e. processing, transmission and propagation. During transmission and propagation of data, secure tunnels are established for providing the security. Data security during processing can be handled using cryptography primitives and protocols. There is always a provision to apply lightweight or full-fledge cryptography primitives and protocols based on the availability of resources.
- Data Monitoring System: In this system, the states of data are maintained that confirm the data availability cycle from generation to destroying. The monitoring system keeps track of the sensors deployed at different locations to minutely observe the importance and accordingly relevance to send it to the required computing center or any other operation.
- Vehicle Monitoring System: This system uses drone-system for monitoring the on-road vehicle and is associated with every other system to have the complete system in operation.
- Data Computing System: In this system, local and global computing centers are developed to have fast processing and availability of nearby users or monitoring centers. In parallel to other computing centers, this computing system is additional to handle unknown computations.
- Data Transmission System: The data transmission system is built to have the establishment of every type of end to end connectivity in the proposed system. This connectivity will provide fully connected and shortest communication provisions for data transmission.
- Federal Government: The complete system will work properly if operations of all systems are under proper control. In any country, this is possible when the federal government has properly right to monitor all activity and interrupt as per the interest of users and application advantages.

Case Study 2: Data Security through Digi-locker in Drone-based Monitoring System

In today's time, there is an increasing urge to secure one's important documents and make them readily available as and when required. As per the need, creating storage that can be used to save and hide important documents is not enough unless there are additional security techniques that can be used to prevent any possible attacks to affect the same. Thus, we need to implement techniques like facial recognition, the functionality of file format independent encryption, email-based login, one-time password approach, and alarm system to ensure efficient protection from any attack or possible theft. The main objective of this case study is to create a secure vault to allow users to upload sensitive information in the form of encrypted data inclusive of endpoint security.

The sub-objectives of this case study are: (i) Create a storage area for keeping the electronic records of data collected through the drone-based system, (ii) The data should be well secured in a way that cryptography primitives and protocols should be integrated to avoid any type of attack, (iii) Implement ID-Password login as the first step to authenticate the users to access any information, (iv) Implement one-time password authentication technique over mail as an alternative way of authentication, (v) Implement facial access control as the second step of authentication, (vi) Implement the alarm set up for unsuccessful login attempts exceeding the value "4", (vii) AES encryption technique is used to encrypt and secure confidential data, and (viii) Separate path creation for downloaded files at administrator and user level.

Figure 2 shows the process of the user accessing the drone's collected data securely. Here, the user has to register for accessing the data. After registration, each user will get the credentials to login and access the database. After the user enters the credentials and tries to log in, user login credentials are verified. If user credentials match with the stored procedure then login is considered to be successful and users are provided with options to access the data. Here, either user can download or upload files. The file is downloaded if other users or drones have stored it. Similarly, files are uploaded if users are interested to send these files to other users with or without the use of the drone system. In case the user fails in successful login then an alarm is generated to trigger the system processes in providing 5 chances. If the user fails in providing correct credentials in five chances then the user is blocked to access any system service. Thereafter, an admin inquiry is passed to invalidate the user and appropriate alerts are generated for users as well.

Figure 3 shows the class diagram for the user authentication process. In this class diagram, the user_interface class has an Enter_Credentials() function to allow the user to enter their credentials. After user credentials, users are authenticated either through a username and password credentials, or biometrics. In the biometric system, the user has to do a facial scan for recognition and authentication in the system. In the authentication process, the user is also given an option to get authenticated using the email system. Here, the user enters its email credentials and he/she has to enter one-time-password (OTP) for integrating the email system with the application. Further, the Alarm System is made to generate an alert if the user is not authenticated after repeated user attempts. This process is followed if the user fails to login with multiple attempts in a short duration.

Figure 4 shows the use case diagram for the user authentication process. In this process, the user login to the Digilocker system with its username and password. The username and password user case is extended with forgot password use case. In

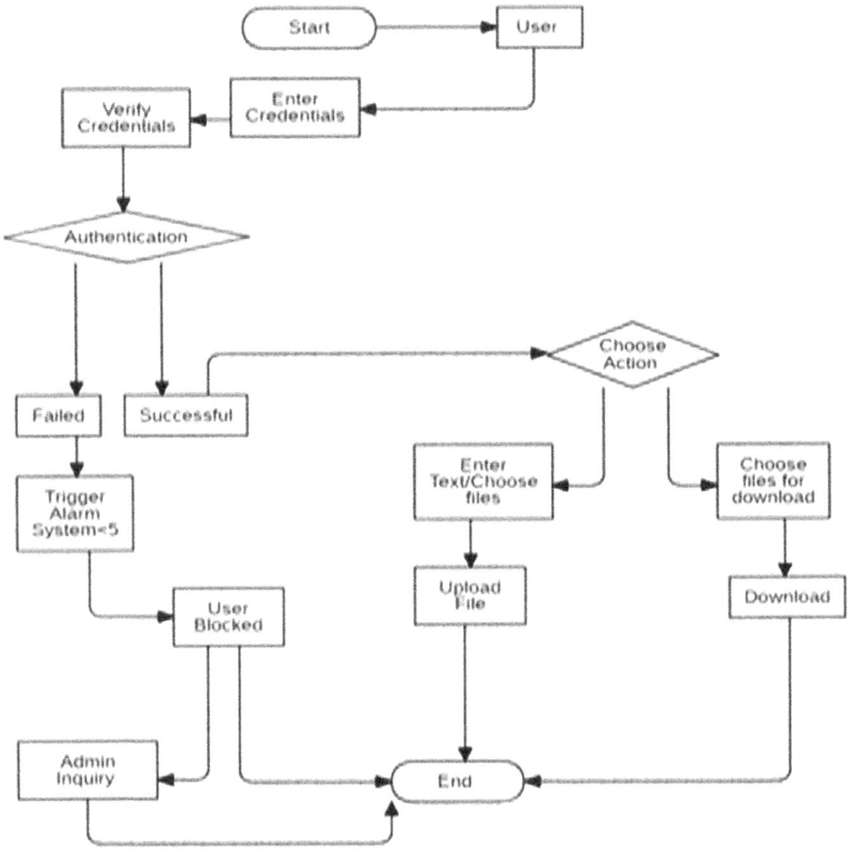

Fig. 2 Data accessibility process in Drone-based monitoring system

this use case, the OTP process is integrated to regenerate the password with proper verification through the backend database process. The backend database process saves the entered data or updates the records in the whole process. Additionally, an encryption/decryption process is applied to securely store the records. The encryption/decryption process uses traditional cryptography primitives and protocols for secure storage. The use case diagram shows that there are mandatory use cases for file download and upload in the system. The file download and upload are default procedures that the user has to opt for access or operating the system.

Figure 5 shows the front-end screen for the user login. Figure 5 shows that the user can enter credentials to access the system. Further, OTP is integrated that sends a temporary password to email id for accessing the system. The complete process is fast enough to secure the system from various attacks. This is a continuous process of ensuring system security through QoS and cryptography primitives and protocols. Thus, the proposed system can be integrated with any application that needs to collect and process data through drone-based system processes.

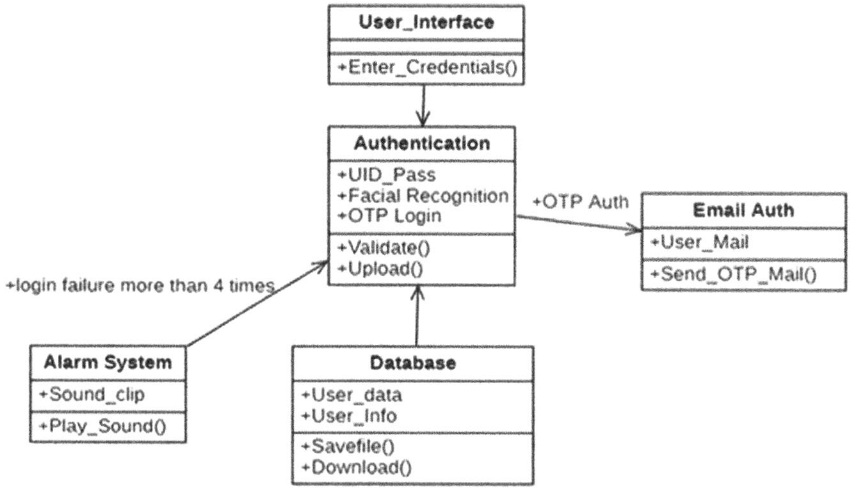

Fig. 3 Class diagram for the user authentication process

Case Study 3: Website security in Drone-based System

This case study is developed to integrate web applications with a drone-based system. The proposed web-application can be integrated with any system that needs the support of the drone system. Here, importance is given to collect, analyze, and visualize the data. Tracking what goes on with a process can help us work a lot faster and more efficiently, and creating reports is an easy way to draw quick conclusions and keep an eye on what's going on. But building reports can be tricky, especially if you're using third-party software or manual spreadsheets. Tracking how portals move through the process is incredibly important. That's why the portals to progress report roll in at#1 on our list. With this report, one can visualize loopholes, if any from end-to-end. The work provides comprehensive inspection of:

- Detection of the open ports on the host (port discovery or enumeration)
- Detection of the vulnerability and security holes (Nmapscripts)
- Detection of the URL vulnerability flaws (Xss scripts)
- Detecting malicious code injections. (SQL injection)
- Various links/hyperlinks attached to the portal. (Link analyzer)
- Get a request associated with the id and password field of the form description of the web page. (Source code Analyzer).

The application progress report eliminates the need to shift through data cells, use manual filtering, or tediously search for incomplete applications. This report cuts down on time spent managing data and leaves the user with a clear understanding of where the applications are headed. Keeping an eye on these metrics will also help you make sure that your websites or web portals can seamlessly move through the process. The tool will give a metric score based on the analysis done at various

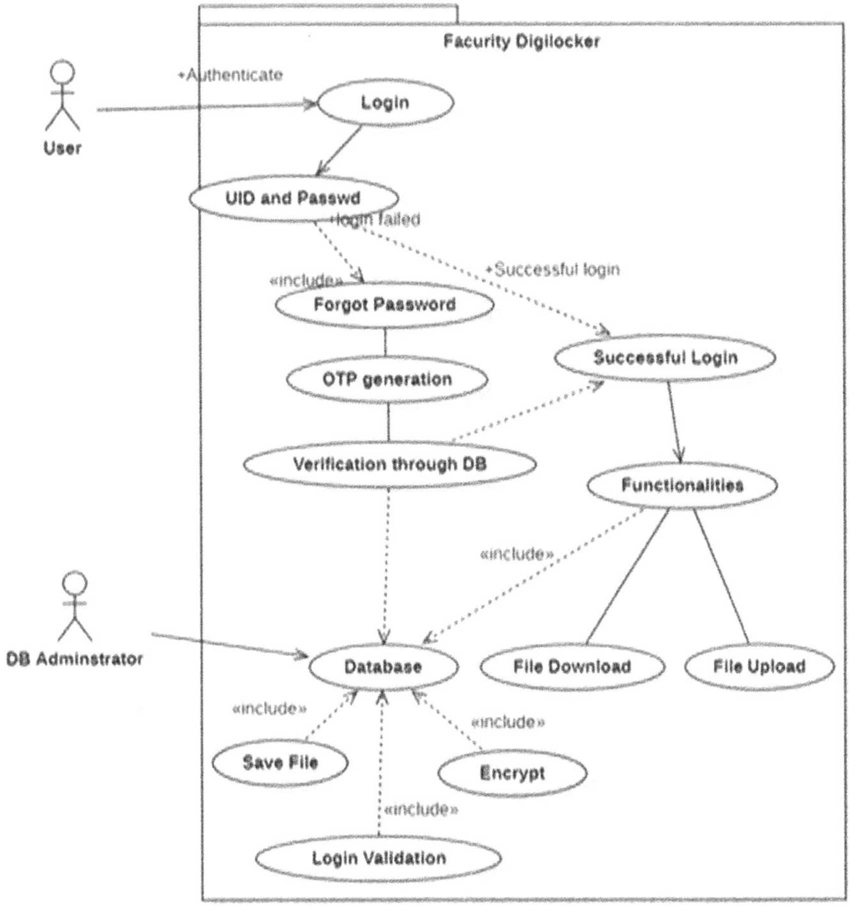

Fig. 4 Use case diagram for the user authentication process

levels and will provide solutions for problems identified during the process. Detect, analyze, and create a report for a portal to depict the safety and security through a score, depicted in the form of a pie chart and thereby generating a report for future reference by the user. What we intend to do is to examine various security points available on the web portal and also check for the loopholes present and alert the user with solutions available to fix them. Through this GUI application, the major drawbacks in existing web-application for the drone-based system are removed. This is made possible with pen-testing using different tools and techniques. In the proposed system, new facilities are provided to users with security and testing. These facilities are different from those which are available in traditional web-application development. The available facilities make the users access and scan the information on a priority-based algorithm. In this priority system, the ethical users with a transparent system accessibility record are allowed to access the services. Thus, an exhaustive

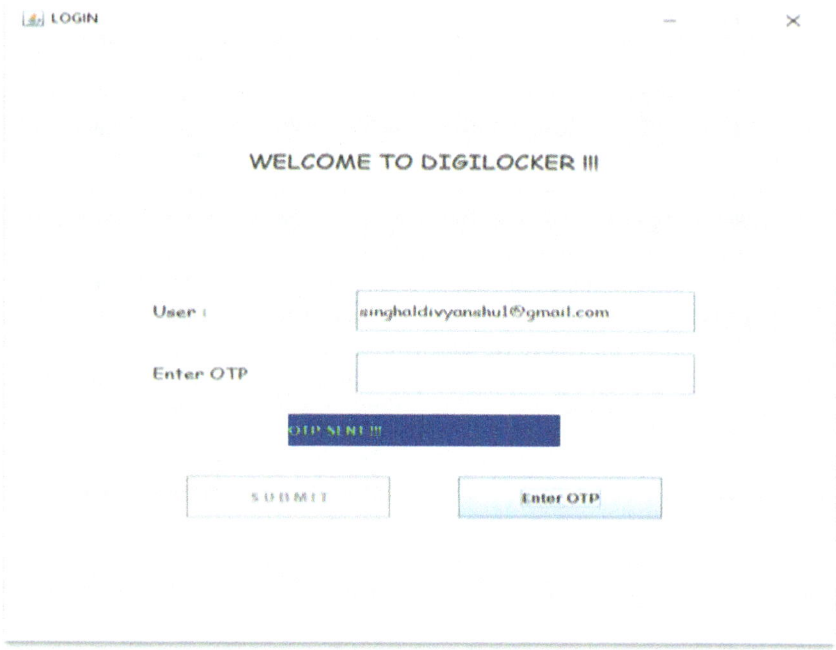

Fig. 5 Front-end for the user authentication process

search mechanism is necessary to scan such users and their intentions to access the data. Accordingly, the system applications are enhanced for optimizing the results and improving the performance of the overall network. The optimized results are displayed to users with information filtered through system choices and preferences. Here, effective searching mechanisms are applied to scan the records and display as per effective searching algorithms. The designed interface is simple and easy to access. The interface is simple and consists of clearly defined attacks vectors. There GUI consists of:

- A block for entering the website's URL that has to be checked.
- Dropbox to select the attacks to be performed.

The application proceeds according to the following steps:

1. The user enters the URL and clicks any of the present attacks, and then that particular url is tested for the same vulnerability.
2. In XSS, the application checks the website for any HTML e.g. <HTML> or any script e.g. <SCRIPT> should not be accepted by the portal. If it is, the portal can be prone to an attack by Cross-Site Scripting. To check the proposed system against the SQL injection attack, the programs check the entered value for a quote (') and parse the data. The parsed data encounter an error or special symbol then

the execution of the query is topped and the process is put in the vulnerable category. A portal is considered to be vulnerable to SQL injection attacks and the system is alerted to take appropriate security measures. The tester checks to verify if the application is passing vital information in the query string. The link finder fetches different hyperlinks associated with the web page without crawling them. When the user selects a source code analyzer from the dropdown, for the inserted URL it readily checks for the Get-requests method handled by the portal. It checks for the Id and password fields present in the form of the respective page. Open port scanning checks for the ports related to the page service. As open ports can be used as communication relays or infiltration vectors into the network the program scans and reports for the open and closed ports.

3. After checking for the respective attacks, a score is to be generated on the parameters of the low, medium, high risk found which will help the user take appropriate actions.

Figure 6 shows the use case diagram for vulnerability assessment in the proposed vulnerable application in a drone-based system. In this use-case diagram, the major

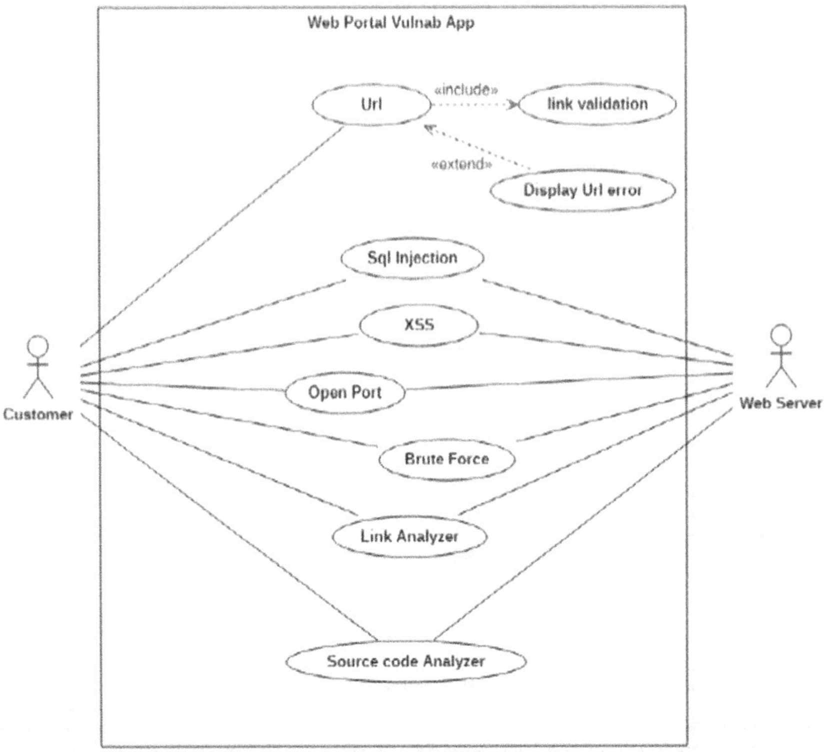

Fig. 6 Use-case for vulnerability assessment in the proposed system

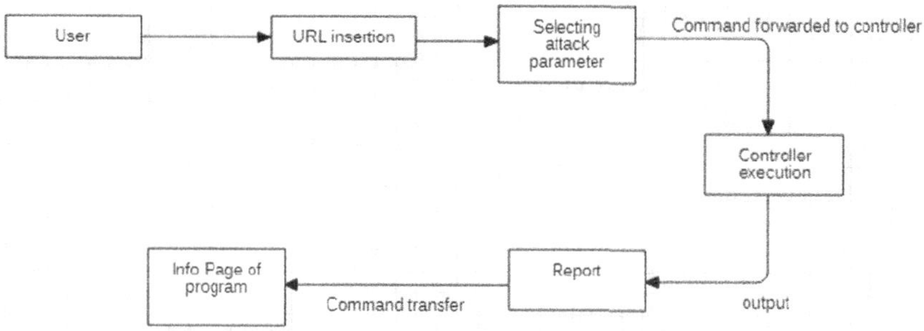

Fig. 7 Vulnerability assessment flow in the proposed system

concentration is drawn towards attacks including SQL injection, XSS, Open port-based attacks, brute force, and link analyzer. In the proposed system, web server and customer are the actors that can interact with the system and perform pre-defined activities. The availability of the URL option ensures that validation processes can be executed that ensure proper links for made available to everyone accessing the system and display system errors rather than redirecting to fake sites.

Figure 7 shows the vulnerability assessment flow in the proposed system. The vulnerability flow shows that the user inserts the URL and the system scans the URL for validation. The inserted URL is scanned against various attacks quickly by selecting the attack parameters. After parameter selection, the controller executes its query for forwarding the command to the controller and operate as per instructions. The controller executable parts generate URL reports on which users are recommended to follow the links and process the page with said program. This way of scanning the system report with advanced user linked activities build user trust and make the system transparent as well.

Figure 8 shows the vulnerability assessment web-application system screenshot. This screenshot shows that the user will have the option to scan the URL against various attacks. If the URL is found to be vulnerable against attacks then the system generates alerts accordingly else system output green-signal to process the URL and complete the transactions. Thus, the proposed system allows the web application to select a particular attack option to scan or scan against all pre-defined attacks by default.

Case Study 4: Drone-based Any object data collector system.

This case study proposes a multi-layered drone-based data collector system. In this system, drones are operated in free-movement directions with collision avoidance using LiDAR. The functionalities of the drones' movement are divided into multiple layers. Figure 9 shows the multi-layered architecture. More details of the layered approach and functionalities are discussed as follows.

Layer-1: Sensor-based information collection system: In this layer, sensors are deployed at different locations. It can be attached to the human body as wearable sensors, attached to objects for live object condition detection, thrown in the air for

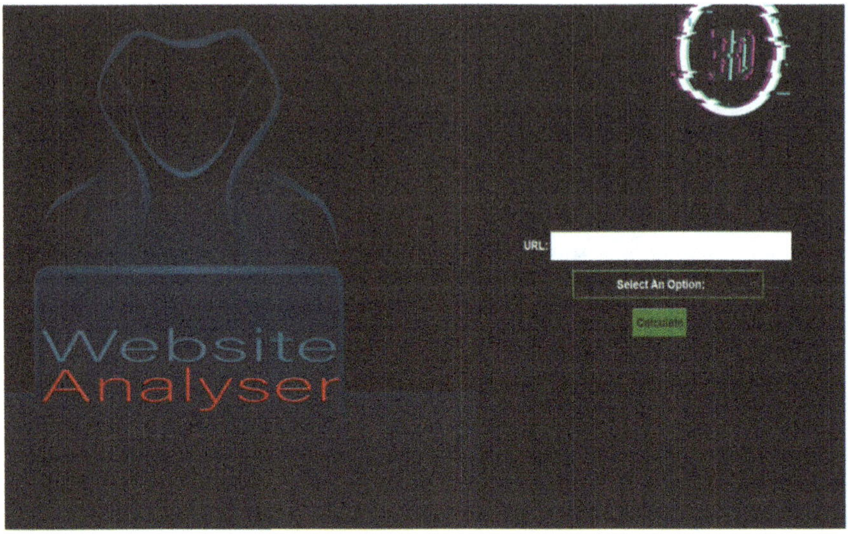

Fig. 8 Vulnerability assessment web-application in the proposed system

Layer-4	AI and Machine Learning	(Profiling, history logger, object identification)
Layer-3	Edge, Fog and Cloud Computing	(pre-processing, analysis, visualization etc.)
Layer-2	Drone Network and Internet of Drones	(Decision Making, Thermal Imaging, Sanitization, Monitoring, Vigilance etc.)
Layer-1	Wearable Sensor Devices Network	(data collection, vehicle identification, thermal scanning etc.)

Fig. 9 Multi-layered architecture for drone-based system

temperature humidity measurement as dust sensors etc. The collected data can be utilized in multiple manners. In one scenario data can be used to build people profiles. In another scenario, it can be used for understanding the people's requirements. Similarly, multiple scenarios could be generated to understand the requirements and accordingly plan the resources.

Layer-2: Internet of Drones (IoDn): Internet of Drones is a network where drones can fly without collision and exchange information as and when it is required. The exchanged information ensures QoS and security but not mandatory. As drones are resource constraint devices. Thus, the need to exchange information in a secure channel is very high. The secure channel could be established with lightweight cryptography primitives and protocols. In the case of the healthcare system, drone-based technology is very helpful in collecting indoor and outdoor data. This data can be used for analyzing the patients. The collected patient's data can be stored in the cloud for the necessary computations. To speed up the analysis process, and improve the QoS, parallel, and distributed computing architecture is divided into the edge, fog, and cloud computing services. The major challenges in IoDn based data collection, analysis, and processing are (i) establishing connections in highly dynamic network, (ii) storage of data over resource constraint drone devices, (iii) exchange mechanisms to frequently push/pop data from storage places to computing places, and (iv) reduced the computational and analysis complexities. The constructed IoDn can be used for various applications like monitoring, alert system, sanitization, and instructions-based data collector system. Industry 4.0 technologies can be integrated to have a better operation, controlling, monitoring, and advanced analysis.

Layer-3: Edge and Cloud Computing Infrastructure: The parallel and distributed computing infrastructure is any major requirement to achieve a zero-time computing response. The efficiency of implementing computing close to data collection and analysis center would generate quick responses. In an example, computing close to healthcare data collection place using edge computing followed by fog and cloud computing services enhances the overall application operations. Thus, the proposed solution will be an efficient solution to make the healthcare system patient-centric rather departmental or specialization centric. Further, the blockchain and industry 4.0 based solution can improve the quality of data and visualization of results. In conclusion, parallel and distributed computing can improve services and operations.

Layer 4: *Advanced Analytics*: In any industry scale application, there is a need for operations at local and global computing places. Local computing places are closer to data usage. Whereas, global places are to analyze a large amount of data at a central and/or remote place. For example, traffic data can be collected using drones and can be sent to local administrative units for processing. Thereafter, all data should be made available to a central place, which is a remote place also, for necessary computations. The central place is comparatively assumed to have large computational resources. Thus, generate a huge database with advanced analytics using the machine, deep-learning, or other AI-based approaches.

Figure 10 shows the drone-based monitoring system using the AnyLogic simulator. The simulation is planned to make free people movement on-ground and count the number of people and their movement through density-based thermal imaging. Figure 11 shows the thermal imaging-based area alert system. In this system, those areas are highlighted (with red color) where there is a large movement. Thus, the chances of people interacting or activities increase as well. To carefully scan these areas and plan necessary services, a drone-based observational system is observed to be more important. Recently, it has been observed that such a system is more effective for the COVID-19 scenario where there is a need to sanitize areas based on usage. This becomes a well-designed system to be useful in the future for remotely cleaning during any form of a pandemic. The available data could be made available to the local and global edge, fog, and cloud servers for necessary computations. These scenarios can be extended to serve in the identification of people, objects, traffic, environmental conditions etc. Such predictions are useful in various applications.

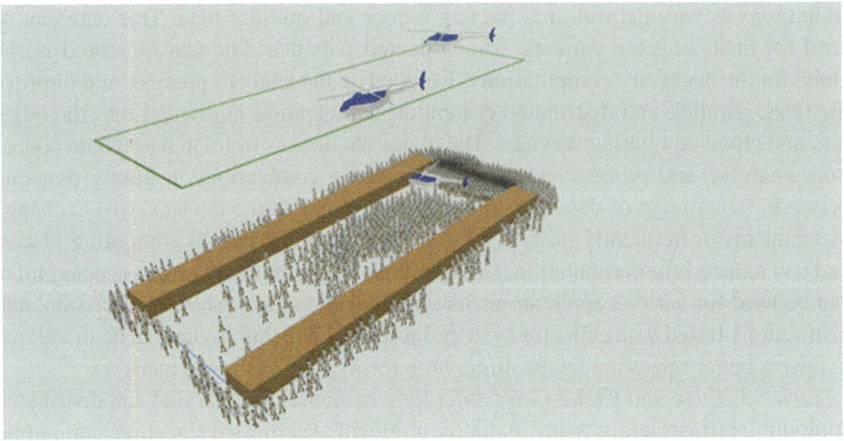

Fig. 10 Drone-based remote monitoring system

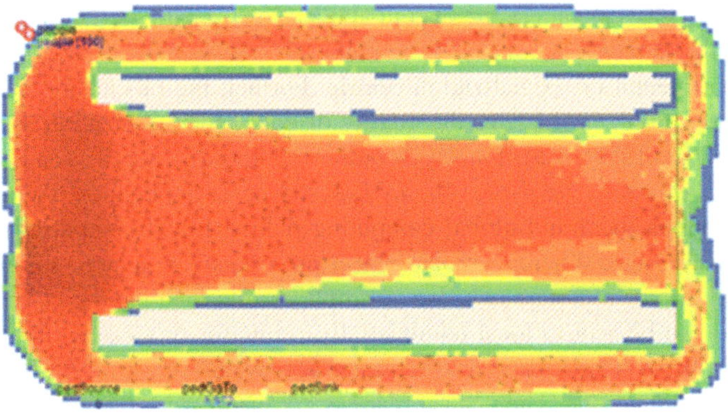

Fig. 11 Density-based thermal image plotting

4 Conclusion

The data collection system having diverse transmission measurements and control activities using sensor devices applies to various applications like human-related monitoring, traffic management, service providing system, product delivery system. If more data is available to analyze then it would make the system more efficient by applying iterative scanning and updating with feedback. In a similar approach, this work has explored a set of case studies that are helpful in drone-based monitoring, ensuring secure data processing, transmission, and propagation, a Digi locker system to securely provide the data to the authenticated system. The wireless connectivity within specified modules and the drone-based system can give multiple scenarios that can be used in various applications. The applications put data in center and give importance to parallel and distributed services especially edge, fog, and cloud services to handle a large amount of data constantly. The present study has identified that multi-layered architecture is important to distinguish the services and distribute the computing functionalities. Further, the wireless interconnection between the drone-based system and application-specific modules is a fail-safe and secure mechanism that needs to consider the critical operations and mechanism for measuring the normal to severe damage to the network and its services. In the future, the present work can be extended to validate the proofs and generate real-time applications with their usage. Highly dynamic channels with wireless connectivity need further importance to ensure that trustworthy data is available for applications. The application scenarios for the drone-based system can be extended for multiple domains. These domains include oil and gas pipeline observations, nuclear devices and place observations, monitoring water supply systems etc. Drone-based systems act as strategic technologies for various critical time analysis. Thus, acceptability and lifetime for such a system are much higher and demanded by various operation measurement and control modules. Thus, a drone-based system can act as a repeater and delivery device/system to place the data at a trustworthy and secure environment irrespective of distance.

References

1. Kumar, A., Rajalakshmi, K., Jain, S., Nayyar, A., Abouhawwash, M.: A novel heuristic simulation-optimization method for critical infrastructure in smart transportation systems. Int. J. Commun. Syst. e4397 (2020)
2. Suzuki, Y., Yoshikawa, S., Yamawaki, A.: A Command List Based Automatic Controlling Method of UAV for Infrastructure Monitoring System with Zero-standby Sensor Node (2018)
3. Kumar, A., Sharma, K., Singh, H., Naugriya, S.G., Gill, S.S., Buyya, R.: A Drone-based Networked System and Methods for Combating Coronavirus Disease (COVID-19) Pandemic (2020). arXiv preprint arXiv:2006.06943
4. Kumar, A., Aggarwal, A., Yadav, D.: A multi-layered outlier detection model for resource constraint hierarchical MANET. In: 2018 5th IEEE Uttar Pradesh Section International Conference on Electrical, Electronics and Computer Engineering (UPCON), pp. 1–7. IEEE
5. Kumar, A., Aggarwal, A.: An efficient simulated annealing based constrained optimization approach for outlier detection mechanism in RFID-sensor integrated MANET. Int. J. Comput. Inf. Syst. Ind. Manage. Appl. **11**, 55–64 (2019)
6. Kumar, A., Sharma, D.K.: An optimized multilayer outlier detection for internet of things (IoT) network as industry 4.0 automation and data exchange. In: International Conference on Innovative Computing and Communications, pp. 571–584. Springer, Singapore
7. Kumar, A., Srikanth, P.: A decision-based multi-layered outlier detection system for resource constraint MANET. In: International Conference on Innovative Computing and Communications, pp. 595–610. Springer, Singapore
8. Colaprico, M., de Ruvo, M.F., Leotta, G., Bizzarri, F., Vergura, S., Marino, F.: DUBIO: a fully automatic drones & cloud based infrared monitoring system for large-scale PV plants. In: 2018 IEEE International Conference on Environment and Electrical Engineering and 2018 IEEE Industrial and Commercial Power Systems Europe (EEEIC/I&CPS Europe), pp. 1–5. IEEE (2018)
9. Cahill, P., Clandestine Development, LLC: Drone Safety Alert Monitoring System and Method. U.S. Patent Application 14/824,011 (2016)

10. Koubâa, A., Qureshi, B., Sriti, M.F., Javed, Y., Tovar, E.: A service-oriented cloud-based management system for the internet-of-drones. In: 2017 IEEE International Conference on Autonomous Robot Systems and Competitions (ICARSC), pp. 329–335. IEEE (2017)
11. Kumar, A., Kumar Sharma, D., Nayyar, A., Singh, S., Yoon, B.: Lightweight proof of game (LPoG): A proof of work (PoW)'s extended lightweight consensus algorithm for wearable kidneys. Sensors **20**(10), 2868 (2020)
12. Kumar, A., Krishnamurthi, R., Nayyar, A., Sharma, K., Grover, V., Hossain, E.: A novel smart healthcare design, simulation, and implementation using healthcare 4.0 processes. IEEE Access **8**, 118433–118471 (2020)
13. Kumar, A., Jain, S.: Proof of game (PoG): a proof of work (PoW)'s extended consensus algorithm for healthcare application. In: International Conference on Innovative Computing and Communications, pp. 23–36. Springer, Singapore (2020)
14. Kumar, A., Jain, S.: Proof of game (PoG): a game theory based consensus model. In: International Conference on Sustainable Communication Networks and Application, pp. 755–764. Springer, Cham (2019)
15. Kumar, A., Gopal, K., Aggarwal, A.: detection and treatment for lightweight mobile ad hoc networks. In: International Conference on Heterogeneous Networking for Quality, Reliability, Security and Robustness, pp. 750–763. Springer, Berlin, Heidelberg (2013)
16. Koubâa, A., Qureshi, B., Sriti, M.F., Allouch, A., Javed, Y., Alajlan, M., Cheikhrouhou, O., Khalgui, M., Tovar, E.: Dronemap planner: a service-oriented cloud-based management system for the internet-of-drones. Ad Hoc Netw. **86**, 46–62 (2019)
17. Rizwan, R., Shehzad, M.N., Awais, M.N.: Quadcopter-based rapid response first-aid unit with live video monitoring. Drones **3**(2), 37 (2019)
18. Haus, M., Krol, J., Ding, A.Y., Ott, J.: Feasibility study of autonomous drone-based IoT device management in indoor environments. In: Proceedings of the ACM SIGCOMM 2019 Workshop on Mobile AirGround Edge Computing, Systems, Networks, and Applications, pp. 1–7 (2019)
19. Kharchenko, V., Torianyk, V.: Cybersecurity of the internet of drones: vulnerabilities analysis and IMECA based assessment. In: 2018 IEEE 9th International Conference on Dependable Systems, Services and Technologies (DESSERT), pp. 364–369. IEEE (2018)
20. Ntumba, P., Bouloukakis, G., Georgantas, N.: Interconnecting and monitoring heterogeneous things in IoT applications. In: International Conference on Web Engineering, pp. 477–481. Springer, Cham (2018)
21. Alharthi, M., Taha, A.E.M., Hassanein, H.S.: An architecture for software defined drone networks. In: ICC 2019–2019 IEEE International Conference on Communications (ICC), pp. 1–5. IEEE (2019)
22. Chemodanov, D., Qu, C., Opeoluwa, O., Wang, S., Calyam, P.: Policy-based function-centric computation offloading for real-time drone video analytics. In: 2019 IEEE International Symposium on Local and Metropolitan Area Networks (LANMAN), pp. 1–6. IEEE (2019)
23. Dahlstrom, R.L.: The emergence of contact based nondestructive testing NDT at height utilizing aerial robotic drone systems. In: Offshore technology conference (2020)
24. Saha, A.K., Saha, J., Ray, R., Sircar, S., Dutta, S., Chattopadhyay, S.P., Saha, H.N.: IOT-based drone for improvement of crop quality in agricultural field. In: 2018 IEEE 8th Annual Computing and Communication Workshop and Conference (CCWC), pp. 612–615. IEEE (2018)
25. Ahmed, A., Nagai, M., Tianen, C., Shibasaki, R.: UAV based monitoring system and object detection technique development for a disaster area. Int. Arch. Photogrammetry Remote Sens. Spat. Inf. Sci. **37**, 373–377 (2008)
26. Jung, N.J., Choi, M.H., Lim, C.W.: Development of drone operation system for diagnosis of transmission facilities. In: 2018 21st International Conference on Electrical Machines and Systems (ICEMS), pp. 2817–2821. IEEE (2018)
27. Maria, G., Baccaglini, E., Brevi, D., Gavelli, M., Scopigno, R.: A drone-based image processing system for car detection in a smart transport infrastructure. In: 2016 18th Mediterranean Electrotechnical Conference (MELECON), pp. 1–5. IEEE (2016)

28. Amer, K., Samy, M., Shaker, M., ElHelw, M.: Deep Convolutional Neural Network-Based Autonomous Drone Navigation (2019). arXiv preprint arXiv:1905.01657
29. Kumar, A., Gopal, K., Aggarwal, A.: A novel lightweight key management scheme for RFID-sensor integrated hierarchical MANET based on internet of things. Int. J. Adv. Intell. Paradigms **9**(2–3), 220–245 (2017)
30. Kumar, A., Aggarwal, A., Gopal, K.: A novel and efficient reader-to-reader and tag-to-tag anti-collision protocol. IETE J. Res. pp. 1–12 (2018)
31. Chugh, N., Kumar, A., Aggarwal, A.: Security aspects of a RFID-sensor integrated low-powered devices for internet-of-things. In: 2016 Fourth International Conference on Parallel, Distributed and Grid Computing (PDGC), pp. 759–763. IEEE (2016)
32. Kumar, A., Aggarwal, A.: Performance analysis of MANET using elliptic curve cryptosystem. In: 2012 14th International Conference on Advanced Communication Technology (ICACT), pp. 201–206. IEEE (2012)
33. Kumar, A., Gopal, K., Aggarwal, A.: Simulation and cost analysis of group authentication protocols. In: 2016 Ninth International Conference on Contemporary Computing (IC3) pp. 1–7. IEEE (2016)
34. Kumar, A., Gopal, K., Aggarwal, A.: Cost and lightweight modeling analysis of RFID authentication protocols in resource constraint internet of things. J. Commun. Softw. Syst. **10**(3), 179–187 (2014)
35. Kumar, A., Goyal, M., Rajalakshmi, K., Aggarwal, A.: A simulation annealing based anti-collision protocol for RFID tag identification. In: 2016 Ninth International Conference on Contemporary Computing (IC3), pp. 1–7. IEEE (2016)

Drone Application in Smart Cities: The General Overview of Security Vulnerabilities and Countermeasures for Data Communication

Huu Phuoc Dai Nguyen and Dinh Dung Nguyen

Abstract With the boost of technologies in computing, the Internet of things (IoT), and Information and Communication Technologies (ICT), the demand for using drones has been increased in real-world applications. Recently, unmanned aerial vehicles (UAVs) or drones can be used to complete a wide range of different tasks from the military to the industry with numerous studies available in the literature. However, the broad use of drones for smart cities also faces several technical, societal issues such as cybersecurity, privacy, and public safety that need to be concerned. This chapter briefly provides a general overview of cybersecurity vulnerabilities and cyber-attacks through a meta-analysis of relevant literature review towards drones such as Wi-Fi security, drone networking security, malicious software, and the like. Moreover, several countermeasures are explained in this chapter including detection methods and defense mechanisms to protect drones from their security vulnerabilities. Importantly, this paper raises cybersecurity awareness for users and presents solutions for future research in this area of interest. In fact, most cybersecurity vulnerabilities are based on sensors, communication links, and privacy via photos. Therefore, to ensure the security of drones, there is a need for a combination of the solutions for multiple sensors, the use of safe communication links instead of Wi-Fi, and the application of the CIA triad concepts.

Keywords Countermeasure · Cybersecurity · Drones · Drone traffic management · IoD · Smart city · UAV · Vulnerabilities

H. P. D. Nguyen (✉)
Doctoral School on Safety and Security Sciences, Óbuda University, Budapest, Hungary
e-mail: phuocdaitt@yahoo.com

D. D. Nguyen
Department of Aeronautics, Naval Architecture and Railway Vehicles, Budapest University of Technology and Economics, Budapest, Hungary
e-mail: dungtbhk@gmail.com

R. Krishnamurthi et al. (eds.), *Development and Future of Internet of Drones (IoD): Insights, Trends and Road Ahead*, Studies in Systems, Decision and Control 332,
https://doi.org/10.1007/978-3-030-63339-4_7

1 Introduction

Originally, drones were established and applied for several specific military aspects. Nevertheless, they have been applied in governmental or civic appliances in many aspects recently; for example, disaster management [1], delivery good and information [2], search and rescue operations [3], observation [4], wildfire managing [5], communication for ad hoc networks [6], weather estimation [7], civilian security [8], agricultural controlling [9], distant sensing (for data, image, or environment) [10], emergency assistance [11], and traffic monitoring [12, 13].

According to Tractica, the commercial drone market was expected to continue overgrowing over the next few years while giving chances to several industry cooperators, reaching a global income of $13.7 billion by 2025 (Fig. 1) [14].

The use of drones has been progressed shortly, which can be skyrocketing. The drones and the Internet of things (IoT) [15–17] offer interactivity and ubiquitous connectivity. In [18], Nayyar and his colleagues addressed the novel evolving the theory of the Internet of Drone Things (IoDT). By introducing this concept, the study has predicted that the technology of drones will create a new scope with new missions, new applications, and faster connectivity. The key technologies such as cloud computing, fog computing, sensors, and IoT protocols play an essential role in the IoDT. These created different applications for organizations from security to logistics, and from agriculture to industry. In addition, the usage of drones in smart cities is one of the most advanced areas of this technology nowadays. Regarding Pratyush Singh, he proposed that "A smart city is an urban region that is highly advanced in terms of overall infrastructure, sustainable real estate, communications, and market viability" [19]. The smart city offers many advantages and it can be applied for citizens, businesses, government, and the environment. Drones can be

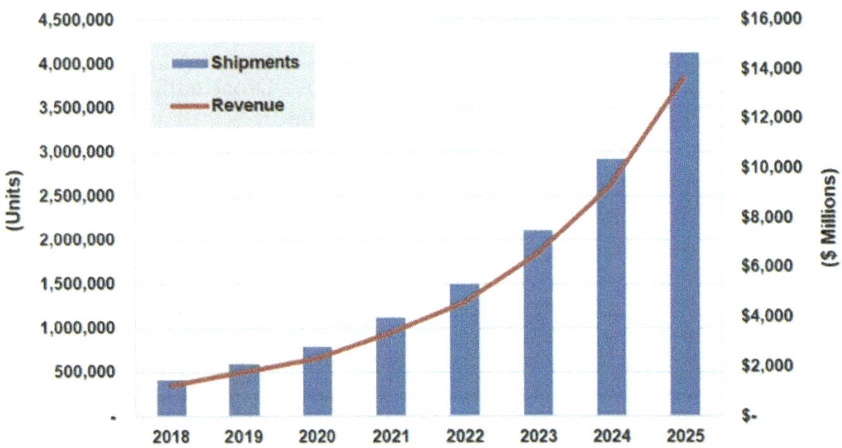

Fig. 1 Commercial drone hardware shipments, and total (hardware + services) revenue, world markets: 2018–2025 [14]

deployed flexibly and fast in many fields in smart cities; for example, monitoring traffic, people, and environment; controlling civilian security; and delivering products" [20]. Moreover, one of the most advantages of drone's usage is more cost-effective than manned planes. In addition, they are more flexible in different locations and situations, as well as hazard cases for people. On the other hand, they can operate closely to destination objects which enhance the accurate method and better actions. Therefore, these drones' features offer benefits for smart city applications. Although drones have a lot of advantages, they also bring several technical and societal issues such as cybersecurity, privacy, and public safety. This chapter shortly describes the comprehensive overview of cybersecurity vulnerabilities and cyber-attacks via a meta-analysis of literature review of related studies towards drones; for instance, Wi-Fi security, drone networking security, malicious software, and the like. Moreover, several countermeasures based on detection and defense mechanisms are summarized in this study to protect drones from their security vulnerabilities. Furthermore, there is a need for addressing the importance of raising cybersecurity awareness for users and indicating several potential solutions for future research. This chapter provides an overview of drone security vulnerabilities and countermeasures for data communication.

The organization of this chapter is as follows: Sect. 1 introduces the usages of drones in many aspects of our daily life. Section 2 describes several literature reviews about smart cities, drones' applications, challenges for drones, and data communication methods. Section 3 indicates several cybersecurity threats and attacks towards drones to identify the current security issues of drones. Section 4 discusses detection measures and defense mechanisms to protect and prevent drones from cyber-attacks and security vulnerabilities. Finally, Sect. 5 presents the conclusion of the chapter.

2 Background

2.1 Smart City

Recently, there has been a rapid urbanization process worldwide. In fact, according to the United Nations Population Fund, it was remarked that human beings lived in metropolitan areas approximately more than 50% in 2010, and this number can be expected to 70% by 2050 (Fig. 2) [21]. Moreover, in Europe, currently, the population stayed in the city about 75% and supposed approximately 80% by 2025 [22]. The importance of city regions is a worldwide circumstance. This is approved by the wide spread of megacities in Asia, Latin America, and Africa 40%, 80%, 40%, respectively.

Today, most resources contributing to economic importance are consumed in cities but contributed to the poor environmental performance. Cities are responsible for significant shares of Global Greenhouse Gas (GHG) emission's total energy approximately from 60 to 80%. However, the energy for electricity and transportation

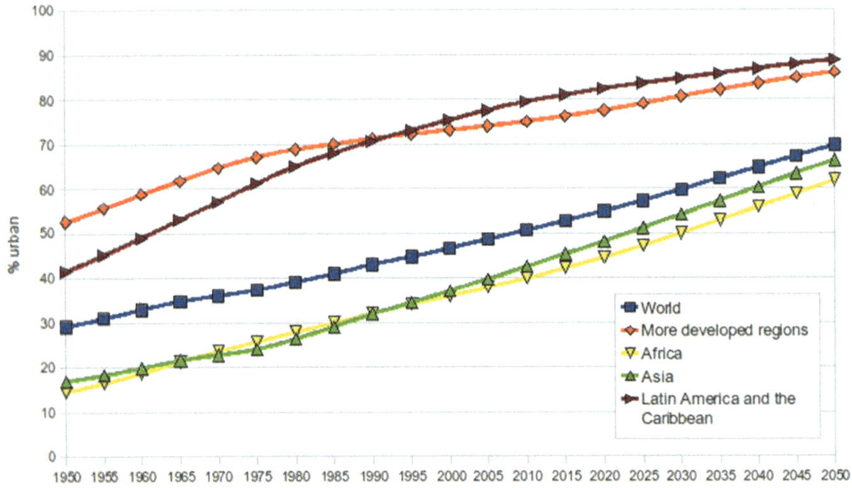

Fig. 2 Percentage of population living in urban areas by region, 1950–2050 [21]

is spent much more when the city density is lower because CO_2 emissions per capita decrease double with the metropolitan area density's growth [23]. Besides, Gartner (2018) suggested that a smart city framework is one of the six technologies that can achieve direct adoption in the next 10 years (Fig. 3) [23]. In addition, the Internet

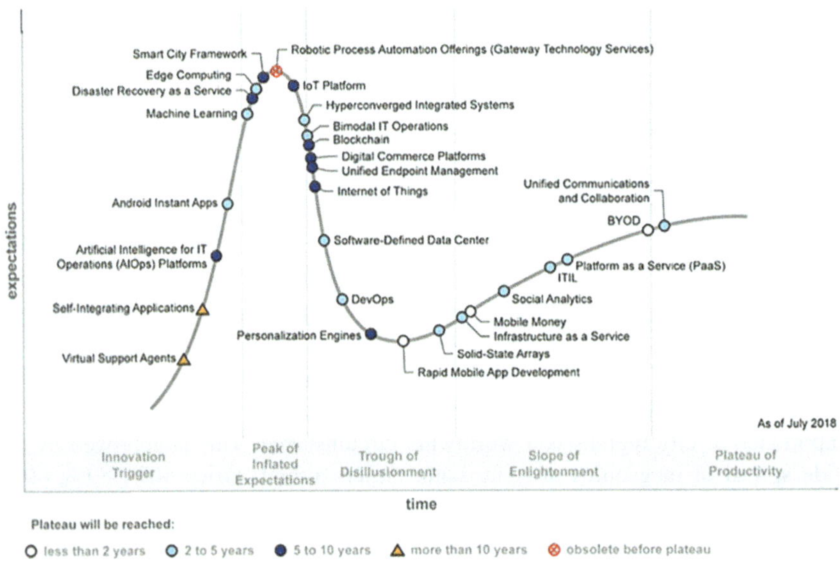

Fig. 3 Hype cycle for IT in GCC, 2018 [23]

of Things (IoT), digital twins, and smart contracts should be focused on growing rapidly.

The urbanization process has dramatically developed the economy and enhanced the social capacity to reconstruct the landscape. Furthermore, it offers a significant improvement to the standard of living. Nonetheless, people also face many new challenges because of this process; for instance, natural resources problems, transport congestions, air, and water pollution. These problematic issues are an impressive assemblage of challenges for future urban planning [24]. Therefore, the smart city concept can execute suitable resolutions to solve these problems better. There are numerous explanations for a smart city in the literature review. The smart city concept has been developed from the initial concept, called information city, and then integrated with the advanced Information and Communication Technology (ICT) [25]—center smart city. There are six main dimensions of a smart city [26, 27]. Besides, a smart city focuses on residents' property and knowledge more than digital or intelligent cities. These dimensions were illustrated in the wheel by Boyd Cohen (Fig. 4) [28]. The author used to benchmark the world's major cities, which is well performed in a forward-looking way in these dimensions.

The first smart cities 'concept and definition were established in the nineties. It has developed gradually over the years. However, there was a remarkably increased number of publications related to this topic after the appearance of European Union projects since 2010 [30]. Moreover, a smart city can be a set of functions provided to a specific group of citizens or administration process, which expresses the reorganize government procedure. However, the knowledge of this concept must be improved and evaluated shortly [31].

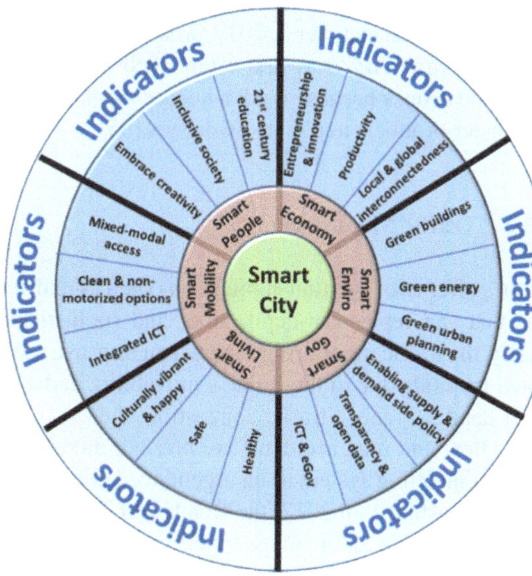

Fig. 4 The smart city wheel [29]

2.2 *Drones*

Several studies have discussed the challenges and opportunities for applying drones in urban areas, especially smart cities [32–34]. In a smart city environment, drones can perform essential tasks such as bringing innovative ideas to life and developing the economy [35]. In fact, drones' ability can change the current applications and it may create new chances.

2.2.1 Package Delivery

Smart City technology creates a new era of delivering product assistance more effectively and accurately for civilians [35]. In this situation, drones have the ability to carry massive packages and transfer autonomously and fast to the clients [35]. By an innovation of the customer waiting time, the authors in references [36, 37] proposed a cooperative truck and drone delivery system, which integrates drones with truck-based delivery operations. This system will enhance the efficiency of a conventional delivery system, as well as more sensitivity and higher computing efficiency compared to the heuristic approach. Several companies like Amazon, Alibaba, and Google have been tested practical trails to manage the parcel delivery of drones [38]. Also, DHL Parcel has been started testing a drone delivery service to deliver medications to one of Germany's North Sea islands [39]. In these practices, the drone flew autonomously but still had to be continuously tracked. In addition, drones will be able to operate autonomously and safely in urban environments thanks to rapidly advanced technologies in obstacle detection and avoidance [40]. For example, UPS company successfully tested a delivery drone in the USA [41] from a company electric van's roof on October 2, 2019. Additionally, a delivery drone project of Amazon namely "Prime Air" was first established in 2016. This leads to a new contest in legalizing drones because it got a license from the U.S. Patent and Trademark Office to transfer products to clients by drones [35].

2.2.2 Traffic Monitoring

Drones offer two potential applications such as real-time traffic monitoring and analysis that can replace in-depth labor and sophisticated observational system [35]. Firstly, the traffic monitoring feature can improve traffic systems instead of the traditional methods due to the drone's movability and efficiency in a large area. On the other hand, to reduce traffic congestion, the exact and real-time information about traffic flow and road accidents are needed; therefore, many big cities require smart traffic controlling system as an urgent need due to the development of traffic volume [42]. For instance, in 2017, the Roads and Transport Authority in Dubai city monitored their traffic and managed their car accidents by using drones [43]. Secondly, in France, the "Elistair Tethered" company in Lyon deployed the "Data From Sky analyze traffic information of various kinds of vehicles by using live transmission data about the traffic flow [44]. Another application of drones in an urban area used a swarm of 10 drones in Athens to record traffic streams in a congested area [45]. A complete dataset can allow an intensive investigation of critical traffic phenomena. This experience will create a unique observatory of traffic congestion that researchers can use to develop and test their models.

2.2.3 Traffic Patrolling

Regarding the transport management system, drones can support and combines with ground vehicles to achieve various traffic patrolling missions such as road patrolling [46], traffic monitoring [47], and enhancing efficiency [48]. This approach can be performed by a framework that vehicle released and recovered the drone during the vehicle's operation. This helps to increase the ability of the far distance driving and mobility of vehicles and control drones in wireless and remote way [49]. However, this method leads to create a new obstacle for the operation between a drone and a vehicle in patrol missions. For example, the Traffic Patrolling Routing Problem with Drones (TPRP-D) was oriented toward complex tasks [50]. In these practices, the vehicles are required to stay and wait for the drones until they finished their operation.

2.2.4 Policing

Smart police systems are recognized as one part of smart cities. These systems are supported by new technologies in order to solve critical and sophisticated security problems in the city. In fact, using drones in police systems is the latest and effective device to detect burglary suspects and investigate crimes in dangerous missions. For instance, in 2015, "Devon and Cornwall police forces" in the U.K. first deployed and tested a police drone unit. Currently, this unit has been completely applied by the full-time staff [51]. Moreover, in England and Wales, there are over a quarter of the 43 forces using drones to investigate crimes [52]. In China, Police used about 1000 drones to assist in tracking suspected crimes and discovering drug fields in 2017 [53].

2.2.5 Extensive Disaster Management

To manage and respond towards crisis situations in extreme natural disasters e.g. earthquakes, floods, volcanos, fires in forests or large infrastructures, and terrorist attacks are in high demand. In these cases, drones can be utilized effectively [54]. Besides, they are adaptable, trustworthy, and safe tools to control and give live information about the situation [35]. Because they can track the location of these disasters and get more information in different places to analyse and evaluate the situation.

Furthermore, drones can be the alternative communication systems for replacing the damaged ones in emergency cases [55]. In addition, drones can be a useful tool to identify survivors 'location, carry medical supplies and equipment, or even transfer the injured people.

2.2.6 Firefighting and Rescue Operation

To enhance firefighting response and rescue operations, drones can be recognized as safer, faster, and efficient technology in comparison to conventional methods. Due to the ability of autonomous operation, accessibilities to hazardous areas, and capability of data-gathering tasks in dangerous cases, drones may be considered as the best solution. For example, drone applications in firefighting and rescue operations were presented by the emergency management organization of Dubai [56], the firefighting department of New York [57].

2.2.7 Wireless Communications

The trustworthy and expandable wireless connection's availability is one of the necessary factors in supporting many services in smart cities [58, 59]. Drones operated based on wireless communication support devices, as a result, they can provide primary or extra services in the smart city for different clients [60]. Likewise, drones can offer wireless connectivity among distant areas to maintain supporting interactive smart city aids [61].

2.3 Challenges for Drones in Smart Cities

Despite the significant advantages of drones discussed above, several serious challenges need to be highlighted [32, 33].

2.3.1 Safety

Drones have been applying in different civilian applications recently. These lead to critical safety problems because of the enormous damage from drones' crashing. There are several reasons for drone accidents such as technical failure, lack of equipment maintenance, mid-air collision, and operation misuse [62]. Other aspects concerning the falling of drones in the public areas are extreme natural statuses such as "turbulence, lightning, and the like" [35]. Furthermore, there is a severe danger of the collisions on the air because of sharing airspace with other commercial planes in urban areas. If the drone's flight path is outside the prohibited areas, it is safe. In contrast, it requires to be taken outside of the forbidden zones [33].

2.3.2 Privacy

In the use of commercial drones, privacy plays an essential role in security concerns. Drones have special accessible ways that discriminate them from other devices. Moreover, they integrate high-accuracy cameras, sensors, and recorders that can perform remotely and precisely [35]. Also, they face several security issues about privacy and personal data protection. In fact, hackers can use malicious applications to exploit drones and get personal data or profile individuals by using the wireless localization method.

2.3.3 Security

The technology inside of drones is the most significant security concerns of using a commercial or civilian drone. Because these technologies could be hacked or destroyed by attackers, which can disrupt services [33]. A drone's navigation system based on GPS can be easily spoofed because GPS signals are unencrypted and unauthenticated [63]. On the other hand, Wi-Fi Jamming attack could lead to out of control of drones' connection systems, as well as the crucial damages for nearby people [35]. The authors in [64] have designed and developed a secure communication protocol for UAVs. This study presented and discussed different communication protocols e.g. UranusLink, UAVCAN, and MAVLink [64]. The authors also provided their critical analysis of the structure and working mechanism of these protocols. Based on these protocols' pros and cons, the authors suggested that the MAVLink protocol can be widely used for UAVs communication. Although it provides better communication than others, it can have serious problems due to the lack of any security mechanism to encrypt messages. As a result, there is an urgent need to figure out a secure communication protocol in order to make sure the security standards of the communication between drones and the Ground Control Station (GCS). On the other hand, this research proposed an artificial intelligence agent, which will get input from a GCS and measure the criticality of the mission and the application security based on obtaining both efficiency and security simultaneously to guarantee the communication protocol.

2.3.4 Drone Operation in Smart Cities

Regarding the Federal Aviation Administration (FAA), Drones can operate in the special areas that are usually forbidden for commercial drones [65]. Nevertheless, based on the way you want to fly and who you are, it should have the permission of the available places and devices. This process is simpler and more flexible for commercial drones' operating. Moreover, to integrate drones with a domestic air system, a "phased-in approach" has been achieved [66].

The fast-growing number of drones may bring new challenges to the aviation industry, especially in air traffic control. Thus, to safeguard normal flights, a specific operating field needs to be established. Indeed, Sandor suggested that UAV traffic management systems (UTM) could provide the flight's achievement to monitor the traffic on the air effectively [67]. This system can separate the operation between the drones and the traditional aircraft. Besides, it can keep the order of the traffic flow with the "low-level" airspace parts [65].

In recent years, the interest in drones has increased among commercial entities and recreational flyers. Therefore, there is a need to ensure safety for people, properties, and other airspace users like helicopters during drone operations. The authors in [68] evaluated the height limit for drone operations in urban airspace with the available technology enablers, categories of drones, and the purpose of the application. Then they introduced drone lanes/tunnels or routes concepts, which are enabled for safe drone operation in urban areas.

2.3.5 Drone Management in Smart Cities

Regarding the FAA regulations, people need to monitor and control drones like airplanes [33], which are classified into four types [65]. The highlights of drone regulations are presented in reference [62], including a global overview and discussion of primary criteria. The author provided the viewpoint of the status of drone regulation' trends in the past, the present, and the future. Moreover, in this study, the authors examined and discussed the privacy, data protection, and public safety in the legal frameworks for operating drones. In our previous researches [65, 69], we presented the regulations of the use of drones and suggested drone-following models in order to control drones' traffic flows. Regarding security and infrastructure, drones may threaten civil aviation and reduce safety in the air. Hence, handling the drones in the sky is vital. The current research proposes a novel model for controlling drones namely the drone-following models, which expresses the one-by-one following the process of drones in the traffic flow [69]. Moreover, this approach is related to the demonstration of the drone acceleration that depends on the variety of velocities and distances between the given drone and its leading one. Based on the numerical simulation results, the safe distance between drones is kept, which means no traffic flow accident. Although the research's results show that these models can help to create the important simulation technologies or a new kind of controls, the equations of motion of drones must be integrated into these models to improve the proposed method [69]. The investigation [70] into the development of UTM for the drones' urban operation has resulted in several significant problems: difficulties in using passive surveillance systems, the complexity of conflict detection and resolution, and the need for a cost-effective solution. The solutions for these problems require full integration of UTM into the urban transport management system and the development of unique methods for managing a large number of drones in the traffic flows.

2.4 Data Communication Methods

Monitoring and managing drones in smart cities are performed through communication systems. The exchange of information between drones and the GCS is solved by several solutions, including wireless communication, Internet of Drone (IoD) solutions. However, the wireless connection must be a reliable, robust, scalable, and fast system for drone applications. Besides, the communication of multiple drones will become more challenging at the same time. Although the multi-drone, sometimes referred to as cooperative drones, drone formation, or groups of drones, has significant advantages compared with single-drone. Still, it can cause considerable challenges regarding the broken link, bandwidth limitation, and power. While frequencies of 2.4–5.8 GHz are applied for the civilian drones, satellites are widely used in common and in drones' military applications. Several existing wireless technologies, such as IEEE 802.x, 3G/4G/LTE, can be deployed for multi-drone applications [71]. In drone communication, several issues should be considered and evaluated, including the speed of the drone, energy limitation, limited on-board storage, and antenna angle. For instance, the antenna's characteristics may cause a lower data rate and reduced radio range.

3 Cybersecurity

Cybersecurity refers to technologies, processes, and practices that address to protect devices, systems, computers, networks, programs, and data from malicious attacks, damages, or unauthorized access [72–75]. Moreover, cybersecurity is known as information technology security or electronic information security. Furthermore, cybersecurity is related to the way of protecting networks, devices, and data from unauthorized access, cyber-threats, or criminal use to guarantee the confidentiality, integrity, and availability of information [76]. There are several elements of cybersecurity; for example, network security, application security, data security, information security, operational security, end-user education, and disaster recovery/business continuity planning [77].

3.1 Threats and Attacks

Although drones offer a lot of benefits, they also face many cyber-threats and cyber-attacks such as global position system jamming and spoofing, Wi-Fi security issues, sensor security concerns, Bluetooth security, UAV network security, malicious software/hardware, and privacy leak of photos. These cyber-threats and cyber-attacks are briefly described in this part to identify the current cybersecurity issues of drones to enhance cybersecurity awareness for users.

3.1.1 Global Positioning System (GPS) Jamming and Spoofing

Drones' navigation depends on the GPS to locate its position and direction. An attacker can break the encryption of this communication channel by using fake signals. Moreover, to operate a drone, it requires the communication links between GPS satellites, signals from drones, and a connection between the base station and the drone" [78]. However, it is easy for hackers to attack GPS signals because of no encryption mechanism. For instance, there was a report in 2011 about a "Lockheed Martin RQ-170 Sentinel drone" of the U.S. Air Force that was targeted by a GPS spoofing attack of Iranian forces [78]. Furthermore, several pieces of research have shown the various level of successful GPS spoofing attacks in media devices even drone is not an exception recently like [79–82]. This attack is very dangerous because it can make a UAV to fly as a set-path or change its velocity and position by the attacker [83–85].

3.1.2 Wi-Fi Security Issues

Wi-Fi connection is one of the main ways for drone communication, it offers several benefits such as flexibility, cost-effectiveness, and ease of installation. However, drone applications also involve Wi-Fi threats and cybersecurity issues [86]. There are many kinds of Wi-Fi security concerns towards a UAV, as follows [87] (Table 1).

3.1.3 Sensor Security Issues

Drone works based on a lot of sensors such as gyroscope, barometer, accelerometer, GPS, magnetometer, rangefinder, Inertial Measurement Unit (IMU), and so on [97]. These sensors can enlarge the abilities of drones, maintain their positions, enhance speed, and avoid obstacles. Like the Wi-Fi security concerns, sensors also have several vulnerabilities and attacks for drones like [98–102]. Almost these vulnerabilities and attacks can be classified into two major types such as active and passive malicious actions from attackers to exploit the sensors. Moreover, according to Amit Kumar Sikder et al., they summarized that there were four primary existing sensor-based threats on their purpose and nature of threats; for example, "information leakage, transmitting malicious sensor patterns or commands, false sensor data injection, and denial of service" [102]. On the other hand, physical damage or malfunction may cause potential threats for drones 'sensors.

3.1.4 Bluetooth and RFID Security

Bluetooth is the main factor in wireless communication. It offers a short-range radio transmission at low-power and low-cost solutions [103]. This technology can be implemented in many IoT devices e.g. mobile phones, laptops, cars, drones, medical

Table 1 Security concerns related to the Wi-Fi connection

Wi-Fi security concerns	Types	Sub-types	References
Wi-Fi threats	Interception, alteration, and disruption	Data interception, rogue apps, misconfigured apps, ad hoc and soft apps, wireless phishing, location tracing	[88–90]
Wi-Fi attacks	Active attacks	Eavesdropping and traffic analysis	[90–92]
	Passive attacks	Masquerading, replay, message notification, denial of service	
		Unauthorized access, rogue access point, man in the middle attack, reply attack, session hijacking, MAC spoofing	[90–94]
Related basic components (radio frequencies, access points, client devices, and users)	Confidentiality attacks	Traffic analysis, eavesdropping, man in the middle attack, evil twin AP	[90, 92, 93, 95, 96, 15]
	Integrity attacks	Session hijacking, replay attack, 802.11 frame injection attack, 802.11 data/802.11X EAP/802.11 radius replay attack, 802.11 data deletion	[90, 92, 93, 96]
	Availability attacks	Denial of service, radio frequency jamming, 902.11 beacon flood, 802.11 associate/authentication flood, 802.11 de-authentication and disassociation, queensland Dos/virtual carrier sense attack, fake SSID, EAPOL flood, AP theft, WPA2 vulnerabilities	[90, 92, 87, 96]
	Authentication attacks	Dictionary & brute force attack	[92, 96]

devices, industrial devices, entertainment devices, and the like [104]. Also, this technology consists of two basic standards; for example, "a legacy authentication procedure and a secure authentication procedure" [105]. These standards allow devices to authenticate each other by using a "long-term key" [104]. Although this technology brings many benefits, it contains vulnerabilities during secure connection establishment—namely Bluetooth impersonation attacks (BIAS) and Key Negotiation of Bluetooth (KNOB) attack [105, 106]. In these papers, the authors showed that the attackers can exploit Bluetooth devices in many ways such as BIAS attacks on legacy, BIAS downgrade attacks, BIAS reflection attacks on secure connections, and KNOB attacks.

RFID technology refers to Radio Frequency Identification technology which uses radio waves to locate and identify a small chip (RFID tag) attached to physical objects including UAV, smart devices, mobile phones, and so on [107]. This technology is integrated into various applications but without any standard security control; therefore, it can be easily targeted by hackers to read, modify, manipulate, or disable. There are many different types of threats towards RFID such as physical attacks, DoS, spoofing, traffic analysis, eavesdropping, cloning, tracking, and the like [107, 108]. On the other hand, security and privacy issues related to RFID are significantly important and need to be deeply considered in order to prevent sensitive information from cyber-attacks.

3.1.5 UAV Networking Security

Denial of service (DoS) attack and Distributed denial of service (DDoS) attack

DoS attack refers to attempts of an attacker to interrupt the service via security vulnerabilities or security holes of device or service to use the desired resources [109] (Fig. 5). On the other hand, DoS attack reduces the speed of exchanging information by sending a lot of requests in order to deny the legitimate users' access to share services or resources and make a machine or network resource unavailable [110]. Regarding David More et al., they clarified two major types of DoS attacks such as logical attacks and resource attacks [111]. Besides, DDoS is a complex format of DoS which uses many botnets to give a lot of attacks to a victim at the same time; as a result, it makes it more difficult to prevent [109]. For example, according to Edwin Vattapparamban et al.'s research, the attacker can use de-authentication flood attacks via wireless network Access Point (AP) to get control of the UAV [112, 113]. This attack can target the MAC address of AP to disconnect the link between the user and the drone, and get full access to the drone.

Man in the Middle

Man in the middle (MITM) attack is a general concept when an attacker used several techniques to put an attacker in the middle of a conversation between a user and an application to eavesdrop or capture the essential information [94, 114] (Fig. 6).

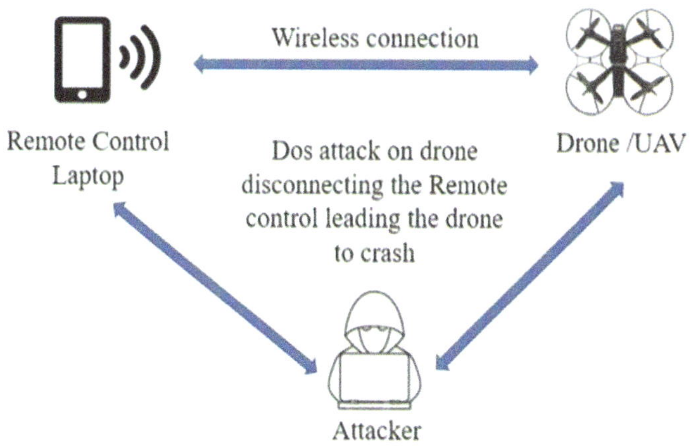

Fig. 5 Denial of service (DoS) attack [78]

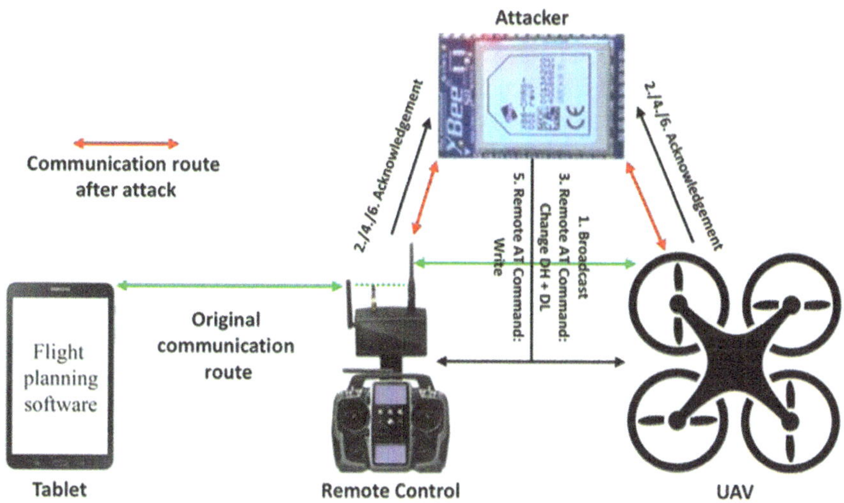

Fig. 6 Man in the middle attack [94]

The main goal of this attack is to steal important data by modifying the communication link between two parties; for example, password, username, and so on. In fact, according to the research of Rodday et al., it showed that a MITM attack could change the parameters for destination high (DH) address and destination low (DL) address for the UAV through radio network in 2016 [115]. Moreover, MITM attack has various forms such as URL manipulation, the rogue domain name server (DNS), Address Resolution Protocol (ARP) spoofing, and duplication of Media Access Control (MAC) [116].

Jamming or spoofing transmissions

The jamming attack is the simplest attack on wireless channels or wireless communication but it is very dangerous. On the other hand, this type of attack is a denial of service attack because this attack can block the communication between two devices on a wireless channel, or even take the control system [117]. Besides, attackers at the ground station can force UAV to switch flight mode into autopilot and make it land down in a pre-picked zone even it is connecting with the control unit by jamming attacks [118]. This attack is very dangerous because it is hard to detect and its impacts are unpredictable.

3.1.6 Malicious Hardware/Software

The flight controller and ground control unit are important elements for UAVs' operating; however, these parts are potentially exploitable with malicious hardware and software [63]. Malicious software such as Trojans, backdoors, and viruses can be designed in the system or third-party software and transferred to drone hardware or software. In fact, in 2013, a trojan was illustrated by Hartmann and Steup when the disclosure of the ground control unit in the Creech U.S. Air Force [119]. In addition, a good example of a virus—namely Maldrone infected civilian drones from the attacker to take control of the UAV [120]. Otherwise, hardware trojans can be implemented purposely inside the UAV's chips to turn off some security functions; as a result, it can have serious effects e.g. a backdoor—Actel ProASIC chip [63, 121].

3.1.7 Privacy Leak by Photos

Drones can be used for taking photos, recording videos with many purposes such as memorizing happy moments, capturing landscapes, military purposes, and so on. However, using drones for taking photos may cause some privacy issues; for instance, flying in forbidden areas and revealing the privacy of legal residents [122]. It not only influences on individuals' privacy but also national security; therefore, several countries take into account making the rules of using drones. For instance, America composed rights for regular individuals to put their home address in a list of no-fly zone database [122]. In addition, in 2019, the European Commission also adopted EU rules for drones to ensure the safety and security for users on the ground and in the air [123].

4 Security Countermeasures

There are diverse ways to protect and prevent the UAV system from cyber-attacks and security vulnerabilities. Every measure at different levels may offer various capabilities towards cyber-attacks. For example, according to Sushee Dahiya and Manik Garg, they clarified several measures for three levels of UAV system such as communication channel level, transceiver level, and control center level [124]. Moreover, several pieces of research and experiments indicated that many methods can be applied to ensure safety and protect UAVs or drones towards cyber-attacks a few years recently. There are two main stages for countering cyber-attacks such as detection and defense.

4.1 Detection Measures

This stage is an essential and proactive thing to recognize the cyber-attacks because it can quickly detect cyber-attacks as soon as they occur. Regarding Sedjelmaci et al., they expressed two methods to detect cyber-attacks that aim the data integrity and network availability for drones; for example, the cyber detection mechanism and threat estimation model [125]. Moreover, like the other network applications, UAV works in a network environment; therefore, an intrusion detection system is a crucial solution to predict and safeguard UAV from active security threats (overload, flash crowds, worms, port scans, jamming attacks, especially DoS/DDoS attack) [116, 118, 126, 127].

4.2 Defense Mechanisms

Based on many types of attacks such as protocol, sensors, components, networks, and physical attacks, Ben Nassi et al. gave several countermeasures for them [128]. Furthermore, several countermeasures for threats and cyber-attacks above are described in Table 2.

Additionally, drones include many sensors and components; as a result, they also have cybersecurity issues related to them. Nevertheless, the CIA (Confidentiality, Integrity, and Availability) triad is the main key to information security for many years. As a consequence, countermeasures for drones or UAVs need to focus on solutions corresponding with this standard [96]. Likewise, RFID technology is integrated into chips on drones but without any standard security controls; hence, it is easily targeted by hackers to modified, read, manipulated, or sabotaged the devices via many types of cyber-attacks such as eavesdropping, spoofing, tracking, cloning attacks, etc. There are possible measures to protect RFID threats; for instance, RFID threat countermeasure framework (cryptographic algorithm and non-cryptographic schemes) [107]. Last but not least, to protect drones from cyber-threats and cyber-attacks effectively, the detection method is one of the most important key factors. Hence, the combination of detection mechanisms, threat estimate model, and intrusion detection system is an essential tool to ensure the security of drones.

Table 2 Countermeasures for cyber-attacks

Threats and attacks	Impacts	Countermeasures
GPS jamming and spoofing	High	RAIM [129], A game-theoretic countermeasure [130], monitor GPS and satellite signal [131, 132], Crowd-GPS-Sec [81], cryptographic authentication signature [85], correlation of encrypted signals [82]
Wi-Fi security concerns	High	Securing UAV-ground and terrestrial communication [133], cryptographic mechanism and security policies [134]
Sensor security issues	High	Enhancing Existing Sensor Management Systems and Location-Privacy Preserving Mechanisms (LPPMs) [102]
Bluetooth or RFID security	Medium	Using long term key, legacy mutual authentication and role switching, and secure connections downgrade [105], Bluetooth firewall [103], legacy and non-legacy compliant countermeasure [106], RFID threat countermeasure framework [107], cryptographic hash functions in the tag [108]
DoS/DDoS attack	High	The ContainerDrone [110], mechanism of source-end, victim-end, and intermediate defense [116]
Man in the middle	High	Strong Wired Equivalent Privacy (WEP)/Wi-Fi Protected Access (WAP) encryption on the access points, HyperText Transfer Protocol Secure (HTTPS), and Public Key-based authentications [94, 116]
Jamming attack	High	Using hardware sandboxing [135], Jam-Me [136]
Malicious software	High	Authorized access, integrity, and availability [63]
Privacy leak by photos	Medium	Detection techniques (drone tracking, speech extraction, sound detection, advanced acoustic cameras, etc.) [123, 137–139]

5 Conclusion

In this chapter, the authors provide a general overview of the current security threats and cyber-attacks of UAVs and their countermeasures via a literature review of related studies. Drone intrusions can be divided into three major categories such as security vulnerabilities of devices integrated into drones like (sensor, Bluetooth, and RFID tags), communication links (UAV network, GPS, satellite), and privacy via photos. Sensors are essential components of drones to get information and make the flight safely; therefore, to prevent drones from cyber-attacks, it requires a combination of solutions for multiple sensors not only for a single one. Besides, communication links like Wi-Fi are unsafe and they are easy to get attacks. To ensure the safety of communication links, countermeasures should be taken more consideration to enhance existing security problems and encourage users to apply a secure communication method like radio instead of Wi-Fi. Privacy issues are also important for users; as a consequence, the solutions for reducing these risks need to be figured out urgently. Furthermore, this chapter expresses that mitigation methods for all drone cyber-threats include detection and defense mechanisms based on the CIA triad concepts. Finally, it is vital to raise cybersecurity awareness of users towards cyber-attacks.

References

1. Erdelj, M., Natalizio, E., Chowdhury, K.R., Akyildiz, I.F.: Help from the sky: leveraging UAVs for disaster management. IEEE Pervasive Comput. **16**(1), 24–32 (2017). https://doi.org/10.1109/MPRV.2017.11
2. Foina, A.G., Sengupta, R., Lerchi, P., Liu, Z., Krainer, C.: Drones in smart cities: overcoming barriers through air traffic control research. In: 2015 Workshop on Research, Education and Development of Unmanned Aerial Systems RED-UAS 2015, pp. 351–359 (2016). https://doi.org/10.1109/red-uas.2015.7441027
3. George, J., Sujit, P.B., Sousa, J.B.: Search strategies for multiple UAV search and destroy missions. J. Intell. Robot. Syst. Theory Appl. **61**(1–4), 355–367 (2011). https://doi.org/10.1007/s10846-010-9486-8
4. Sun, Z., Wang, P., Vuran, M.C., Al-Rodhaan, M.A., Al-Dhelaan, A.M., Akyildiz, I.F.: BorderSense: border patrol through advanced wireless sensor networks. Ad Hoc Netw. **9**(3), 468–477 (2011). https://doi.org/10.1016/j.adhoc.2010.09.008
5. Barrado, C., Meseguer, R., López, J., Pastor, E., Santamaria, E., Royo, P.: Wildfire monitoring using a mixed air-ground mobile network. IEEE Pervasive Comput. **9**(4), 24–32 (2010). https://doi.org/10.1109/MPRV.2010.54
6. De Freitas, E.P. et al.: UAV relay network to support WSN connectivity. In: 2010 International Congress on Ultra Modern Telecommunications and Control Systems, ICUMT 2010, pp. 309–314 (2010). https://doi.org/10.1109/icumt.2010.5676621
7. Cho, A., Kim, J., Lee, S., Kee, C.: Wind estimation and airspeed calibration using a UAV with a single-antenna GPS receiver and pitot tube. IEEE Trans. Aerosp. Electron. Syst. **47**(1), 109–117 (2011). https://doi.org/10.1109/TAES.2011.5705663
8. Maza, I., Caballero, F., Capitán, J., Martínez-De-Dios, J.R., Ollero, A.: Experimental results in multi-UAV coordination for disaster management and civil security applications. J. Intell. Robot. Syst. Theory Appl. **61**(1–4), 563–585 (2011). https://doi.org/10.1007/s10846-010-9497-5
9. Puri, V., Nayyar, A., Raja, L.: Agriculture drones: a modern breakthrough in precision agriculture. J. Stat. Manag. Syst. **20**(4), 507–518 (2017). https://doi.org/10.1080/09720510.2017.1395171
10. Xiang, H., Tian, L.: Development of a low-cost agricultural remote sensing system based on an autonomous unmanned aerial vehicle (UAV). Biosyst. Eng. **108**(2), 174–190 (2011). https://doi.org/10.1016/j.biosystemseng.2010.11.010
11. Rajesh Ganesan, A.H.E., Mercilin Raajini, X., Nayyar, A., Sanjeevikumar, P., Hossain, E.: BOLD: Bio-Inspired Optimized Leader Election for Multiple Drones (2020)
12. Semsch, E., Jakob, M., Pavlíček, D., Pěchouček, M.: Autonomous UAV surveillance in complex urban environments. In: Proceedings—2009 IEEE/WIC/ACM International Joint Conference on Web Intelligence and Intelligent Agent Technology, IAT 2009, vol. 2, pp. 82–85 (2009). https://doi.org/10.1109/wi-iat.2009.132
13. Khan, N.A., Jhanjhi, N.Z., Brohi, S.N., Usmani, R.S.A., Nayyar, A.: Smart traffic monitoring system using Unmanned Aerial Vehicles (UAVs). Comput. Commun. **157**(April), 434–443 (2020). https://doi.org/10.1016/j.comcom.2020.04.049
14. Tractica: The Commercial Drone Market Is Experiencing Steady, Sustained Growth and Consolidation, with Global Revenue Expected to Reach $13.7 Billion by 2025 (2019) [online].

Available at: The Commercial Drone Market Is Experiencing Steady, Sustained Growth and Consolidation, with Global Revenue Expected to Reach $13.7 Billion by 2025. Accessed 02 Aug 2020
15. Singh, S.P., Nayyar, A., Kumar, R., Sharma, A.: Fog computing: from architecture to edge computing and big data processing. J. Supercomput. **75**(4), 2070–2105 (2019). https://doi.org/10.1007/s11227-018-2701-2
16. Krishnamurthi, A., Nayyar, R., Solanki, A.: Innovation opportunities through Internet of Things (IoT) for smart cities. Green Smart Technology for Smart Cities, pp. 261–292. CRC Press, Boca Raton, FL, USA (2019)
17. Kumar, A., Nayyar, A.: si 3-Industry: a sustainable, intelligent, innovative, internet of things industry. A Roadmap to Industry 4.0: Smart Production, Sharp Business and Sustainable Development, pp. 1–21. Springer, Cham (2020)
18. Nayyar, A., Nguyen, B.L., Nguyen, N.G.: The Internet of Drone Things (IoDT): future envision of smart drones. In: First International Conference on Sustainable Technologies for Computational Intelligence, vol. 1045, pp. 563–580. Springer, Singapore (2020). https://doi.org/10.1007/978-981-15-0029-9_45
19. Singh, P.: Smart cities: concept of new urbanization in India. Int. J. Acad. Res. Dev. **2**(5), 127–130 (2017)
20. Mohamed, N., Al-Jaroodi, J., Jawhar, I., Idries, A., Mohammed, F.: Unmanned aerial vehicles applications in future smart cities. Technol. Forecast. Soc. Change **153**, 119293 (2020). https://doi.org/10.1016/j.techfore.2018.05.004
21. Mandl, C., Kirschner, T.: European Smart City Initiative Assessment of Best Practices to Stimulate Market-/Demand-Pull (2014). https://doi.org/10.13140/2.1.4559.7449
22. Robert, J.: Urban Europe (2018)
23. Gartner: Gartner 2018 Hype Cycle for IT in GCC Identifies Six Technologies That Will Reach Mainstream Adoption in Five to 10 Years (2018) [online]. Available at: https://www.gartner.com/en/newsroom/press-releases/2018-12-13-gartner-2018-hype-cycle-for-it-in-gcc-identifies-six-technologies-that-will-reach-mainstream-adoption-in-five-to-10-years. Accessed 13 Jul 2020
24. Alusi, A., Eccles, R.G., Edmondson, A.C., Zuzul, T.: Sustainable cities: oxymoron or the shape of the future? SSRN Electron. J. (2012). https://doi.org/10.2139/ssrn.1726484
25. Solanki, A., Nayyar, A.: Green Internet of things (G-IoT): ICT technologies, principles, applications, projects, and challenges. Handbook of Research on Big Data and the IoT, pp. 379–405. IGI Global (2019)
26. Review, A.S.A.: Smart Cities—Six Dimensions (2016). https://doi.org/10.3850/978-981-07-8859-9
27. Lee, J.H., Phaal, R., Lee, S.H.: An integrated service-device-technology roadmap for smart city development. Technol. Forecast. Soc. Change **80**(2), 286–306 (2013). https://doi.org/10.1016/j.techfore.2012.09.020
28. Smart Circle: Boyd Cohen: 'The Smart City Wheel' (2014) [online]. Available at: https://www.smart-circle.org/smart-city/boyd-cohen-smart-city-wheel/. Accessed 13 Jul 2020
29. Defensenetworks: The Smart City Imperative—Why IT Matters (2017) [online]. Available at: http://www.densenetworks.com/smart-cities/the-smart-city-imperative-why-it-matters. Accessed 13 Jul 2020
30. Jucevičius, R., Patašienė, I., Patašius, M.: Digital dimension of smart city: critical analysis. Procedia—Soc. Behav. Sci. **156**(April), 146–150 (2014). https://doi.org/10.1016/j.sbspro.2014.11.137
31. Marsal-Llacuna, M.L., Colomer-Llinàs, J., Meléndez-Frigola, J.: Lessons in urban monitoring taken from sustainable and livable cities to better address the Smart Cities initiative. Technol. Forecast. Soc. Change **90**(PB), 611–622 (2015). https://doi.org/10.1016/j.techfore.2014.01.012
32. Mohammed, F., Idries, A., Mohamed, N., Al-Jaroodi, J., Jawhar, I.: UAVs for smart cities: opportunities and challenges. In: 2014 International Conference on Unmanned Aircraft Systems, ICUAS 2014—Conference Proceedings, pp. 267–273 (2014). https://doi.org/10.1109/icuas.2014.6842265

33. Vattapparamban, E., Güvenç, I., Yurekli, A.I., Akkaya, K., Uluağaç, S.: Drones for smart cities: issues in cybersecurity, privacy, and public safety. In: 2016 International Wireless Communications and Mobile Computing Conference, IWCMC 2016, pp. 216–221 (2016). https://doi.org/10.1109/IWCMC.2016.7577060
34. Menouar, H., Guvenc, I., Akkaya, K., Uluagac, A.S., Kadri, A., Tuncer, A.: UAV-enabled intelligent transportation systems for the smart city: applications and challenges. IEEE Commun. Mag. **55**(3), 22–28 (2017). https://doi.org/10.1109/MCOM.2017.1600238CM
35. Khan, M.A., Safi, E.A., Khan, I.U., Alvi, B.A.: Drones for good in smart cities: a review. In: International Conference on Electrical, Electronics, Computers, Communication, Mechanical and Computing, Jan 2018, p. 8
36. Moshref-Javadi, M., Lee, S., Winkenbach, M.: Design and evaluation of a multi-trip delivery model with truck and drones. Transp. Res. Part E Logist. Transp. Rev. **136**, 101887 (2020). https://doi.org/10.1016/j.tre.2020.101887
37. Crişan, G.C., Nechita, E.: On a cooperative truck-and-drone delivery system. Procedia Comput. Sci. **159**, 38–47 (2019). https://doi.org/10.1016/j.procs.2019.09.158
38. Popper, B.: UPS researching delivery drones that could compete with Amazon's Prime Air (2013) [online]. Available at: https://www.theverge.com/2013/12/3/5169878/ups-is-researching-its-own-delivery-drones-to-compete-with-amazons. Accessed 13 Jul 2020
39. DHL: DHL Parcelcopter (2020) [online]. Available at: https://www.dpdhl.com/en/media-relations/specials/dhl-parcelcopter.html. Accessed 13 Jul 2020
40. Bensinger, G., Nicas, J.: Delivery Drones Hit Bumps on Path to Doorstep (2015) [online]. Available at: https://www.wsj.com/articles/technical-hurdles-delay-drone-deliveries-1426867441. Accessed 13 Jul 2020
41. Jee, C.: UPS Has Won Approval to Run the First Drone Delivery Airline in the US (2020) [online]. Available at: https://www.technologyreview.com/2019/10/02/102587/ups-has-won-approval-to-run-the-first-drone-delivery-airline-in-the-us/. Accessed 04 Apr 2020
42. Kanistras, K., Martins, G., Rutherford, M.J., Valavanis, K.P.: Survey of unmanned aerial vehicles (UAVs) for traffic monitoring. Handbook of Unmanned Aerial Vehicles, pp. 2643–2666 (2015). https://doi.org/10.1007/978-90-481-9707-1_122
43. Tesorero, A.: Drones to monitor Dubai roads in 2017 (2016) [online]. Available: https://www.khaleejtimes.com/nation/transport/drones-to-monitor-dubai-roads-in-2017. Accessed 13 Jul 2020
44. Press: Elistair Tethered Traffic Drone Monitors Lyon Rush Hour (2017) [online]. Available at: https://www.suasnews.com/2017/01/elistair-tethered-traffic-drone-monitors-lyon-rush-hour/. Accessed 04 Apr 2020
45. Barmpounakis, E., Geroliminis, N.: On the new era of urban traffic monitoring with massive drone data: the pNEUMA large-scale field experiment. Transp. Res. Part C Emerg. Technol. **111**, 50–71 (2020). https://doi.org/10.1016/j.trc.2019.11.023
46. Pal, B.B., Chakraborti, D., Biswas, P., Mukhopadhyay, A.: An application of genetic algorithm method for solving patrol manpower deployment problems through fuzzy goal programming in traffic management system: a case study. Int. J. Bio-Inspired Comput. **4**(1), 47–60 (2012). https://doi.org/10.1504/IJBIC.2012.044930
47. Savkin, A.V., Huang, H.: Asymptotically optimal deployment of drones for surveillance and monitoring. Sensors (Switzerland) **19**(9) (2019). https://doi.org/10.3390/s19092068
48. Liu, M., Liu, X., Zhu, M., Zheng, F.: Stochastic drone fleet deployment and planning problem considering multiple-type delivery service. Sustainability **11**(14), 1–18 (2019). https://doi.org/10.3390/su11143871
49. Saleem, Y., Rehmani, M.H., Zeadally, S.: Integration of Cognitive Radio Technology with unmanned aerial vehicles: Issues, opportunities, and future research challenges. J. Netw. Comput. Appl. **50**, 15–31 (2015). https://doi.org/10.1016/j.jnca.2014.12.002
50. Luo, H., Zhang, P., Wang, J., Wang, G., Meng, F.: Traffic patrolling routing problem with drones in an urban road system. Sensors (Switzerland) **19**(23), 1–20 (2019). https://doi.org/10.3390/s19235164

51. Loughran, J.: Fully operational drone units have been launched by two police forces for the first time in the UK (2017) [online]. Available at: https://eandt.theiet.org/content/articles/2017/07/two-police-forces-introduce-drone-units-in-historic-first-for-law-enforcement/. Accessed 13 Jul 2020
52. Ward, V.: Police to use drones to aid criminal investigations (2016) [online]. Available at: https://www.telegraph.co.uk/news/uknews/crime/12081915/Police-to-use-drones-to-aid-criminal-investigations.html. Accessed 04 Apr 2020
53. Lei, Z.: 1,000 drones used by police across country (2017) [online]. Available at: https://www.chinadaily.com.cn/china/2017-06/19/content_29792454.htm. Accessed 04 Apr 2020
54. Mishra, B., Garg, D., Narang, P., Mishra, V.: Drone-surveillance for search and rescue in natural disaster. Comput. Commun. **156**, 1–10 (2020). https://doi.org/10.1016/j.comcom.2020.03.012
55. Tuna, G., Nefzi, B., Conte, G.: Unmanned aerial vehicle-aided communications system for disaster recovery. J. Netw. Comput. Appl. **41**(1), 27–36 (2014). https://doi.org/10.1016/j.jnca.2013.10.002
56. Estes, A.C.: Dubai's Turning Drones into Firefighters [online]. Available at: https://gizmodo.com/dubais-turning-drones-into-firefighters-1505685714. Accessed 13 Jul 2020
57. Rojas, R.: New York City's Firefighting Arsenal Will Soon Include Drones (2016) [online]. Available at: https://www.nytimes.com/2016/09/09/nyregion/new-york-city-fire-department-drones.html. Accessed 13 Jul 2020
58. Jawhar, I., Mohamed, N., Al-Jaroodi, J.: Networking and communication for smart city systems. In: 2017 IEEE SmartWorld, Ubiquitous Intelligence & Computing, Advanced & Trusted Computed, Scalable Computing & Communications, Cloud & Big Data Computing, Internet of People and Smart City Innovation (SmartWorld/SCALCOM/UIC/ATC/CBDCom/IOP/SCI), pp. 1–7 (2018). https://doi.org/10.1109/uic-atc.2017.8397563
59. Ullah, F., Al-turjman, F., Nayyar, A.: IoT-based green city architecture using secured and sustainable android services. Environ. Technol. Innov. **20** (2020). https://doi.org/10.1016/j.eti.2020.101091
60. Zeng, Y., Zhang, R., Lim, T.J.: Wireless communications with unmanned aerial vehicles: opportunities and challenges. IEEE Commun. Mag. **54**(5), 36–42 (2016). https://doi.org/10.1109/MCOM.2016.7470933
61. Alsamhi, S., Ma, O., Ansari, M., Gupta, S.: Collaboration of drone and internet of public safety things in smart cities: an overview of QoS and network performance optimization. Drones **3**(1), 13 (2019). https://doi.org/10.3390/drones3010013
62. Stöcker, C., Bennett, R., Nex, F., Gerke, M., Zevenbergen, J.: Review of the current state of UAV regulations. Remote Sens. **9**(5), 33–35 (2017). https://doi.org/10.3390/rs9050459
63. Altawy, R., Youssef, A.M.: Security, privacy, and safety aspects of civilian drones: a survey. ACM Trans. Cyber-Phys. Syst. **1**(2), 1–25 (2017). https://doi.org/10.1145/3001836
64. Khan, N.A., Jhanjhi, N.Z., Brohi, S.N., Nayyar, A.: Emerging use of UAV's: secure communication protocol issues and challenges. Elsevier Inc. (2020)
65. Dung, N.D.: Developing models for managing drones in the transportation system in smart cities. Electr. Control Commun. Eng. **15**(2), 71–78 (2020). https://doi.org/10.2478/ecce-2019-0010
66. Federal Aviation Administration: Integration of Civil Unmanned Aircraft Systems (UAS) in the National Airspace System (NAS) Roadmap, 2nd edn., p. 61 (2018)
67. Sándor, Z.: Challenges caused by the unmanned aerial vehicle in the air traffic management. Period. Polytech. Transp. Eng. **47**(2), 96–105 (2019). https://doi.org/10.3311/PPtr.11204
68. Pathiyil, L., Low, K.H., Soon, B.H., Mao, S.: Enabling safe operations of unmanned aircraft systems in an urban environment: a preliminary study. Atmri. Ntu. Edu. Sg, pp. 1–10 (2016)
69. Dung, N.D., Rohacs, J.: The drone-following models in smart cities. In: 2018 IEEE 59th International Scientific Conference on Power and Electrical Engineering of Riga Technical University (RTUCON)—Proceedings (2018). https://doi.org/10.1109/rtucon.2018.8659813

70. Szullo, A., Seller, R., Rohacs, D., Renner, P.: Multilateration based UAV detection and localization. Proceedings of the International Radar Symposium (2017). https://doi.org/10.23919/IRS.2017.8008235
71. Hayat, S., Yanmaz, E., Muzaffar, R.: Survey on unmanned aerial vehicle networks for civil applications: a communications viewpoint. IEEE Commun. Surv. Tutorials **18**(4), 2624–2661 (2016). https://doi.org/10.1109/COMST.2016.2560343
72. Lord, N.: What is Cyber Security? Definition, Best Practices & More (2019) [online]. Available at: https://digitalguardian.com/blog/what-cyber-security
73. Comodo one: What is Cyber Security?, 2018. [Online]. Available: https://one.comodo.com/blog/cyber-security/what-is-cyber-security.php
74. Kaspersky: What is Cyber Security? (2020) [online]. Available at: https://www.kaspersky.com/resource-center/definitions/what-is-cyber-security
75. Craigen, D., Diakun-Thibault, N., Purse, R.: Defining cybersecurity. Technol. Innov. Manag. Rev. **4**(10), 13–21 (2014). https://doi.org/10.22215/timreview835
76. CISA: What is Cybersecurity? (2009) [online]. Available at: https://www.us-cert.gov/ncas/tips/ST04-001
77. Rouse, M.: What is Cybersecurity? Everything You Need to Know (2020) [online]. Available at: https://searchsecurity.techtarget.com/definition/cybersecurity
78. Uluağaç, S., Vattapparamban, E., Güvenç, İ., Yurekli, A.I., Akkaya, K.: Drones for Smart Cities: Issues in Cybersecurity, Privacy, and Public Safety, pp. 216–221 (2016)
79. Wang, K., Chen, S., Pan, A.: Time and Position Spoofing with Open Source Projects (2015)
80. Horton, E., Ranganathan, P.: Development of a GPS spoofing apparatus to attack a DJI Matrice 100 Quadcopter. J. Glob. Position. Syst. **16**(1) (2018). https://doi.org/10.1186/s41445-018-0018-3
81. Jansen, K., Schafer, M., Moser, D., Lenders, V., Popper, C., Schmitt, J.: Crowd-GPS-Sec: leveraging crowdsourcing to detect and localize GPS spoofing attacks. In: Proceedings—IEEE Symposium on Security and Privacy, May 2018, pp. 1018–1031 (2018). https://doi.org/10.1109/sp.2018.00012
82. O'Hanlon, B.W., Psiaki, M.L., Bhatti, J.A., Shepard, D.P., Humphreys, T.E.: Real-time GPS spoofing detection via correlation of encrypted signals. Navig. J. Inst. Navig. **60**(4), 267–278 (2013). https://doi.org/10.1002/navi.44
83. Arteaga, S.P., Hernandez, L.A.M., Perez, G.S., Orozco, A.L.S., Villalba, L.J.G.: Analysis of the GPS spoofing vulnerability in the drone 3DR solo. IEEE Access **7**, 51782–51789 (2019). https://doi.org/10.1109/ACCESS.2019.2911526
84. Kerns, A.J., Shepard, D.P., Bhatti, J.A., Humphreys, T.E.: Unmanned aircraft capture and control via GPS spoofing. J. Field Robot. **31**(4), 617–636 (2014). https://doi.org/10.1109/ICIF.2008.4632328
85. Shepard, D.P., Bhatti, J.A., Humphreys, T.E., Fansler, A.A.: Evaluation of smart grid and civilian UAV vulnerability to GPS spoofing attacks. In: 25th International Technical Meeting of the Satellite Division of The Institute of Navigation 2012, ION GNSS 2012, vol. 5, pp. 3591–3605 (2012)
86. Nayyar, A., Jain, R., Mahapatra, B., Singh, A.: Cyber security challenges for smart cities. Driving the Development, Management, and Sustainability of Cognitive Cities, pp. 27–54 (2019)
87. Arana, P.: Benefits and Vulnerabilities of Wi-Fi Protected Access 2 (WPA2). Infs **612**, 1–6 (2006)
88. Choi, M.K., Robles, R.J., Hong, C.H., Kim, T.H.: Wireless network security: vulnerabilities, threats and countermeasures. Int. J. Multimed. Ubiquitous Eng. **3**(3), 77–86 (2008)
89. Thangaraj, P., Geethanjali, N., Kathiresan, K., Madhumathi, R.: Wifi infrastructure security system from vulnerable attacks. Int. J. Comput. Sci. Netw. Secur. **13**(12), 2013 (2013)
90. Choudhary, G., Sharma, V., Gupta, T., Kim, J., You, I.: Internet of Drones (IoD): Threats, Vulnerability, and Security Perspectives, vol. 37, pp. 1–13 (2018)
91. Arinze Sunday, N.: Wireless Local Area Network (WLAN): Security Risk Assessment and Countermeasures, pp. 1–52. Blekinge Inst. Technol. Sch. (2008)

92. Waliullah, M., Gan, D.: Wireless LAN security threats & vulnerabilities: a literature review. Int. J. Adv. Comput. Sci. Appl. **5**(1), 176–183 (2014). https://doi.org/10.1017/CBO9781107415324.004
93. Farrukh Khan, M.: WLAN security: issues and solutions. Handbook of Wireless Local Area Networks: Applications, Technology, Security, vol. 16, no. 1, pp. 449–457 (2005)
94. Rodday, N.M., De Schmidt, R.O., Pras, A.: Exploring security vulnerabilities of unmanned aerial vehicles. In: Proceedings of NOMS 2016—2016 IEEE/IFIP Network Operations and Management Symposium, Apr, pp. 993–994 (2016). https://doi.org/10.1109/noms.2016.7502939
95. Choudhary, G., Sharma, V., Gupta, T., Kim, J., You, I.: Internet of Drones (IoD): Threats, Vulnerability, and Security Perspectives (2018)
96. Javaid, A.Y.: Cyber Security Threat Analysis and Attack Simulation for Unmanned Aerial Vehicle Network (2015)
97. Flynt, J.: What Sensors Do Drones Use? (2019)
98. Petracca, G., Sun, Y., Jaeger, T., Atamli, A.: AuDroid: preventing attacks on audio channels in mobile devices. In: ACM International Conference Proceeding Series, 7–11 Dec, pp. 181–190 (2015). https://doi.org/10.1145/2818000.2818005
99. Petracca, G., Marvel, L.M., Swami, A., Jaeger, T.: Agility maneuvers to mitigate inference attacks on sensed location data. In: Proceedings of MILCOM 2016—2016 IEEE Military Communications Conference, pp. 259–264 (2016). https://doi.org/10.1109/milcom.2016.7795336
100. Petracca, G., Reineh, A.A., Sun, Y., Grossklags, J., Jaeger, T.: Aware: preventing abuse of privacy-sensitive sensors via operation bindings. In: Proceedings of the 26th USENIX Conference on Security Symposium, pp. 379–396 (2017)
101. Petracca, G., Atamli, A., Sun, Y., Grossklags, J., Jaeger, T.: Aware: Controlling App Access to I/O Devices on Mobile Platforms (2016)
102. Sikder, A.K., Petracca, G., Aksu, H., Jaeger, T., Uluagac, A.S.: A Survey on Sensor-Based Threats to Internet-of-Things (IoT) Devices and Applications (2018)
103. Lonzetta, A.M., Cope, P., Campbell, J., Mohd, B.J., Hayajneh, T.: Security vulnerabilities in bluetooth technology as used in IoT. J. Sens. Actuator Netw. **7**(3), 1–26 (2018). https://doi.org/10.3390/jsan7030028
104. Antonioli, D., Tippenhauer, N.O., Rasmussen, K.: BIAS: Bluetooth Impersonation AttackS, pp. 549–562 (2020). https://doi.org/10.1109/sp40000.2020.00093
105. Antonioli, D., Tippenhauer, N.O., Rasmussen, K.: BIAS: Bluetooth Impersonation AttackS. In: Proceedings of the IEEE Symposium on Security and Privacy, pp. 549–562 (2020). https://doi.org/10.1109/sp40000.2020.00093
106. Antonioli, D., Tippenhauer, N.O., Rasmussen, K.: The knob is broken: exploiting low entropy in the encryption key negotiation of Bluetooth BR/EDR. In: SEC'19: Proceedings of the 28th USENIX Conference on Security Symposium, pp. 1047–1061 (2019)
107. Spruit, M., Wester, W.: RFID Security and Privacy: Threats and Countermeasures (2013)
108. Peris-lopez, P., Hernandez-castro, J.C., Estevez-tapiador, J.M.: RFID system: a survey on security threats and proposed solutions, vol. 4217 (2006). https://doi.org/10.1007/11872153
109. Lau, F., Rubin, S.H., Smith, M.H., Trajković, L.: Distributed denial of service attacks. Proceedings of IEEE International Conference on Systems, Man and Cybernetics vol. 3, pp. 2275–2280 (2000). https://doi.org/10.1109/icsmc.2000.886455
110. Chen, J., Feng, Z., Wen, J.Y., Liu, B., Sha, L.: A container-based DoS attack-resilient control framework for real-time UAV systems. In: Proceedings of 2019 Design, Automation & Test in Europe Conference & Exhibition, DATE 2019, pp. 1222–1227 (2019). https://doi.org/10.23919/date.2019.8714888
111. D. Moore, G. M. Voelker, and S. Savage, "Inferring internet denial-of-service activity, Proc. 10th USENIX Secur. Symp., vol. 24, no. 2, pp. 115–139, 2001
112. Gudla, C., Rana, S., Sung, A.H.: Defense Techniques Against Cyber Attacks on Unmanned Aerial Vehicles, pp. 110–116 (2018)
113. Compton, S.: 802. 11 Denial of Service Attacks and Mitigation (2019)

114. Imperva: Man in the middle (MITM) attack (2020) [online]. Available at: https://www.imperva.com/learn/application-security/man-in-the-middle-attack-mitm/
115. Dahlman, E., Lagrelius, K.: A Game of Drones: Cyber Security in (2019)
116. Rani, C., Modares, H., Sriram, R., Mikulski, D., Lewis, F.L.: Security of unmanned aerial vehicle systems against cyber-physical attacks. J. Def. Model. Simul. **13**(3), 331–342 (2016). https://doi.org/10.1177/1548512915617252
117. Mpitziopoulos, A., Gavalas, D., Konstantopoulos, C., Pantziou, G.: A survey on jamming attacks and countermeasures in WSNs. IEEE Commun. Surv. Tutor. **11**(4), 42–56 (2009). https://doi.org/10.1109/SURV.2009.090404
118. Arthur, M.P.: Detecting signal spoofing and jamming attacks in UAV networks using a lightweight IDS. In: CITS 2019—Proceeding of the 2019 International Conference on Computer, Information and Telecommunication Systems, pp. 1–5 (2019). https://doi.org/10.1109/cits.2019.8862148
119. Hartmann, K., Steup, C.: The vulnerability of UAVs to cyber attacks—an approach to the risk assessment. In: Proceedings of the 5th International Conference on Cyber Conflict, pp. 1–23. IEEE (2013)
120. Paganini, P.: A Hacker Developed Maldrone, the First Malware for Drones (2015) [online]. Available at: http://securityaffairs.co/wordpress/32767/hacking/maldrone-malware-for-drones.html
121. Gil Casals, S., Owezarski, P., Descargues, G.: Generic and autonomous system for airborne networks cyber-threat detection. In: AIAA/IEEE Digital Avionics Systems Conference—Proceedings, pp. 1–14 (2013). https://doi.org/10.1109/dasc.2013.6712578
122. Zhi, Y., Fu, Z., Sun, X., Yu, J.: Security and privacy issues of UAV: a survey. Mob. Netw. Appl. **25**(1), 95–101 (2020). https://doi.org/10.1007/s11036-018-1193-x
123. European Commission: European Commission Adopts Rules on Operating Drones (2019) [online]. Available at: https://ec.europa.eu/transport/modes/air/news/2019-05-24-rules-operating-drones_en
124. Dahiya, S., Garg, M.: Proceedings of UASG 2019—Unmanned Aerial Vehicles: Vulnerability to Cyber Attacks, vol. 51. Springer (2020)
125. Sedjelmaci, H., Senouci, S.M., Messous, M.A.: How to detect cyber-attacks in unmanned aerial vehicles network? In: 2016 IEEE Global Communications Conference, GLOBECOM 2016—Proceedings (2016). https://doi.org/10.1109/glocom.2016.7841878
126. Condomines, J., Zhang, R., Larrieu, N.: Ad Hoc Networks Network intrusion detection system for UAV ad-hoc communication: from methodology design to real test validation. Ad Hoc Netw. **90**, 101759 (2019). https://doi.org/10.1016/j.adhoc.2018.09.004
127. Choudhary, G. et al.: Intrusion Detection Systems for Networked Unmanned Aerial Vehicles: A Survey
128. Nassi, B., Shabtai, A., Masuoka, R., Elovici, Y.: SoK—Security and Privacy in the Age of Drones: Threats, Challenges, Solution Mechanisms, and Scientific Gaps, pp. 1–17 (2019)
129. Javaid, A.Y., Jahan, F., Sun, W.: Analysis of Global Positioning System-based attacks and a novel Global Positioning System spoofing detection/mitigation algorithm for unmanned aerial vehicle simulation. Simulation **93**(5), 427–441 (2017). https://doi.org/10.1177/0037549716685874
130. Eldosouky, A., Ferdowsi, A., Saad, W.: Drones in Distress : A Game-Theoretic Countermeasure for Protecting UAVs Against GPS Spoofing (2003)
131. Warner, J.S., Ph, D., Johnston, R.G., Ph, D.: GPS Spoofing Countermeasures (2003)
132. Camtepe, S.A., Scientific, T.C., Foo, E.: A Survey and Analysis of the GNSS Spoofing Threat and Countermeasures (2016). https://doi.org/10.1145/2897166
133. Wu, Q., Mei, W., Zhang, R.: Safeguarding Wireless Network with UAVs: A Physical Layer Security Perspective, vol. 117583, pp. 1–15
134. Sharma, V.: Three-tier neural model for service provisioning over collaborative flying ad hoc networks. Neural Comput. Appl. (2016). https://doi.org/10.1007/s00521-016-2584-1
135. Mead, J.: Prevention of Drone Jamming Using Hardware Sandboxing (2016)

136. Di Pietro, R., Oligeri, G., Tedeschi, P.: JAM-ME: Exploiting Jamming to Accomplish Drone Mission (2019). https://doi.org/10.1109/cns.2019.8802717
137. Birnbach, S., Baker, R., Martinovic, I.: Wi-Fly?: Detecting Privacy Invasion Attacks by Consumer Drones (2017). https://doi.org/10.14722/ndss.2017.23335
138. Busset, J. et al.: Detection and tracking of drones using advanced acoustic cameras. Unmanned/Unattended Sensors Sens. Networks XI; Advanced Free-Space Optical Communication Techniques and Applications, vol. 9647, p. 96470F (2015). https://doi.org/10.1117/12.2194309
139. Jeon, S., Shin, J.W., Lee, Y.J., Kim, W.H., Kwon, Y.H., Yang, H.Y.: Empirical study of drone sound detection in real-life environment with deep neural networks. In: 25th European Signal Processing Conference, EUSIPCO 2017, pp. 1858–1862 (2017). https://doi.org/10.23919/eusipco.2017.8081531

Cloud-Based Drone Management System in Smart Cities

Dinh-Dung Nguyen

Abstract Recently, unmanned aerial vehicles (UAVs), or drones, can be used to complete several different military tasks to the industry with numerous studies available in the literature. With the accelerated development of technologies, especially computing, sensing, the Internet of things (IoT), and Information and Communication Technologies (ICT), the demand for using drones has been increased in real-world applications. However, there will be more accidents when more drones are active in the sky. Therefore, it is essential to manage drones in operation areas, especially the urban environment. This research introduces an approach, called a cloud-based approach for managing drones in a smart city. This approach is based on the cloud devices and services such as computation, storage, and web services. A ground control station controls and monitors drones, allowing users to define path planning and achieve the information from drone's sensors. Users, or remoted pilots, can create paths or missions for drones, saved, and transferred to a connected drone. This approach lets users control and monitor drones as connected objects in a real-time environment. An experimental study of monitoring and controlling drones via the Internet (4G D-com Viettel) has been carried out, aiming to evaluate the real-time performance of monitoring and controlling drones. The experimental results have illustrated that the proposed method is a cloud solution that enables to manage and control drones in a real-time environment.

Keywords Drones · Internet of things · Internet of Drones · Cloud computing · Drone managing system · Smart city

D.-D. Nguyen (✉)
Department of Aeronautics, Naval Architecture and Railway Vehicles, Budapest University of Technology and Economics, Budapest, Hungary
e-mail: ddnguyen@vrht.bme.hu

Department of Aircraft System Design, Faculty of Aerospace Engineering, Le Quy Don Technical University, Hanoi, Vietnam

R. Krishnamurthi et al. (eds.), *Development and Future of Internet of Drones (IoD): Insights, Trends and Road Ahead*, Studies in Systems, Decision and Control 332,
https://doi.org/10.1007/978-3-030-63339-4_8

1 Introduction

Drones, commonly term of unmanned aerial vehicles (UAVs), are remotely piloted aircraft with vital roles in protection and commercial sectors. Drones can direct themselves automatically without any human control. A drone can equip various IoT devices, including sensors and payloads, to do several specific tasks [1]. Recently, drones are used in several applications like traffic monitoring [2], surveillance and mapping in agriculture applications [3], cargo delivery, etc.

In 2019, the UAV market was achieved USD 19.3 billion and was expected to touch USD 45.8 billion by 2025 [4]. The growing use of drones for numerous purposes, including monitor, surveillance, precision agriculture, and product delivery, contributes to the UAV market's growth. However, the most vital sectors predicted to drive the UAV market's growth are the rise in the military UAVs application by protection authorities global.

The UAV business has been segmented into various regions, such as North America, Europe, Asia Pacific, the Middle East, Latin America, and Africa. In 2019, UAV's largest market was North America, where it is estimated to be continuously the largest market in the next few years (see Fig. 1). The adoption of drones for edge and marine surveillance projects has been increasing in nations like the US and Canada, which drives the UAV market's increase in North America.

Besides, business drones' requirements are being directed by before-mentioned circumstances due to spreading marketing use events, affordable tools and assistance, high definition (HD) imaging and sensing, and combined data statistics. According to Tractica, it is expected that the business drone demand will stay overgrowing while giving vital chances to several manufacturing cooperators, leading a global income of USD 13.7 billion by 2025 (see Fig. 2) [5].

The business drone market is concentrating on center resolutions that drones exceed. The capacity to illustrate such useful answers is the essential determinant of which corporations can obtain long-term administration and reduction by the wayside.

By introducing the concept of "Internet of Drone Things (IoDT)," Nayyar et al. have forecasted that the current drone technology will broaden in various new fields

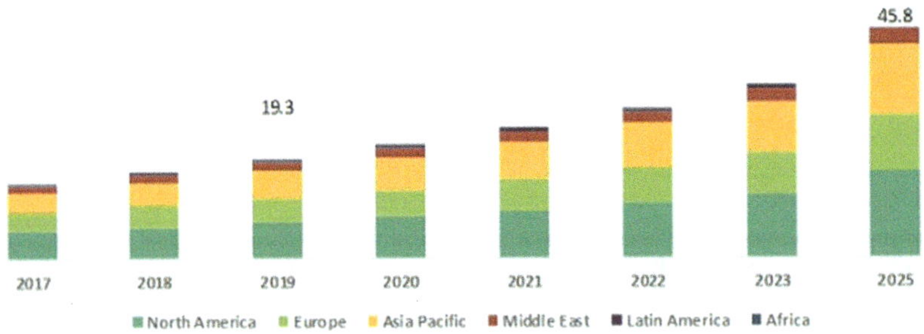

Fig. 1 The UAV market by region (USD billion) [4]

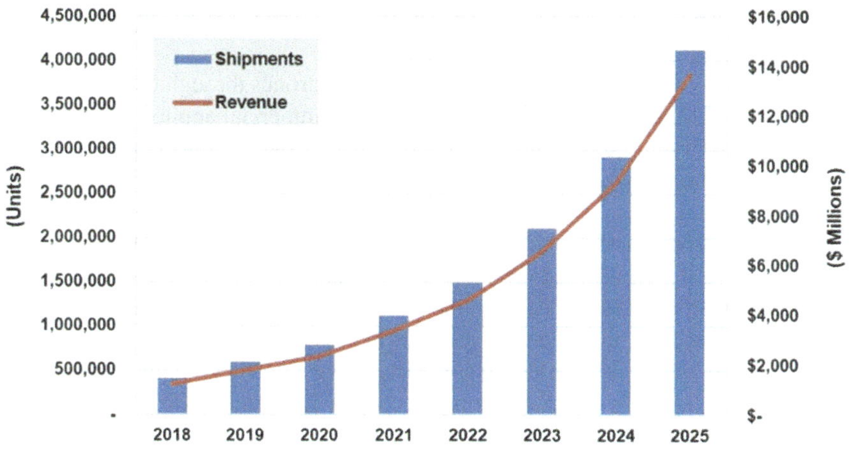

Fig. 2 Commercial drone hardware shipments, and total (hardware + services) revenue [5]

with missions, applications, and faster connectivity [6]. Besides, this study presented the technologies, applying, and safety problems related to the concept of IoDT and the experimental of IoDT.

Drones can be equipped with sensors to obtain information and get procedures similar to an actuator [7]. These drones can deliver packages, patrol areas, monitor infrastructure, search, and secure. In these platforms, drones were used as tele-operated vehicles through the Internet, based on low-level co-operations straight associated with the primary drone flights. However, controlling and managing drones through the Internet poses new challenges, which means that a large number of drone applications in particular airspace raises the need for drone traffic management or, in general, unmanned aircraft vehicle traffic management (UTM). Many studies have been proposed potential methods and technologies as well as system architecture for the UTM [8–12].

The investigations [13, 14] into the development of UTM for the drones' urban operation and analysis of the possible solutions have resulted in the identification of several significant problems, including

- difficulties in using passive surveillance systems (due to the low flight altitudes and large buildings),
- the complexity of conflict/obstacle detection and resolution (due to high traffic intensity and a lot of built obstacles such as houses), and
- need for a cost-effective solution (low-cost UTM, due to the very low operational cost of drones).

The solutions for these problems require full integration of UTM into the urban transport management systems and the development of unique methods for managing a large number of drones in formation flight. Such approaches include the management of dynamically variable groups of drones [15, 16], swarm optimization [17,

18] and drone-following models for individual vehicles [23, 24] moving with similar trajectories; being in the same "trajectory tunnels".

For outdoor environments, managing connective drones through the Internet was conducted in several research projects as well as commercial applications [19–21]. Besides, an IoT program for controlling cooperated drones in indoor conditions was introduced in the study [22], where indoor flight plans were used to control multiple connected drones.

This study proposes an approach called a cloud-based method for managing drones in a smart city, which is based on cloud devices and services such as computation, storage, and web services. In Sect. 2, some related works will be presented. The cloud-based connected drone management system is demonstrated in Sect. 3. Section 4 will show elements of a drone, including software, hardware, and communication units. The experimental study and evaluation process are illustrated in Sect. 5. This paper is ended by concluding Sect. 6.

2 Related Works

Various studies related to managing drones in smart cities are presented in this section.

In our previous researches [23, 24], we presented the regulations of the use of drones and proposed the drone-following models to manage drones in the urban environment. More accidents will occur when more drones are active in the sky, threatening urban air transport such as infrastructure (buildings, public areas) and a safe environment. Thus, the need for managing drones in urban areas is necessary. These studies proposed a novel approach to managing drones called the drone-following models, describing the one-by-one following drones' vehicles in urban air transport. This approach was based on defining the drone acceleration that depends on the variations in velocities and gaps from the given drone to its front one. Based on the numerical simulation results, the safe distance between drones is kept, which means no traffic flow accident. Although the numerically simulated results illustrate that proposed models can improve the powerful simulation technologies or a new prototype of controls, drones' equations of motion must be integrated into these models to improve the proposed method.

In an IoT environment, many proposal platforms are proposed to use drones as connected objects based on low-level system communication. The authors in [22] proposed an IoT program to control and monitor connected drones in indoor situations. This platform consists of three significant parts, as follows:

- the web application used to define flight plans, and it is allowed for sending and accomplishing these trajectories by continuous communication with the drones,
- a local Wi-Fi network where the server and the drones are connected,
- one or more connected drones that transfer and get information from the server.

In this study, users in a web application used indoor flight plans to control and monitor connected drones. This method can control multiple drones because the trajectories do not ought to be designed and taken simultaneously. With sensors equipped on drones, data can be shown in monitoring tasks by automatic communication stream among the server and the chosen drone. The proposed platform is validated based on user events and the investigation procedure. However, this platform is not designed for high-level services.

For high-level services, a net-drone using connected drones can be managed and monitored by IoT systems [21]. Such a system can operate many drones at the same time. The formation of drones will be deployed on-demand network co-operation in an emergency network infrastructure, like an immediate rise in community or emergencies. Moreover, such a group of drones is used by a service provider to improve network quality. Because net-drones were used to collect data about links, traffic, and neighboring drones, collected information can be employed to increase network entrance to users nearby. In this study, net-drones are managed by checking their altitude and the gap between drones. Several platforms based on IoT were presented in the literature [25–28].

Because of the drones' extensive application and safety vulnerability, especially a secure connection between drones and the ground control station (GCS) or between the GCS and end-users, they have performed them an engaging objective for hackers and enemies. Simultaneously, there were a few research available to discuss UAVs' protection resolutions, the author in [29] have designed and developed a secure communication protocol for UAVs. This study presented and discussed different communication protocols, including UranusLink, UAVCAN, MAVLink. The authors also provided their critical analysis of the structure and working mechanism of these protocols. Based on these protocols' pros and cons, the authors have distinguished that MAVLink is the standard generally accepted protocol for UAVs connection.

Nevertheless, MAVLink requires a security tool to encrypt information even it provides better communication than others. Therefore, there is a requirement for a safe transmission protocol to guarantee the needed protection standard positions for interaction between drones and the GCS. This study proposed a secure communication protocol with an artificial intelligence agent's aid, taking the input from a GCS and measuring the mission's criticality and application safety to obtain capability and protection simultaneously.

Regarding a group of drones, the authors in [31] presented a novel and feasible path planning technique based on a particle swarm optimization algorithm used to generate the desired trajectory sent to a single track for each drone based on its established location. This method is not only simple implemented for the group of drones but also generated the optimum paths for each drone in the formation.

Because drones can fly near people and property, safety regulations play an essential role in the drone managing system. All drones must then be controlled and regulated as aircraft according to the Federal Aviation Administration (FAA) laws [32]. Several cases of studies concerning the proper usage of drones are introduced. For example, in the use case of drones in the Slovak Republic [33], the application of drones in transportation must ensure protection and power effectiveness [34, 35].

The authors in [36] highlighted the significance, influence, and variety of UAV regulations. Most studies did not mention specific rules, even the legal provisions, which applied to their information gathering flights. Some studies presented national and international regulatory and short introductions, presenting only short summaries and overly generalized findings. However, secrecy, information protection, and public protection are concerned, investigated, and presented by lawyers and social scientists regarding societal aspects. Such researchers also determined the gap between current regulatory frameworks. This study also provided three critical aspects on which current UAV regulations focused, including

- the goal of the law is to manage the use of airspace by UAVs,
- establishing operational conditions to ensure suitable flights,
- undertaking regulatory schemes of flight confirmations, pilot permissions, and information gathering signature.

Through the above research discussions, it can be remarked that drones' management is a vital concern in the city transportation system and the urban air traffic management system. Firstly, the transport system is a whole complex system, effective operation of which becomes the essential tasks, which can be widely observed, analysed and managed by using an extensive distribution network of sensors and actuators integrated into a system communicating through the internet [37]. Secondly, aerial transportation will continue to increase and face new challenges such as a need for more capacity, more efficiency, and more safety. Thus, the hottest topic of the integration of air traffic management (ATM) with a total transport-managing system is the management of drones in urban areas. The primary identified problems are the difficulties of passive surveillance, possible very high traffic intensity, and conflict detection and resolution, including conflicts with built obstacles. The solutions for these problems require the full integration of ATM into the urban transport-management systems and the development of unique methods for managing a large number of vehicles in formation flight. Therefore, this study has proposed a new approach, called cloud-based drone managing system, for managing and controlling the dynamically variable groups of drones or a single drone. This research has been considered as a part of the project aimed to develop and integrate drones with the urban transport-managing system.

3 Cloud-Based Drone Management System

The Cloud-based drone managing system (CbDMS) motivated by IoT and the Internet of Drones (IoD) technologies have shown exemplary performance in trading with complicated and active traffic flows. This platform has three main layers as follows (see Fig. 3):

- physical layer, including connected drones, and fundamental infrastructure;
- cloud layer, including storage, computation and interfaces, is based on the wireless system, using the Internet, and

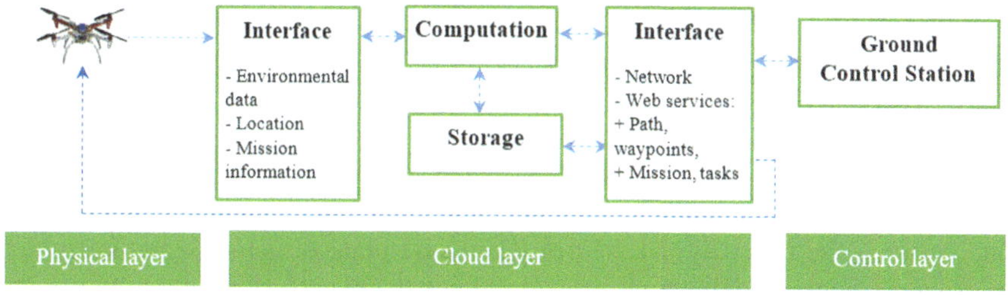

Fig. 3 The cloud-based drone managing system

- control layer is a hierarchically organized software set used to control and manage drones in the urban air transport system.

Because of the increasing number of drone applications in smart cities, employing CbDMS in air transport in smart cities cannot be denied. In the following subsections, each layer of the CbDMS is shed light.

3.1 Physical Layer

The physical layer is referred to as drones. Drones can take their tasks, including surveillance, traffic monitoring, wireless network applications, delivery [38, 39].

This layer attaches multiple network components and co-operations, such as drone-to-drone (D2D) or drone-to-target (D2T). The physical layer is connected with the IoT cloud by wireless technologies, including two categories: short and long-range [40]. The drones receive control signals and information about traffic situations from the cloud layer to guide their responses due to the desired of the ground control station (GCS) in the control layer. In the drone era, this transition is simple due to integrating drones with the IoT world.

3.2 Cloud Layer

The cloud layer has three main components, including storage, computation, and interface components. Providing storage can be used for streams of data about the environment, location, and mission information from drones and captured by this layer. Depending on the application's requirements, this data is stored in a regular Structured Query Language (SQL) database or shared data model, which supports the display of large-scale group processing on saved information. Data process on the cloud has two types: (i) real-time stream processing, using to detect critical events or possible threats; (ii) batch processing, applied to see critical situations by storing incoming data. Several computation algorithms, such as image processing,

Map/Reduce jobs, and data analytics, are executed in the cloud, which may reduce the processing time and improve performance. Interface components include network and web services interfaces, which are used to communicate between the physical layer and the control layer.

The cloud layer is also acknowledged as the center of the CbDMS, which is the core administration section of system performances. The layer aims to transfer the data between the physical and control layers and manage network administration and source allocation. The control layer defines the necessary policy to the center controller in the cloud layer, and the controller transfers those demands to the drones in the physical layer. This operation will be done with the interface components' cooperation based on Internet protocols [41, 42]. Furthermore, communication plays an essential role in transferring data from drones to the control layer, which provides higher efficiencies than conventional communication such as telemetry wireless communication. Besides, the cloud layer can be equipped with advanced sensors or controllers that can manage time consciousness and heterogeneity, providing higher efficiencies.

Interfaces in this layer can use a variety of different communication protocols such as cellular, wireless local area network (WLAN), low-power wide-area network (LPWAN), and wireless personal area network (WPAN) [43]. Depending on each drone application, goal, and overall need, more than one connection can be used for more reliable information, sincerity, and quality. A Wi-Fi transmission system is used for commercial drone operation, where drones communicate directly with the central station without an entrance location. Besides, long-range wide area networks (LoRaWAN) and long-term evolution (LTE) provide better reliability and low latency communication systems compared to Wi-Fi [44].

3.3 Control Layer

The control layer, referred to as the GCS, uses to control and manage drones in the traffic flows. The GCS is essential for monitoring drones from a place near to or inside the flying field. The GCS collects, progresses, and transforms the information from drones and gives it to different users in a similar network.

This layer consists of application software that can send the control signal to drones and receive data from drones. The users can record drones, set and modify task parameters thorough data analysis implemented by the cloud-based on such software. This application also enables remoted pilot monitoring and controlling the drones and their tasks remotely by connecting/disconnecting available drones.

4 Components of Drone

The drones are a complicated part of technology, including hardware and software that ensures a sustained and continuous flight. Today, the autonomous control system, such as self-driving vehicles, is possible because of integrating new productions of computational and mechanical methods. The IoT devices play the most critical role in transferring between offline location (hardware) and online cyberspace (software).

4.1 Software

The software part is recognized as the system's mind, while the drone's brain is the flight controller unit (FCU). This component is needed to ensure the drone and the user's specific and stable performance, including multiple layers such as firmware, middleware, and operating system. These layers connect users to the drone hardware, also called flight controllers.

Nowadays, several flight controllers can be used in a variety of drones because of their open-sourced packages, including Ardupilot [45], Pixhawk [46], iNav [47], Paparazzi [48], and LibrePilot [49].

Generally, the performance of the software is associated with the hardware platform. The software components are critical because more robust hardware cannot hide the software element's incompetence.

4.2 Hardware

The hardware has several components, which are demonstrated in Fig. 4. A drone can connect to the cyber world through this hardware. Two essential components of a drone are sensors and electronic speed controllers (working as an actuator). Sensors can be classified into [43]:

- proprioceptive sensors, measuring information internal for self-monitoring;
- exteroceptive sensors, covering data outside such as height and distance, and
- exproprioceptive sensors, relating the inner and outer performance of the drone.

Such sensors will continuously collect data about monitoring, movements, and senses. Based on that data, the actuators will decide to respond equally to the differences in the scene. Flight controllers (FC) can equip with sensors like accelerometers, gyroscope, and magnetometer that frequently get data about the drone, such as height, motion performance, and velocity changing. It could be impossible to test and control the drone operation without FC. A particular position can be navigated and hovered by the FC decision made with the aid of the Inertial Navigation System (INS) and Electronic Speed Controller (ESC). The drone's motion and the flying

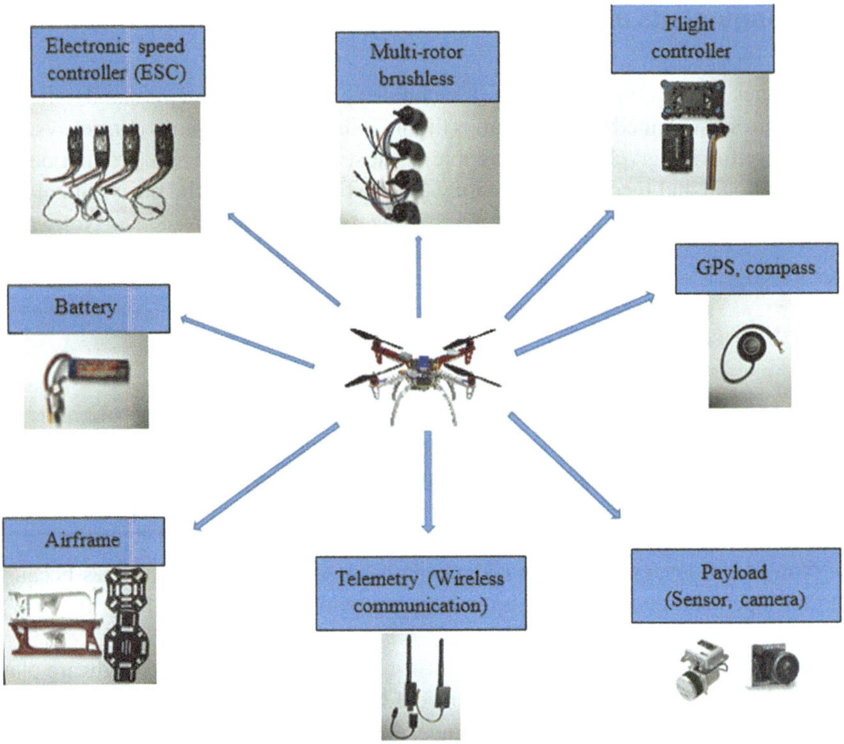

Fig. 4 Components of a drone

progress are monitored by exchanging data between the drone and the GCS. This exchange of information is taken via telemetry information flow, causing challenges for the long-range flights. To overcome this problem, satellites and Internet devices are the options for a case of a constant, uninterrupted telemetry information flow. Several payloads such as sensors or cameras can be equipped on the drone due to its tasks. In commercial drones, a First Person View (FPV) camera is mounted to record video during flights.

Another essential component of a drone is a battery that causes the drone's weight, affecting it more challenging to fly. Due to the limitation of the battery, drone tasks must be analyzed and carefully understood. An accident of the drone will occur if there is any small battery error, which leads to frustration in a wireless connection, data processing, and different essential parts in the drone. The battery's dimensions and the recharging time are the two essential criteria in drone batteries. Before and after performing their missions, drones need to find the charging stations to recharge their battery. Kuhn-Munkres algorithm is carried out to get the drones' optimum power to land in smart cities' suitable locations [30].

4.3 Communication Method

Checking and managing drones in smart cities are illustrated by transmission operations. The exchange of information between drones and the GCS is solved by several solutions, including wireless communication and the Internet of drone (IoD) solutions. However, the wireless connection must be a reliable, robust, scalable, and fast system for drone applications.

Besides, the communication of multiple drones will become more challenging at the same time. Although the multi-drone, sometimes referred to as cooperative drones, drone formation, or groups of drones, has significant advantages compared with single-drone. Still, it can cause considerable challenges regarding the broken link, bandwidth limitation, and power. While 2.4–5.8 GHz frequencies are assigned to inappropriate guiding of the civilian drones, satellites are used in large-scale and military applications. Several existing wireless technologies, such as IEEE 802, 3G/4G/LTE, can be deployed for multi-drone applications [50].

In drone communication, several issues should be considered and evaluated, including the speed of the drone, energy limitation, limited onboard storage, and antennas angle. For instant, the antenna's characteristics may cause a lower data rate and reduced radio range.

5 Experimental Study and Evaluation

This section shows an experimental investigation of monitoring and controlling drones via the Internet (4G D-com Viettel). This exploratory study aims to assess the real-time display of monitoring and controlling drones. This framework is built on hardware on the drone (Pixhawk PX4) and software on the ground (Mission planner).

The drone used in the experiments was built from airframe Z450 and processor Pixhawk PX4 that is responsible for its low-level motor control and body stabilization. This drone equipped with a companion computer (Raspberry Pi 3B) that is used for streaming data between the drone and GCS [51]. This drone also has a camera (Raspberry camera V1) that is used for ground observation and management. The video stream from this camera can be obtained via a Wi-Fi connection to be processed on the companion computer.

Besides, this testbed platform is to validate the proposed CbDMS, which provides the management and control of drone applications for delivery, surveys, security, ambulance, and emergency response.

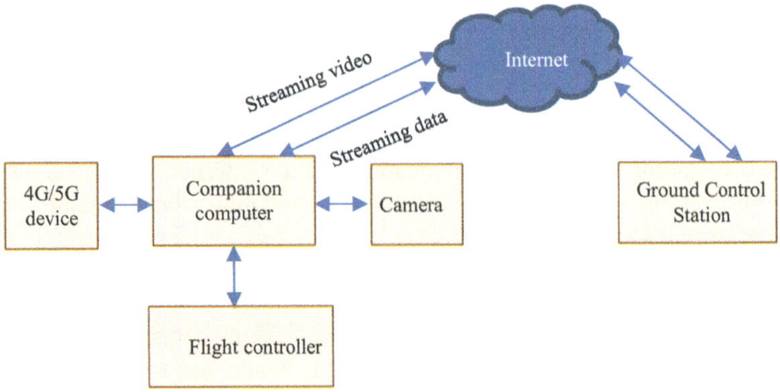

Fig. 5 The diagram of the testbed platform

5.1 Experimental Setup

We carried out tests with a drone to validate the proposed approach and assess its achievement—the test situation of following a set of waypoints with a real drone in a particular area.

The materials to setup hardware are demonstrated as the following:

- Flight controller: Pixhawk PX4
- Companion computers: Raspberry Pi 3B
- Micro SD card: 16 GB
- Camera: Raspberry camera V1
- D-Com: Viettel
- UBEC: HobbyKing™ HKU5 5 V/3A UBEC
- Direct Cable for connecting companion computer to the flight controller.

The diagram of this testbed platform is illustrated in Fig. 5.

The connection between hardware components is shown in Fig. 6.

Using a direct cable to connect the Pixhawk's TELEM2 port to the RPi's Ground, TX and RX pins as shown in Fig. 7.

Ground control station (Mission planner) connects to drone through 4G Wi-Fi router (D-com Viettel), which allows monitoring and controlling the drone. Besides, the drone will receive mission and command through MAVLink messages.

5.2 Experimental Results

The experimental result is demonstrated in Fig. 8. Originally, the drone was placed at a home position. When it received the command from the GCS, it take-off and do mission, visiting the created waypoints. It is seen that the desired trajectory and

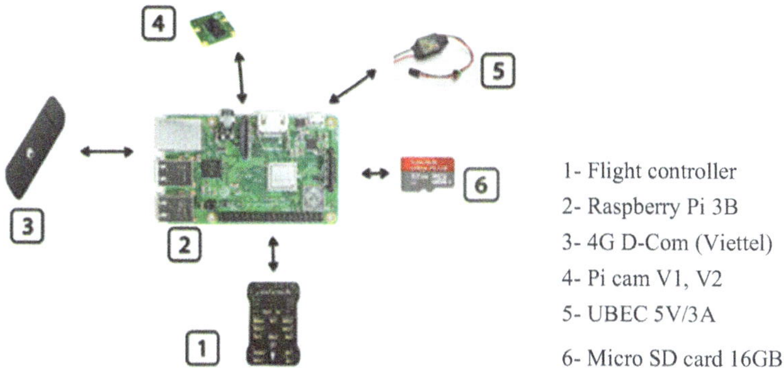

Fig. 6 The connection between hardware components

Fig. 7 The connection between companion computer and flight controller

actual trajectory are correlated. The gap between the two trajectories represents GPS location because the drone receives the GPS location.

The attitude information, including roll, pitch and yaw angles are shown in Figs. 9, 10, 11 and 12.

It can be observed that proportional-integral-derivative (PID) control is entirely sufficient to follow set position at low speeds. Thus, the drone aerodynamic changes slightly (such as no wind disturbance). For initial test flights, proper tracking was achieved, even providing the following errors on the scale of 2–40 (Figs. 9, 10 and 11).

It can be seen that the propellers are susceptible to the pitch and roll dynamics. The FCU has tried to control the drone moving accordingly to the required pitch.

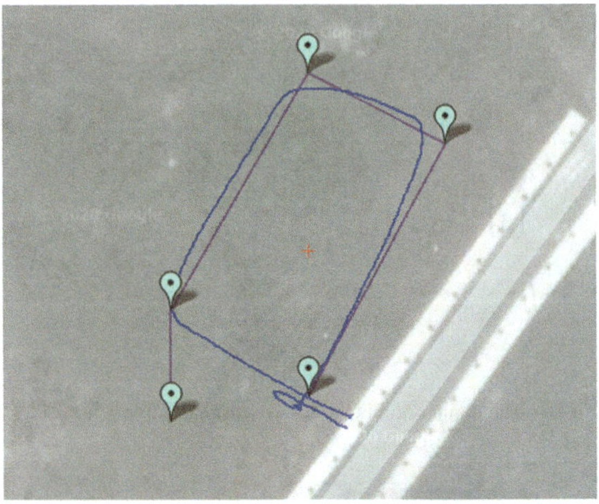

Fig. 8 The difference between desired and real trajectories (pink line—desired trajectory, blue line—real trajectory)

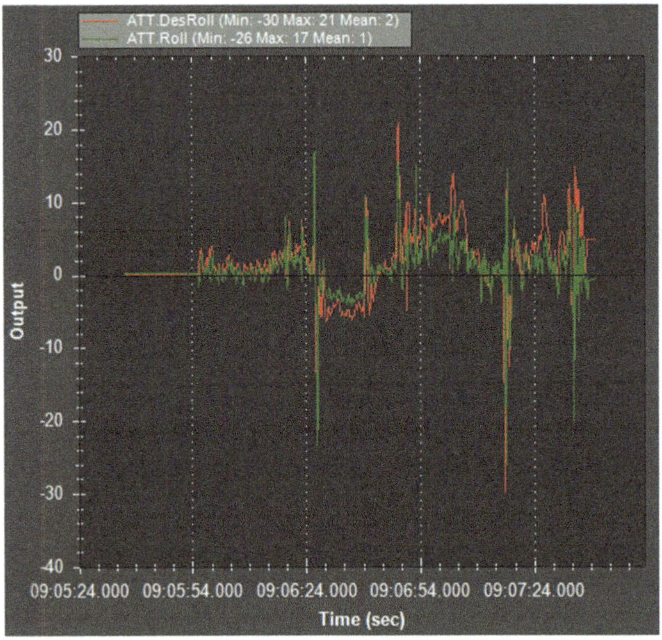

Fig. 9 The pilot's desired roll angle in degrees—red line, the drone's actual roll in degrees—green line

It is illustrated in Fig. 12 that a linear controller achieved the altitude flight control with the increase in the vertical speedup. The FCU has performed to be extremely useful in height control despite a minor gap between the desired and actual altitudes.

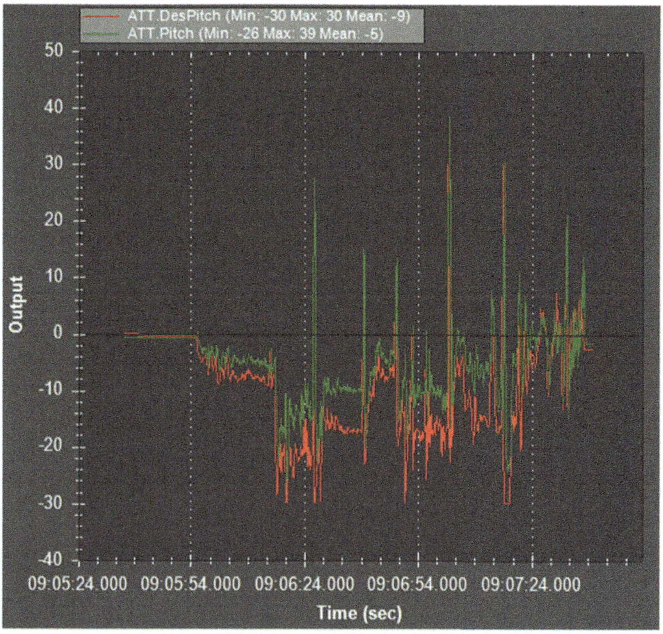

Fig. 10 The pilot's desired pitch angle in degrees—red line, the drone's actual pitch in degrees—green line

During the experiments, the video streamed by the drone downwards facing camera, which can observe and control the drone in a real-time environment.

These experimental results have demonstrated that the proposed CbDMS is a cloud solution that enables to manage and control drones in a real-time environment. The monitoring efficiency can be increased by raising the regularity of refreshing GPS coordinates or adding filtering techniques (Kalman filters).

5.3 Discussion and Future Research Directions

We have presented a CbDMS to manage and control a drone in a real-time environment. Experimental results using real video captured by a drone demonstrate that the proposed approach can determine the primary position and orientation of the desired trajectory.

The CbDMS is an advanced approach for managing drones to meet critical features. The CbDMS associates to real-time streaming, Cloud computing, regularly refreshed information, and intelligent acknowledgment to dynamically varying situations. With CbDMS, complicated missions can be taken with efficiency, improving safety and applicability.

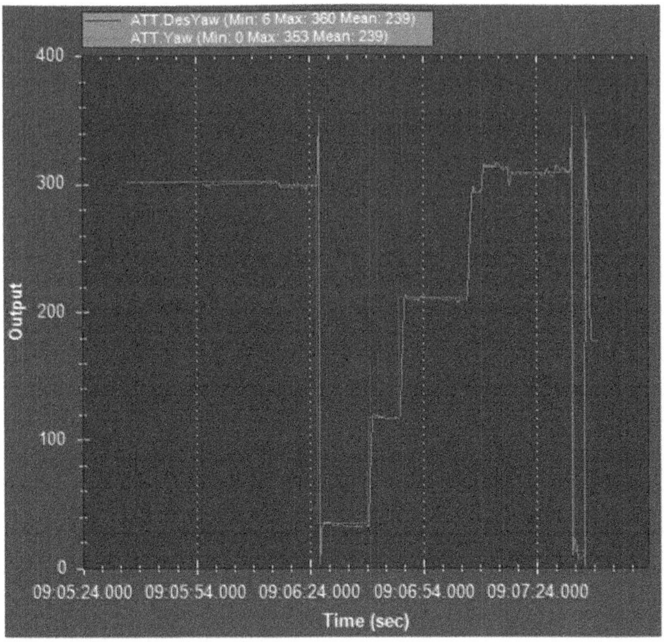

Fig. 11 The pilot's desired yaw angle in degrees—red line, the drone's actual yaw in degrees—green line

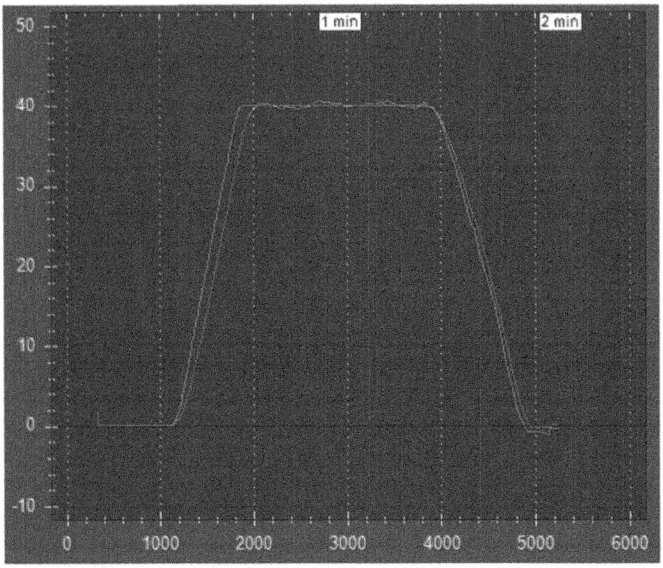

Fig. 12 The difference between desired and actual altitude of drone, green line—desired altitude, red line—actual altitude

It is necessary to examine the limitations of the proposed method. Firstly, it can be noted that the performance of controlling and managing real-time drones over the network is highly dependent on a guaranteed quality of assistance. Controlling drones over the Internet has two typical constraints, including hard real-time and soft real-time controls, which impose safe and extraordinary quality of co-operation networks. For instance, operating a drone through the Internet may cause harm or crash to drone because of a missing authority or request delay. An intelligent onboard device is the best solution to avoid crashes if a command is not received to overcome this problem.

In our experimental study, a drone autonomously followed a list of waypoints, which was sent to drone through the Internet. This constraint is a soft real-time, which means that the Internet is used to deliver offline commands to the drone.

Secondly, drones can detect obstacles and plan their paths by using onboard sensors that receive information in real-time. It means that drones can survey and gather environmental information. Keeping this information up to date enables online managing and controlling drones, which is one of the most advantages of drone applications. However, one drawback may occur in online trajectory creating and collision avoidance, such as less accurate due to insufficient input data. Regarding path planning and trajectories, several criteria, including whole journey length, fulfillment time, coverage field, and maneuvers, are applied to assess the execution of drone applications.

Finally, a drone used in our experiment is one of the most current testbeds for small drone evolution. However, its aerodynamics is sophisticated and requires to be correctly modeled to allow accurate trajectory checks. Several suitable FCUs have been developed and announced in the literature, which concentrated on smoothly trajectory, attitude, and altitude control, in controlled outdoor environments.

This work is part of a research that aims to integrate drones with the urban transport system. As a direct next step, we aim to investigate the multiple drone traffic by introducing drone following models; the effect of the following models on drone energy use during flight is investigated.

We also plan to extend this experiment with a group of drones that one can be a leader and others as followers. Other improvements can be carried out in future research that developing the proposed method to manage and control drones more accurately in a real-time environment. Furthermore, several possible extensions of this method are countless. We are planning to design and test applications based on the Leader-Follower formation of drones.

6 Conclusion

In this work, the proposed approach, cloud-based drone managing system, illustrates a drone managing system consisting of a physical layer, cloud layer, and control layer to guarantee reliable and effective drone performance. Because of the increasing number of drones in the urban environment, employing CbDMS in the smart city cannot be denied.

The experimental study demonstrates that the proposed CbDMS enables management and control drones in a real-time environment. The performance correctness can be increased by raising the regularity of refreshing GPS coordinates or adding filtering techniques such as Kalman filters. Besides, controlling a drone through the Internet may cause harm or crash to drone because of a missing authority or request delay. An intelligent onboard device is the best solution to solve this problem, avoiding crashes if a command is not received.

This study is part of the research that aims to integrate drones with the urban transport system. In the next step, we intend to investigate the multiple drone traffic by introducing drone following models; the effect of the following models on drone energy use during flight is investigated.

Besides, several extensions of this study can be conducted in future research, including improvement of the proposed method to ensure more accuracy operation, and experiment with a group of drones that one can be a leader and others as followers.

References

1. Cuffari, B.: Using Sensors in Drones (2018). https://www.azosensors.com/article.aspx?ArticleID=1149. Accessed 20 May 2020
2. Khan, N.A., Jhanjhi, N.Z., Brohi, S.N., Usmani, R.S.A., Nayyar, A.: Smart traffic monitoring system using Unmanned Aerial Vehicles (UAVs). Comput. Commun. (2020)
3. Puri, V., Nayyar, A., Raja, L.: Agriculture drones: a modern breakthrough in precision agriculture. J. Stat. Manag. Syst. **20**(4), 507–518 (2017)
4. Unmanned Aerial Vehicle (UAV) Market. https://www.marketsandmarkets.com/Market-Reports/unmanned-aerial-vehicles-uav-market-662.html. Accessed 8 Feb 2020
5. The Commercial Drone Market Is Experiencing Steady, Sustained Growth and Consolidation, with Global Revenue Expected to Reach $13.7 Billion by 2025. https://tractica.omdia.com/newsroom/press-releases/the-commercial-drone-market-is-experiencing-steady-sustained-growth-and-consolidation-with-global-revenue-expected-to-reach-13-7-billion-by-2025/. Accessed 8 Feb 2020
6. Nayyar, A., Nguyen, B.L., Nguyen, N.G.: The Internet of Drone Things (IoDT): future envision of smart drones. In: First International Conference on Sustainable Technologies for Computational Intelligence, pp. 563–580. Springer, Singapore (2020)
7. Bruzzone, A. et al.: Disasters and emergency management in chemical and industrial plants: drones simulation for education and training. In: International Workshop on Modelling and Simulation for Autonomous Systems, pp. 301–308. Springer (2016)
8. Kopardekar, P.H.: Unmanned aircraft systems traffic management (UTM) safely enabling UAS operations in low-altitude airspace. Presented at the UAV Industry Conference—Drones for Business Success, Auckland, New Zealand (2017)
9. Prevot, T., Rios, J., Kopardekar, P., Robinson, J.E., III, Johnson, M., Jung, J.: UAS traffic management (UTM) concept of operations to safely enable low altitude flight operations. In: 16th AIAA Aviation Technology, Integration, and Operations Conference, Washington, DC, United States (2017)
10. Pathiyil, L., Low, K., Soon, B.H., Mao, S.: Enabling safe operations of unmanned aircraft systems in an urban environment: a preliminary study. In: The International Symposium on Enhanced Solutions for Aircraft and Vehicle Surveillance Applications—ESAVS (2016)

11. Jiang, T., Geller, J., Ni, D., Collura, J.: Unmanned aircraft system traffic management: concept of operation and system architecture. Int. J. Transp. Sci. Technol. **5**(3), 123–135 (2016)
12. Foina, A.G., Krainer, C., Sengupta, R.: An unmanned aerial traffic management solution for cities using an air parcel model. In: 2015 International Conference on Unmanned Aircraft Systems (ICUAS), pp. 1295–1300. IEEE (2015)
13. Szullo, A., Seller, R., Rohacs, D., Renner, P.: Multilateration based UAV detection and localization. In: Proceedings of the 18th International Radar Symposium (IRS), 28–30 June, Praha, p. 10. IEEE. https://doi.org/10.23919/irs.2017.8008235
14. Dobi, S., Rohács, D.: HungaroControl nemzetközi szerepvállalás az UTM környezetben: A USIS projekt. Proceedings of the XII. Innováció és Fenntartható Felszíni Közlekedés, 29–31 Aug 2018, Budapest, Paper 42, p. 5. kitt.uni-buda.hu/mmaws/2018/pages/program/papers/Paper_42_Dobi-Rohács_D_IFFK_2018.pdf
15. Wang, X., Yadav, V., Balakrishnan, S.: Cooperative UAV formation flying with obstacle/collision avoidance. IEEE Trans. Control Syst. Technol. **15**(4), 672–679 (2007)
16. IEEE International Conference on Robotics and Automation (ICRA), 21–25 May 2018, Brisbane, Australia, 2018, pp. 6365–6372
17. Alfeo, A.L., Cimino, M.G.C.A., De Francesco, N., Lazzeri, A., Lega, M., Vaglini, G.: Swarm coordination of mini-UAVs for target search using imperfect sensors. Intell. Decis. Technol. **12**(2), 149–162 (2018)
18. Nguyen, D.D.: A developed particles swarm optimisation algorithm for managing drones in smart cities. Presented at the International Symposium on Sustainable Aviation 2019, 26–29 May 2019, Budapest, Hungary, Proceedings, p. 4
19. Erdelj, M., Król, M., Natalizio, E.: Wireless sensor networks and multi-UAV systems for natural disaster management. Comput. Netw. **124**, 72–86 (2017)
20. Chowdhury, S., Emelogu, A., Marufuzzaman, M., Nurre, S.G., Bian, L.: Drones for disaster response and relief operations: a continuous approximation model. Int. J. Prod. Econ. **188**, 167–184 (2017)
21. Park, K.-N., Kang, J.-H., Cho, B.-M., Park, K.-J., Kim, H.: Handover management of net-drones for future Internet platforms. Int. J. Distrib. Sens. Netw. **12**(3) (2016)
22. Pereira, A.A., Espada, J.P., Crespo, R.G., Aguilar, S.R.: Platform for controlling and getting data from network connected drones in indoor environments. Future Gen. Comput. Syst. **92**, 656–662 (2019)
23. Nguyen, D.D., Rohacs, J.: The drone-following models in smart cities. In: 2018 IEEE 59th International Scientific Conference on Power and Electrical Engineering of Riga Technical University (RTUCON), pp. 1–6. IEEE (2018)
24. Nguyen, D.D.: Developing models for managing drones in the transportation system in smart cities. Electr. Control Commun. Eng. **15**(2), 71–78 (2019)
25. Chuang, H., Hou, K.L., Rho, S., Chen, B.W.: Cooperative comodule discovery for swarm-intelligent drone arrays. Comput. Commun. **154**, 528–533 (2020)
26. Bera, B., Chattaraj, D., Das, A.K.: Designing secure blockchain-based access control scheme in IoT-enabled Internet of Drones deployment. Comput. Commun. **153**, 229–249 (2020)
27. Lin, X., et al.: Mobile network-connected drones: field trials, simulations, and design insights. IEEE Veh. Technol. Mag. **14**(3), 115–125 (2019)
28. Kishore, S., Rogan, N., Aman Pandey, S., Nagasudhan, N., Vikram Raj, K.: IoT based unmanned aerial vehicle for mobile monitoring and management of municipal solid waste (MSW) landfill sites and air quality index (AQI). Int. J. Sci. Technol. Res. **8**(11), 3671–3678 (2019)
29. Khan, N.A., Jhanjhi, N.Z., Brohi, S.N., Nayyar, A.: Emerging use of UAV's: secure communication protocol issues and challenges. In: Drones in Smart-Cities, pp. 37–55. Elsevier (2020)
30. Mirzaeinia, A., Bradley, S., Hassanalian, M.: Drone-station matching in smart cities through Hungarian algorithm: power minimization and management. In: AIAA Propulsion and Energy Forum (2019)
31. Hoang, V.T., Phung, M.D., Dinh, T.H., Ha, Q.P.: Angle-encoded swarm optimization for UAV formation path planning. In: 2018 IEEE/RSJ International Conference on Intelligent Robots and Systems (IROS), pp. 5239–5244. IEEE (2018)

32. Vattapparamban, E., Güvenç, İ., Yurekli, A.İ., Akkaya, K., Uluağaç, S.: Drones for smart cities: issues in cybersecurity, privacy, and public safety. In: 2016 International Wireless Communications and Mobile Computing Conference (IWCMC), pp. 216–221 (2016)
33. Fedorko, G., Žofčinová, V., Molnár, V.: Legal aspects concerning use of drones in the conditions of the slovak republic within the sphere of intra-logistics. Period. Polytech. Transp. Eng. **46**(4), 17–84 (2018)
34. Barmpounakis, E.N., Vlahogianni, E.I., Golias, J.C.: Unmanned aerial aircraft systems for transportation engineering: current practice and future challenges. Int. J. Transp. Sci. Technol. **5**(3), 111–122 (2016)
35. Péter, T., Szabó, K.: A new network model for the analysis of air traffic networks. Period. Polytech. Transp. Eng. **40**(1), 39–44 (2012)
36. Stöcker, C., Bennett, R., Nex, F., Gerke, M., Zevenbergen, J.: Review of the current state of UAV regulations. Remote Sens. **9**(5) (2017)
37. Nguyen, D.D., Rohacs, J.: Smart city total transport-managing system. In: International Conference on Industrial Networks and Intelligent Systems. pp. 74–85. Springer, Cham (2018)
38. Gupta, Lav, Jain, Raj, Vaszkun, Gabor: Survey of important issues in UAV communication networks. IEEE Commun. Surv. Tutor. **18**(2), 1123–1152 (2015)
39. Mohamed, N., Al-Jaroodi, J., Jawhar, I., Idries, A., Mohammed, F.: Unmanned aerial vehicles applications in future smart cities. Technol. Forecast. Soc. Change (2018)
40. Marchese, M., Moheddine, A., Patrone, F.: IoT and UAV integration in 5G hybrid terrestrial-satellite networks. Sensors (Basel) **19**(17) (2019)
41. Zhou, Z., Feng, J., Zhang, C., Chang, Z., Zhang, Y., Huq, K.M.S.: SAGECELL: software-defined space-air-ground integrated moving cells. IEEE Commun. Mag. **56**(8), 92–99 (2018)
42. Athanasiadou, G.E., Batistatos, M.C., Zarbouti, D.A., Tsoulos, G.V.: LTE ground-to-air field measurements in the context of flying relays. IEEE Wirel. Commun. **26**(1), 12–17 (2019)
43. Al-Turjman, Fadi, Abujubbeh, Mohammad, Malekloo, Arman, Mostarda, Leonardo: UAVs assessment in software-defined IoT networks: an overview. Comput. Commun. **150**, 519–536 (2020)
44. Yuan, Z., Jin, J., Sun, L., Chin, K., Muntean, G.: Ultra-reliable IoT communications with UAVs: a swarm use case. IEEE Commun. Mag. **56**(12), 90–96 (2018)
45. Ardupilot Open Source Autopilot (2020) [online]. Available at: http://ardupilot.org/. Accessed 27 May 2020
46. Open Source for Drones PX4 Open Source Autopilot (2020) [online]. Available at: https://px4io/. Accessed 27 May 2020
47. INAV: Navigation-Enabled Flight Control Software (2020) [online]. Available at: https://github.com/iNavFlight/inav. Accessed 27 May 2020
48. PaparazziUAV [online]. Available at: https://wiki.paparazziuav.org/wiki/Main_Page. Accessed 27 May 2020
49. LibrePilot—Open—Collaborative—Free. Available at: https://www.librepilot.org/site/index.html. Accessed 27 May 2020
50. Hayat, Samira, Yanmaz, Evşen, Muzaffar, Raheeb: Survey on unmanned aerial vehicle networks for civil applications: a communications viewpoint. IEEE Commun. Surv. Tutor. **18**(4), 2624–2661 (2016)
51. Nayyar, A., Puri, V.: Raspberry Pi—a small, powerful, cost effective and efficient form factor computer: a review. Int. J. Adv. Res. Comput. Sci. Softw. Eng. **5**(12), 720–737 (2015)

Smart Agriculture: IoD Based Disease Diagnosis Using WPCA and ANN

K. Subhadra and N. Kavitha

Abstract With technological improvement, there has been smart development has been widely used with a lot of sensors for different smart applications. The sensor data has been collected and process for the different sensor-based IoT development. Most of the sensors have to provide decision support in the real-time field. The agriculture field monitor by the drone camera and connected to the IoT framework is called the internet of Drone (IoD), The proposed research, IoT server timely decision on agriculture parameter and disease control is done using a drone image-based image processing for machine learning techniques, also analysis the irrigation parameters for agriculture monitor to control the water utility of the agricultural field. This proposed research applied wavelet-based statistical features to increase the accuracy of diseases and remedies for classification. Wavelet decomposition with PCA (WPCA) has been used to extract geometric features that have been extracted resulting in a set of 11 features. All these features have been applied as input for Enhancing the artificial neural network in the classification of diseases occurring in plant parts and suggest the remedies for final classified diseases. The combined result of WPCA based statistical features shows the accuracy of 97.460, hence, the proposed work turns out to be a better method to diagnose the pathological issues for crops with different plants also effectively deal with irrigation parameters.

Keywords Adaptive histogram (AHE) · Enhanced convolutional neural network (ECNN) · Wavelet PCA features · Enhanced artificial neural network

---

K. Subhadra (✉) · N. Kavitha
Nehru Arts and Science College, Thirumalayampalayam, Coimbatore, India
e-mail: subha.badra@gmail.com

N. Kavitha
e-mail: nkavee1@gmail.com

R. Krishnamurthi et al. (eds.), *Development and Future of Internet of Drones (IoD): Insights, Trends and Road Ahead*, Studies in Systems, Decision and Control 332,
https://doi.org/10.1007/978-3-030-63339-4_9

1 Introduction

The statistical report shows India has been secured the second rank in agriculture production. Agriculture is one of the leading fields and plays a key role in the socio-economic advancement of the nation. India is having approximately 210 million acres of land meant for only agriculture. The Wheat, Rice, Jowar, Corn, Nuts, Grams, Cereals, Sunflower are the most important crops and Mango, Orange, Banana, Grapes, Apple, Pomegranate, Watermelon, Berries, Sapota, Guava are the most familiar fruits grown. Vanilla, Dragon, Cotton, Silk, Tea, Coffee beans, Spices are the foremost commercial crops [1]. The agriculture output depends on rainfall, soil profile, and climatic condition. Variation in any of these leads to minimizing the yield. To maximize the overall comes under good monitoring. To enhance the agricultural environment monitoring the sensors and IoT technology has been used as smart technology for smart agriculture. IoT technology provides an immediate update to the user. The pathological issues occurring in these crops have been the major source for the decline in yield of the crop and hence it is the key challenge to control these diseases in a smart way. These plant pathogens occur mainly due to attacks in Virus, Parasites, Bacteria, Fungi, etc. These pathogens or the infectious agents may be either autotrophic (also called parasites) or saprophytes (also called heterotrophy). These saprophytes are going to survive on the dead tissues while the parasites are going to survive on the living tissues. Facultative are also there which are going to survive on both [2].

In ancient times, the diagnosis of the diseases in plants was manual which turned out to be quite expensive and difficult to consult due to lack of knowledge in cultivators and scarcity of experts. In such situations, various alternatives have been discovered to maintain the yield of the crop, preserve the environment, and reduce the practice of harmful chemicals and sprays. Hence the key apprehension for the agriculture experts, scientists, and researchers are the disease detection, recognition, classification and to suggest control measures. Thus early detection of the disease and its prevention is the only way to ensure higher yield. So it is inevitable to apply the latest technologies like information science, artificial intelligence, digital image, and processing to get the higher yield thereby help the development of the national economy [3].

Computation procedures have been used for computerization in diverse applications of the farming field. IoT outlines for smart farming can bring numerous benefits, counting dipping the risk of vendor lock-in, approving machinery, and sensing/automation systems [4]. The drone-based agriculture monitoring can aid to find easy and efficient timely decisions about crop conservation with smart technology.

The finding of plant disease has completed by human visual awareness. from diverse concerns, as these could convert simply interoperable in the general farm's smart arrangement, easier data argument among diverse, heterogeneous components, improved computerization with fewer exertion by engaging internet standards, etc.

To Project and progress a Robust programmed disease diagnosis organization for crops using digital image processing and Machine learning techniques, which senses, recognizes, and categorizes the pathological problems in plant crops and provides appropriate advice and information to control the disease, thereby prevent from losses. Therefore, techniques of machine learning and image processing are most suitable for this purpose [5].

Image processing is a area that includes of numerous techniques. It includes multiple processing phases such as the acquisition of image data, pre- processing, segmentation, analysis, interpretation of images, recognition, and classification from the real world. IoT based Image processing and Machine learning is an developing domain startled to the phenomena specially in scheming an artificial and a smart system that digs up the data and material using the images. Commonly, crop hurts from numerous diseases which embrace Powdery Mildew (Oidium mangiferae), Anthracnose (Colletotrichum gloeosporioides), Die Back (Botryodiplodia (Lasiodiplodia), Phoma Blight, Bacterial canker, Red rust, sooty mould etc. [6]. Hence, the progression of these types of diversity of arrangements regarding the key claims has been improved with the widespread use of methods of information system and knowledge along with computer science such as image processing, artificial intelligence, neural networks, machine learning algorithms, fuzzy logic, genetic algorithms, digital signal processing, etc. used to excerpt leaf material and discover the diseases [7]. The machine learning techniques in the mobile claim field are progressively being hired in the field of agriculture and the related fields

The Drone in agriculture plays a vital role has been used to optimize the farm based on a large range of images and their condition of crops, fields, and applying pesticides. by using a multispectral and hyperspectral sensor the farmer can identify the nutrient deficiencies, pest damage, fertilizer needs, water quality, and also Used to analyze plant nutrients, plant diseases, water quality, and mineral and surface chemical composition. Figure 1 represents the drone image in agriculture.

Figure 1 shows farming carried out in traditional ways and the modern-day farmers with the use of sophisticated equipment. The ultimate goal is to employ the automatic monitoring and diagnosis of the diseases occurring in the field with advanced IoT tools and equipment with mobile vision techniques. Early detection of crop diseases using image processing and machine learning techniques will use in the agriculture/horticulture field as well as in the national development. Figure 2 represents the traditional and modern agriculture.

Proposed system WPCA features with the use of the ANN are used. This paper is planned as follows. Section 2 describes the proposed work-related events. Section 3 defines the anticipated WPCA features with an improved ANN-based leaf disease classification overview. Sections 4 and 5 deals with the WPCA features with ANN experimental results and analysis. Section 6 describes the Conclusions and future work.

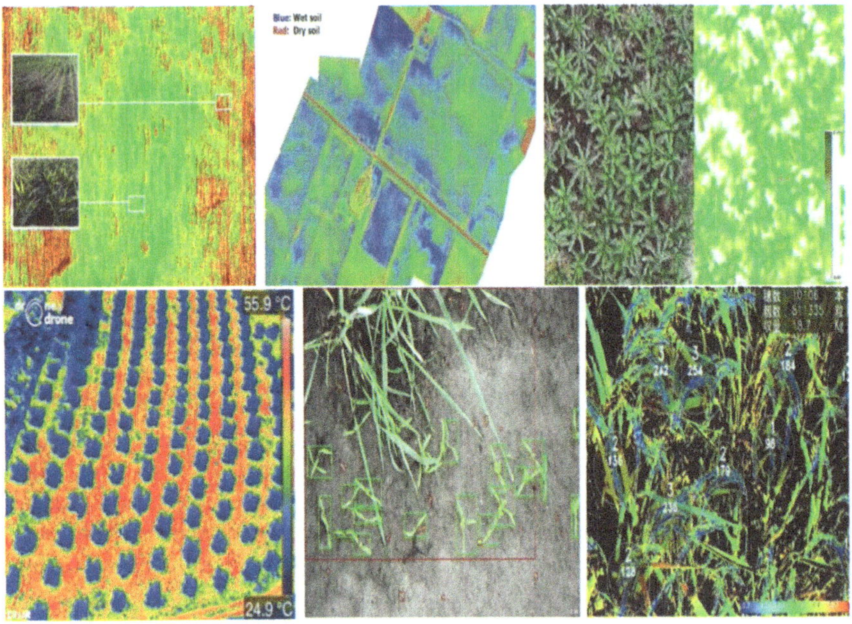

Fig. 1 Various drone image in agriculture field

2 Literature Review

The pest occurrence increased on food yields owed to altering environmental situations which are principal to a rapid surge in the arrival of sicknesses on crops. These sicknesses source overwhelming economic, Social, and ecological losses; this amazing issue grabs Consideration to diagnose diseases in an accurate and timely way. The farmer desires appropriate and consistent foundations of material as input for captivating suitable conclusions. Recent research is implemented in disease detection and parameter monitoring.

Ali et al. (2017) detected the infections on cotton leaf utilizing Principle Component Analysis (PCA), Nearest Neighborhood Classifier (KNN). The insufficiency of the cotton plant may identified by the leaf with the examination shade changing often happen by the pathogen or any other insects [8]. The cotton plant insufficiency may lead to deduce the yield from the plant.

Akila et al. (2018) deals with the Gray scale transformation for discovering the plant malady and histogram has been used for expansion of Plame with the lucitity. both method play the vital role in the remote image detection and various other applications. The RGB pictured are often used image in all image related applications. The histogram adjustment can lead to effective feature extraction [9].

Detection in grape leaf is done by Padol et al. (2016) via the Support Vector Machine (SVM) classifier. Digital clicks of leaves are taken as input and fed into

Fig. 2 Traditional versus modern farming

training and testing. Further image quality is enhanced. Resizing the image to size 300 × 300 is performed and components of green color are attained using thresholding. Gaussian filtering eliminates the noise. K-means clustering identifies the segmented region of diseased and extraction of features of color and texture are done. Hence, the sort of leaf infections are detected via Linear Support Vector Machine (LSVM) classification technique and achieves high accuracy of 88.89% [10].

The recognition and organization of plant leaf and stem sicknesses are done by an approach developed by Bashish et al. (2010). Initially, in the image-processing- based process, the K-Means technique performs the segmentation [11] and is transmitted to a pre-trained neural network. Al-Ghor area in Jordan is taken as input data in a testbed. High accuracy is achieved here significantly and detects leaf diseases automatically. Accuracy and detection automatically are achieved by the experimental results. The statistical classification based approach of the developed Neural Network classifier attains 93% precision in detection and classification [12].

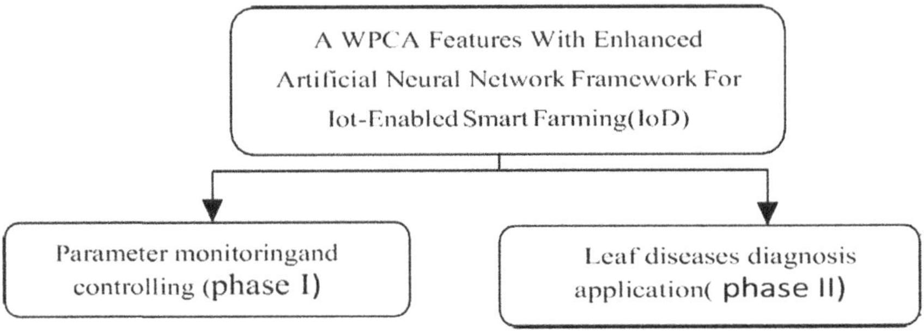

Fig. 3 Flow diagram of the proposed system

3 Proposed System Design

3.1 The IoD Based System Architecture

The recognition and organization of sickness signs using machine learning and image processing techniques include two phases. The fundamental concept of this paper is to provide a highly enabled monitoring of crop field with diseases diagnosis (Fig. 3).

The first phase in the proposed system is based on agricultural parameter monitoring with drone-based sensors in paddy crop field area monitoring and controlling without human interaction [13]. Here we have using temperature humidity, PH sensors interfacing with microcontroller. Sensors fixed to soil sample values of sensor transmit into microcontroller this controller displays the values of each sensor. Also, the system displays suitable crops for that soil sample. Soil parameter monitoring continuously controls the crop area motor devices using the mobile application through the IoT web server. Figure 4 represents the drone parameter for agricultural monitoring [14].

In the second phase, the crop leaf disease has been detected using WPCA- EANN machine learning algorithms by image processing applications. This phase is implemented KNV filter for pre-processing [15], adaptive K-means segmentation, WPCA feature extraction, and EANN classification methods (Fig. 5).

Capture the leaf image using a drone camera and the applied the above- mentioned methods using IoT server-based mobile application to diagnosis the diseases. From the following Fig. 6, the description for implementation of IoT based leaf diseases diagnosis with color classification for dead stressed and healthy leaf from the image identification.

3.1.1 Preprocessing Using KNV Filtering

Once the noise portion for each pixel is resolute apply the linear grouping between the present pixel and the mean of the non-noisy neighbours of the current space to

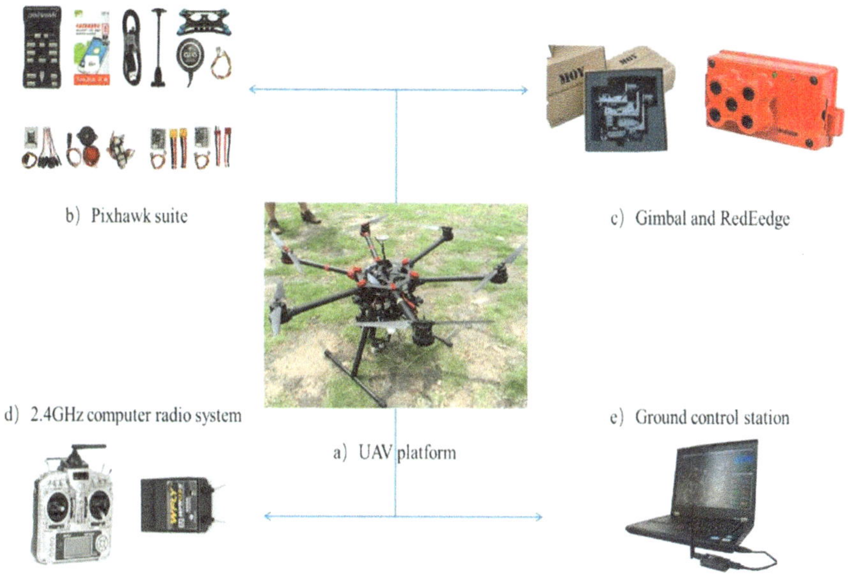

Fig. 4 Drone parameter for monitoring

give back the sensed noisy pixel. If the designated space comprises all the elements as noisy, the size of the space is augmented to the next space size, and the process is repeated till the window size reaches a predefined maximum window size. KNV filter values replace the pixel value by summing the values of neighboring pixels. Color perception is attained from the Hue Saturation Value (HSV). All values of pixels from the surrounding area summed up and result in median value and middle pixel value replaces the pixel. The color space parameter is obtained from the RGB colors of the filtered images (Fig. 7).

3.1.2 Segmentation Using Adaptive K-Means Clustering Algorithm

The segmentation has been compared with original K-mean method and adaptive k-mean techniques. The proposed method gives good result for segmentation. The conventional method ought of k-centroids for initialization makes huge demerits for the location-based centroid change the output can also be changed [16]. The random number-based cluster and location placement may lead to improper segment with degraded result.

This problem has been overcome with proposed adaptive k-mean clustering with, local minimum and local maximum ideals are added from the image. In an adaptive scheme, an iterative aspect method is applied by minimization of the given objective function to produce the optimal value of initial k- centroids. Following algorithm has been used to segment the output region from the input drone image data set.

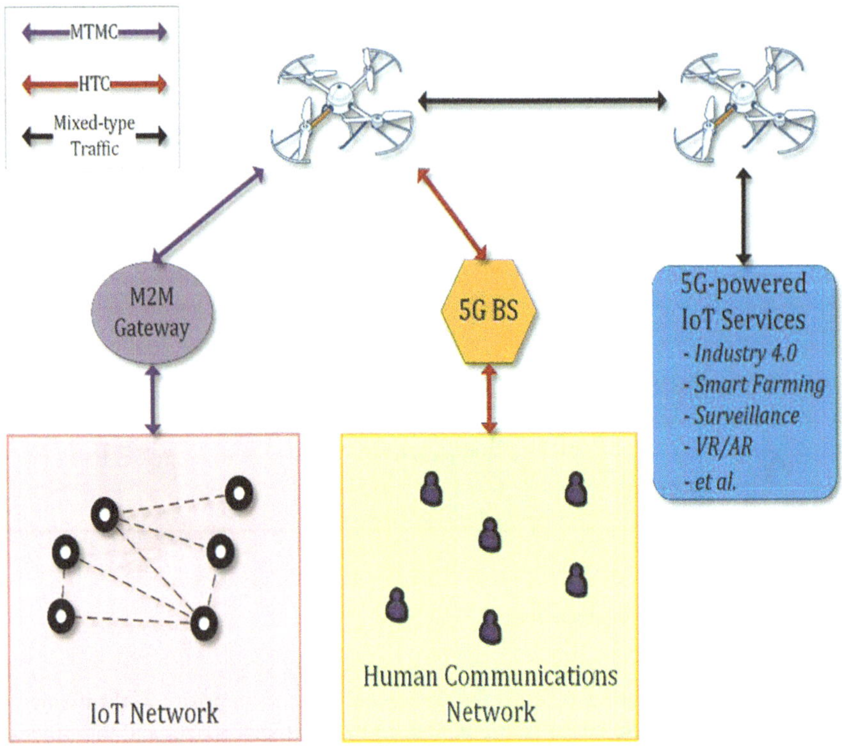

Fig. 5 IoD system architecture

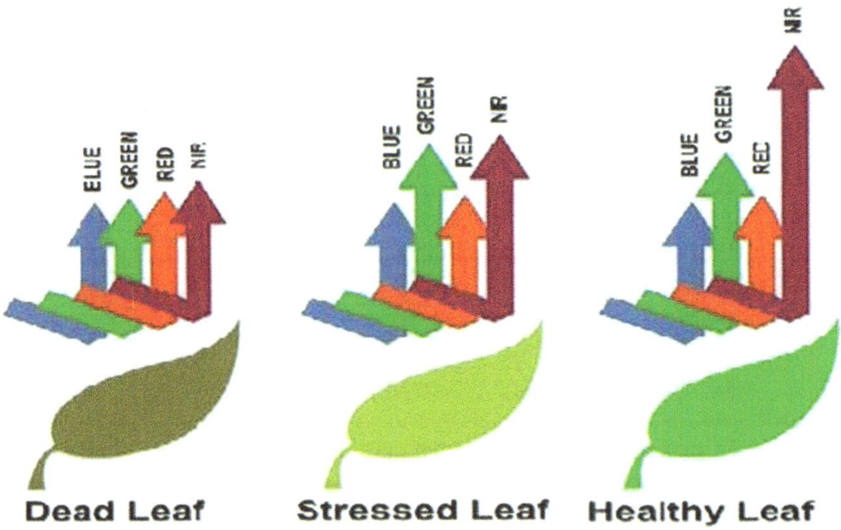

Fig. 6 Color based leaf diseases diagnoses

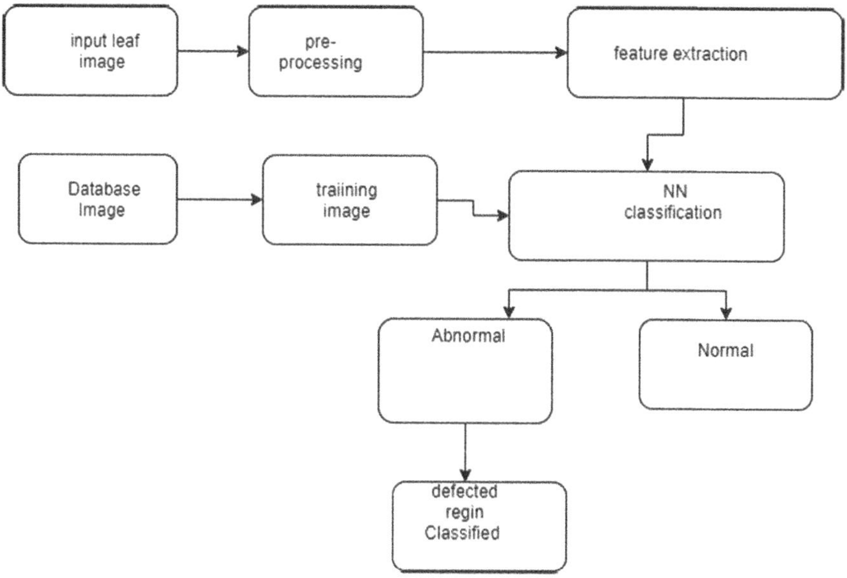

Fig. 7 Block diagram of proposed architecture

Input: Input drone images of dataset

Output: Segmented region

Step 1. Initialize a sub-window size, W=3 and maximum window size, Wmax = 15.

Step 2. Select a sub-window W×W with center pixel .

Step 3. If is not equal to 0 or 255, shift the window and go to Step 1 (noise free pixel).

Step 4. Collect the set of pixels ( ) from the sub-window ignoring the pixel with 0 or 255.

Step 5. If the size of ,

Replacewith
wienerof
pixelsin. Shift
thewindow

Go to Step1

Step 6. Set W=W+2,

Step 7. If W ≤ Wmax, go to Step 2.

Else replace the center pixel by mean of the pixels in sub-window of size Wmax End

center pixel by mean of the pixels in sub-window of

size Wmax End

3.1.3 Feature Extraction

The extraction of features is obtained from the segmentation completion. The detection of plant diseases is done by texture and shape features. The Wavelet-PCA based Statistical texture features, it is calculated from Contrast, Correlation, Energy, Homogeneity, Entropy (Randomness), Root Mean Square (RMS), Skewness, Kurtosis, Variance, Smoothness, Inverse Difference Moment, Mean, Median, Standard Deviation, Max Value, and Min Value.

Wavelet-PCA based Statistical Feature Extraction

At this stage of research, images with a resolution of 256 × 256 pre- processed images apply a level 'Haar' wavelet transform [17] to get sub-bands of resolution of different levels of resolution. These sub-band images further used in PCA parameter values are obtained. This PCA matrix is further used to calculate twenty statistical features which are formulated below. Let's first divide all images into three groups namely Fruit (Fr), Flower (Fl), and Leaf (Lf). The researcher explains here the process of feature-extraction for a set of fruit images. Similar steps can be used to extract features for a set of flowers and leaf images. Now, take a set of fruit images for analysis of research, which is represented as—

$$\text{IMP} = \{\text{Fr1},\ \text{Fr2},\ \text{Fr3},\ \ldots,\ \text{From}\} \tag{1}$$

Where Fr represents an image from the Fruit group and m representing the number of images of fruit used for analysis. The key objective of this wavelet-based PCA feature extraction implementation is to enhance the accuracy of recognition of pathological problems in general plant crops. The set of images used for classification here is represented as—

$$\text{SFr} = \{\text{sfr1},\ \text{sfr2}, \text{sfr3}, \ldots, \text{sfrn}\} \tag{2}$$

Where n represents the number of class-sets and a particular class is represented by sfr for the respective image. Here, we are going to extract 20 wavelet-based features that have been used for an efficient classification scheme to diagnose the disease in plants. Let, the obtained feature-values set be represented as

$$\text{FV} = \{\text{fv1},\ \text{fv2},\ \text{fv3}, \ldots,\ \text{fv20}\} \tag{3}$$

The crops and plant disease has been the major problem for good yield in the agricultural field. The problem in agricultural yield can also possible to struggle with the economy. Efficient disease monitoring has been important for agricultural economic growth. To prevent and control disease, there has been a need to fix preventive measures and find lesion areas. The black dot finds in the leaf and fruit measures the Anthracnose disease, this can be identified using the directional set feature extraction technique of MRKT with an artificial neural network to categorize a plant portion image whichever well or unhealthy.

The enhancement of this work has been made to find Powdery Mildew disease occurring to lower plants. The enhancement carried out with wavelet-based geometric features have been used to surge the exactness for both Anthracnose and Powdery Mildew illnesses. In this extended work, Twenty Wavelet-PCA based statistical features are extracted addition to twelve MRKT based directional features, resulting in a set of a total of 32 features. All these features are applied as input to the artificial neural network for the classification of diseases occurring in plant parts. The result analysis of MRKT and wavelet-PCA feature extraction is discussed elaborately. Leaves are identified as healthy or unhealthy by utilizing texture and shape features of the leaf images.

Features termed texture-features include homogeneity, contrast, Energy, cluster prominence and entropy are computed for the H image as given in Table 1.

Table 1 Equation feature extraction

S. No.	Shape features	Formula
1.	Contrast	$\sum_{i,j-0}^{n-1} (i,j)2C(i,j)$
2.	Energy (C)	$\sum_{i,j-0}^{n-1} C(i,j)$
3.	Homogeneity	$\sum_{i,j-0}^{n-1} (i,j)2C(i,j)/(1+(i-j)2)$
4.	Entropy (Randomness)	$\sum_{i,j-0}^{n-1} (i,j)2C(i,j)\log c(i,j)$

3.1.4 Classification Approach Using Enhanced Artificial Neural Networks (ECNN)

Once the desired features are selected and extracted, the immediate step is to classify the image according to their respective features. Two or more classifiers or techniques can be further combined to obtain better results [18]. Further, we can also employ a learning method to enhance the methodology. Hence, the classification techniques and classifiers are having a huge impact on the diagnosis of pathological problems in crops. However, the probability, similarity, and decision region boundaries are the keynote concepts they are going to build the classifiers and hence have an impact on the accuracy of classification too [17].

In this work, an improved ANN has been used for the input-layer having 11 nodes, stated by a total 11 appreciated feature-set containing of different Wavelet-PCA based statistical features and twelve-dimensional MRKT directional features mined as explained in previous sections. In this work, four hidden layers have been used. Each hidden layer is consisting of four nodes or neurons. The selection of an optimum number of hidden layers and units depends on several factors such as - the numbers of input and output units, amount of noise in the trained dataset, the complexity of the classification, size of the training dataset, distribution of training dataset pattern regions, activation function, type of hidden unit connectivity and training algorithm employed [19].

A small number of hidden units will lead to extremely large training and generalization errors because of under-fitting. But a large number of hidden units will provide very small training errors at the extra cost of an unnecessary slowness of training algorithm. This slowness and use of the number of hidden units also result in poor generalization until and unless any other regularization methodology has been utilized to avoid overfitting.

EANNs are proficient in learning. This learning activity is achieved by alterations in their weight values. Most of the EANNs enclose some form of learning rule which is further going to modify the weights of the connections by the input patterns which is going to be presented. Figure 8 shows a simple example of an EANN. During the training process, the neuron trained to fire (or not), for specific and subsequent input-data-set patterns. While using operation mode, when an already accomplished input pattern is experimental at the input, its corresponding output is assigned to the current output. If the input pattern does not correspond to any of the data in a trained list of input patterns, the firing rule is considered to determine whether to fire or not to fire. The operation of a single neuron is shown in Fig. 9.

The single-layer architecture of the neural network is one in which all units are individually connected, and hence comprising of the maximum overall situation. So, it has improved possible computational competence than the hierarchically designed and arranged multi-layer construction. In multi- layer networks, units are

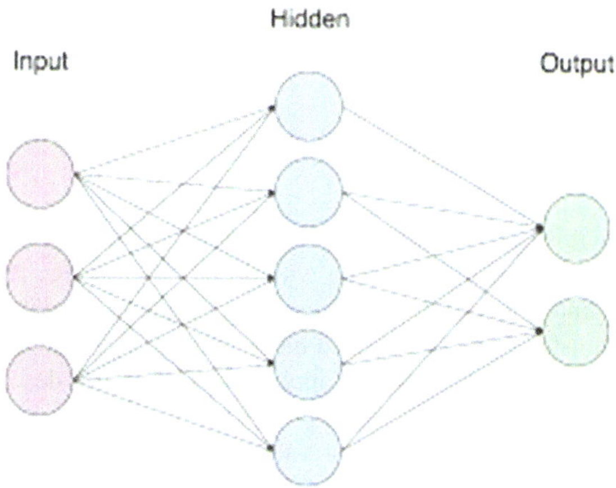

Fig. 8 Simplified artificial neural network

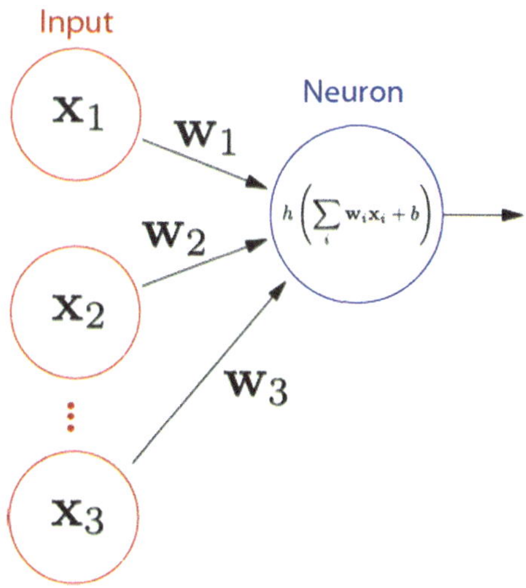

Fig. 9 Operations of single neuran

more precisely designated by their respective layers, instead of following a global numbering system.

Whenever training of neural networks is performed, it results in a diverse solution because of different initial weights, validation, bias values, several divisions of data in training, and sample test sets every time. Similarly, several neural networks trained for the same difficulty provide several different outputs with the same input. To guarantee the correctness of a neural network in training it has to be proficient several times. To boost the accuracy further, one can attempt given methods:

Reset the initial network weights and biases to new values and perform the Training of the network again.
Rise the amount of training vectors. Rise the number of hidden nodes.
Rise the amount of input, if more applicable data is accessible.
Try a diverse training process.

4 Result and Discussion

In this research, the first phase android application used IoT based irrigation parameters, disease monitoring with drone-based image and control strategies are implemented. various kinds of sensors temperature light depending resister, PH sensor and moisture sensors are used monitoring the crop field corresponding crop fields with IoT server-based mobile application from Fig. 7 describe the continuous observation sensor result.

The monitored sensor values exist to identify the user and turn on or turn off the devices with the help of the application. Also, users know the details of the cultivation period present crop, a suggestion for the next crop based on soil PH values, and updating the crop. In Fig. 10 describe the crop cultivation period and updating the current crop. In the second phase, For analysis, the proposed system has prepared an IoT database with 32-dimensional feature vectors (Twelve MRKT features and twenty Wavelet-PCA based features) for all 500 sets of captured images for each part of plant i.e. leaf, fruit, and flower. Out of this prepared database, 30% of the database images are used for testing purposes while the remaining 70% of database images are used for training of the network. Each captured mobile camera image is having an initial resolution of 1080 × 720.

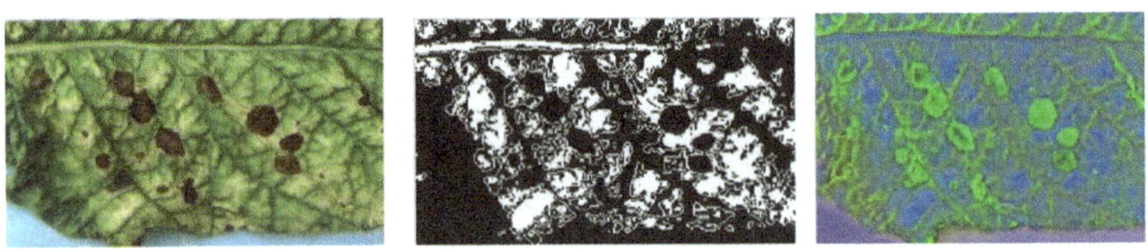

Fig. 10 **a** Input image, **b** pre-processing image, **c** pre-processing colored image

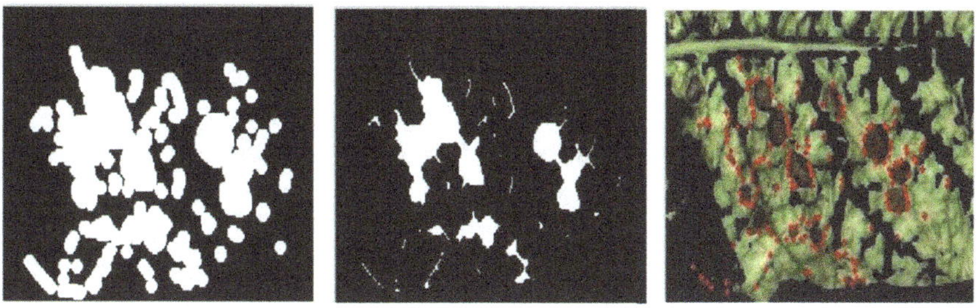

Fig. 11 Segmented image

Figure 10 the proposed segmentation results of the Agricultural leaf input image. The images are of Multiplanar Reconstruction (MPR) type. The diseases region find as (a) input image (b) pre-processing image (c) colored image separated and final image for exact final merged pre-processed image can be viewed. Partial segmentation of GM and WM regions is done by the proposed technique

The overall imagery from the agricultural of iterations for adaptive histogram defined values. In Fig. 10a–c as an input image there has been no clear idea about VM and WM these have been resolved with the proposed modified histogram algorithm and verified in the output part. The proposed work carried out the Diseases identification and tissue segmentation in the critical case of work

Figure 11 shows the efficiency in the proposed methodology for removing critical cases in agriculture drone images. The automated drone-based diagnosis is broadly chosen by the diagnosis of the disease. Segmentation of the affected region and unaffected portions are shown in Fig. 11c. Improper segmentation of tissue regions is also observable. This reduces the count of true negative value.

Figure 9a–c represents the disease identification and tissue segmentation In this proposed work, WPCA features are extracted from segmented diseased regions. According to the extracted features, the multiclass leaf diseases are predicted with the help of an enhanced ANN approach. The Consistent values have been listed in Table 2. The 11 diverse types of features have been finding also used to classify the effect of high grade in the leaves. It is also detected that the continuous intensity differences present in the input image do not have any effect on the working of the WPCA method. The proposed WPCA algorithm succeeded in Feature extraction the features delivered perfection in disease classification based on WPCA features.

Through comparator circuit in the Raspberry Pi connected different sensors for agricultural field monitoring (Soil moisture, humidity, temperature detection). This monitoring circuit has also involved a drone camera for capturing a sequence of images for crop field identification. Soil moisture sensor gives a resistance variation at the output. That signal is applied to the comparator and signal conditioning circuit. The sensor values are continuously sent through the cloud server using the MySQL server. The proposed android application displays the values. The user efficiently controls the motor and water levels from the webserver. The proposed IoT application also intimate the crop cultivation period through mail and Motor status. Based on

Table 2 Feature extraction values

Contrast	0.256411790514938
Correlation	0.920350034297844
Energy	0.139596184544799
Entropy	7.310705136113330
Homogeneity	0.885857687975673
Kurtosis	2.734068936510878
Mean	1.118500745156483e + 02
Skewness	0.166067980041149
Smoothness	0.999999822344007
RMS	15.948602088876783
Variance	1.465163327819722e + 03

the sensor values suggest the next crop to farmers. Figure 10 sensor values and the remaining cultivation period are displayed.

5 Performance Analysis

Further for the performance analysis, the author has calculated the Confusion Matrix. This Confusion Matrix is generally a 2×2 matrix. The elements of this matrix are—False-Negative (FN), False-Positive (FP), True-Negative (TN), and True-Positive (TP). Where FN and FP are the parameters for the number of pictures in the challenging database with the incorrect classification while TN & TP are the parameters for the number of images in the testing database with the correct classification. Further, few more statistical values such as Sensitivity, Specificity, Miss Rate, and Accuracy can be calculated. These statistical values can be calculated using a set of equations

$$\text{Sensitivity (TPR)} = \text{TP}/\text{P} \quad \text{where, } \text{P} = \text{TP} + \text{FN} \tag{4}$$

$$\text{Specificity (TNR)} = \text{TP}/\text{N} \quad \text{where, } \text{N} = \text{TN} + \text{FP} \tag{5}$$

$$\text{Fall Out (FPR)} = \text{FP}/\text{N} \tag{6}$$

$$\text{Miss Rate (FNR)} = \text{FN}/\text{P} \tag{7}$$

The classification of the various diseases of the crop is performed by IoT server training the classifiers with the total of 11 extracted features of MRKT and wavelet-PCA. In this research work, the Enhanced method for using artificial neural network machine learning are used as classifiers, the result analysis shows the proposed EANN

Table 3 Classification results

Diseases	Alternaria alternate
Accuracy	97.6400
Remedies	Copper fungicides

Table 4 Performance analysis

Classifier	Accuracy
Proposed system	97.6400
CNN	93.8
ECNN	96

is most suitable than the conventional classifier and also proposed implementation suggest he suitable remedies for identified diseases.

Table 3 final results obtained given input images are detected the Alternaria Alternate with the 97.460 accuracy results also suggest the suitable remedies for corresponding plant diseases with help of IoT server. The proposed system obtained accuracy parameter is compared among developed wavelet decomposition PCA (WPCA) with EANN algorithm, Convolutional Neural Network (CNN) and Enhanced Convolution Neural Network (ECNN) based classification approaches and tabulated in Table 4.

The developed wavelet decomposition PCA (WPCA) features based EANN algorithm disease classification technique is analyzed for accuracy parameters against prevailing methods of enhanced deep Convolutional Neural Network (CNN) and Convolutional Neural Network (CNN) based classification approaches. Images are represented in the x-axis and its accuracy on the y-axis. Thence this developed work uses the adaptive k-means clustering algorithm for reflecting more exact diseased regions of leaf images. The accurate segmentation of the diseased region improves the accuracy rate. Figure 9 represents the details of the validated results that the developed decomposition PCA (WPCA) features based EANN algorithm-based classification scheme achieves better accuracy performance compared with the existing systems. The designed classification scheme achieves a higher true positive rate for CNN and ECNN disease prediction. Also from Table 3 proposed system result shows the classification result of Alternaria Alternate diseases result enhanced to suggest the copper fungicides making a solution to control Alternaria Alternate diseases (Fig. 12).

6 Conclusion

The Proposed android application for IoT based leaf disease identification and irrigation parameters monitoring and controlling process is carried out. The proposed system carried out two phases. The first phase effectively sensor data carried and control the irrigation water. Second phases Wavelet decomposition with PCA

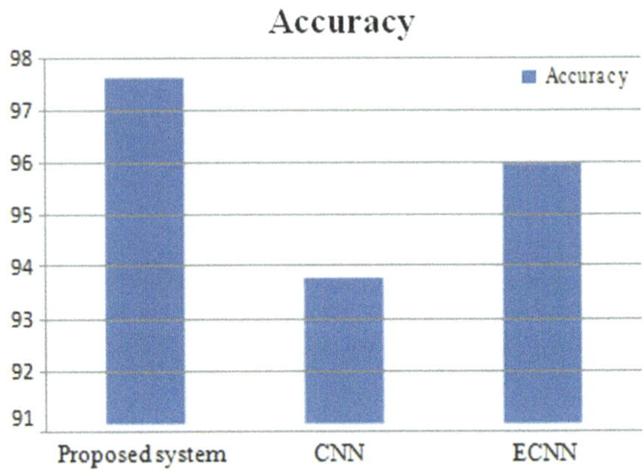

Fig. 12 Performance evolution graph results

(WPCA) based statistical features are extracted resulting in a set of 11 features and enhanced artificial neural network methods are used to find leaf disease classification. The performance of the proposed method and the conventional method compression process is improved through introducing proposed technology. Hybrid Fractal with WPCA-EANN Function-based image disease detection is found to be better when compared to conventional compression methods and achieved higher accuracy 97.460%. Based on the leaf image datasets, the results of the experiments demonstrate that the proposed classifiers are effective, and significantly outperforms when compared with other existing systems in multi-label leaf diseases classification. Also, IoT sensor data are received controlled effectively without any interruption. Future studies could explore whether accuracy could be improved by increasing the training data set, or by adjusting the parameters of the algorithms by implementing a new predictive model to improve accuracy and more IoT irrigation sensor enhanced within this system implement the automation in agricultural devices.

References

1. Hay, S.I.: Remote sensing and disease control: past, present, and future. Trans. R. Soc. Trop. Med. Hyg. **91**(2), 105–106 (1997)
2. Raut, S., Fulsunge, A.: Plant disease detection in image processing using MATLAB. IJIRSET **6**(6), 10373–10381 (2017)
3. Hong, S., Rim, S., Lee, J., Kim, J.: Remote sensing for estimating chlorophyll amount in rice canopies. In: Geoscience and Remote Sensing, 1997. IGARSS'97. Remote Sensing-A Scientific Vision for Sustainable Development, 1997 IEEE International, vol. 1, pp. 89–91. IEEE (1997)
4. Shibayama, M., Takahashi, W., Morinaga, S., Akiyama, T.: Canopy water deficit detection in paddy rice using a high-resolution field spectroradiometer. Remote Sens. Environ. **45**(2), 117–126 (1993)
5. Zhang, M., Qin, Z., Liu, X., Ustin, S.L.: Detection of stress in tomatoes induced by late blight disease in California, USA, using hyperspectral remote sensing. Int. J. Appl. Earth Observ. Geoinf. **4**(4), 295–310 (2003)

6. Barbedo, J.G.A.: Digital image processing techniques for detecting, quantifying, and classifying plant diseases. SpringerPlus **2**(1), 660 (2013)
7. Cline, B.L.: New eyes for epidemiologists: aerial photography and other remote sensing techniques. Am. J. Epidemiol. **92**(2), 85–89 (1970)
8. Ali, H., Lali, M.I., Nawaz, M.Z., Sharif, M., Saleem, B.A.: Symptom-based automated detection of citrus diseases using the color histogram and textural descriptors. Comput. Electron. Agric. **138**, 92–104 (2017)
9. Akila, M., Deepan, P.: Detection and Classification of Plant Leaf Diseases by using Deep Learning Algorithm. CONNECT–2018, vol. 6, issue 7 (2018)
10. Padol, P.B., Yadav, A.A.: SVM classifier based grape leaf disease detection. In: 2016 Conference on advances in signal processing (CASP), pp. 175–179. IEEE (2016)
11. Al Bashish, D., Braik, M., Bani-Ahmad, S.: A framework for detection and classification of plant leaf and stem diseases. In: 2010 International Conference on Signal and Image Processing, pp. 113–118. IEEE (2010)
12. Thangadurai, K., Padmavathi, K.: Computer vision image enhancement for plant leaves disease detection. In: 2014 World Congress on Computing and Communication Technologies (WCCCT), pp. 173–175. IEEE (2014)
13. Puri, V., Nayyar, A., Raja, L.: Agriculture drones: a modern breakthrough in precision agriculture. J. Stat. Manage. Syst. **20**(4), 507–518 (2017)
14. Ganesan, R., Raajini, X.M., Nayyar, A., Sanjeevikumar, P., Hossain, E., Ertas, A.H.: BOLD: bio-inspired optimized leader election for multiple drones. Sensors **20**(11), 3134 (2020)
15. Singh, V., Misra, A.K.: Detection of plant leaf diseases using image segmentation and soft computing techniques. Inf. Process. Agric. **4**(1), 41–49 (2017)
16. Jhuria, M., Kumar, A., Borse, R.: Image processing for smart farming: detection of disease and fruit grading. In: Proceedings of the 2013 IEEE Second International Conference on Image Information Processing (2013)
17. Huang, W., Guan, Q., Luo, J., Zhang, J., Zhao, J., Liang, D., Huang, L., Zhang, D.: New optimized spectral indices for identifying and monitoring winter wheat diseases. IEEE J. Sel. Top. Appl. Earth Observ. Remote Sens. **7**(6), 2516–2524 (2014)
18. Nayyar, A., Nguyen, B.L., Nguyen, N.G.: The internet of drone things (IoDT): future envision of smart drones. In: First international conference on sustainable technologies for computational intelligence, pp. 563–580. Springer, Singapore (2020)
19. Ramakrishnan, M., Nisha, A.S.A.: Groundnut leaf disease detection and classification by using back propagation algorithm. In: IEEE ICCSP Conference, 978–1-4 799-8081-9/15 (2015)

DroneMap: An IoT Network Security in Internet of Drones

Rajani Reddy Gorrepati and Sitaramanjaneya Reddy Guntur

Abstract Internet of Drones (IoD) is a major role for the central military, agriculture and IoT applications that requires critical information to be processed. It ensures that security and network privacy issues in the Internet of Drones (IoD) have malware/vulnerable attacks and Distributed Denial of Service (DDoS) attacks are highly energy-constrained, which are a direct standard of cryptography protocols and secured key IoD algorithms. IoD is a capable of enhanced state-of-the-art of Drones while providing services from an existing cellular networks. IoD is vulnerable to malicious attacks over radio waves frequency space due to the increasing number of attacks and threats to a wide range of security measures for IoD networks. Low cost of Unmanned Aerial Vehicles (UAV) known as Drones for enabling various IoT applications. UAV are also used in several applications in surveillance, disaster, environment and management search and rescue monitoring solutions that are limited to point-to-point communication patterns, and are not suitable for distributed applications in multi-UAV scenarios. UAV has limited processing and storage capabilities with massive computations requirements for certain applications. In this book chapter, we represent the Drone-map planner that are service-oriented fog-based drone management system that controls, monitors and communicates with Drones over the network. Drone-map planner that allows to communicate with multiple Drones over the internet, which are enables to control anywhere and anytime without any long distance restrictions. Drone-Map planner provides access to fog computing resources for drones to heavy load computations. To classify the attacks based on the threats and vulnerabilities associated with the networking of drone and their incorporation into the existing cellular setups. This chapter of book summarizes the challenges and research directions to be followed for the security of IoD.

Keywords Internet of Drones · Security · UAV · Internet of things · Drone map

R. R. Gorrepati
Department of Computer Science and Engineering, Vignan's Foundation for Science, Technology, and Research, Vadlamudi, Guntur 522013, India

S. R. Guntur (✉)
Department of Electronics and Communication Engineering, Vignan's Foundation for Science, Technology, and Research, Vadlamudi, Guntur, India
e-mail: drgsr_ece@vignan.ac.in

R. Krishnamurthi et al. (eds.), *Development and Future of Internet of Drones (IoD): Insights, Trends and Road Ahead*, Studies in Systems, Decision and Control 332,
https://doi.org/10.1007/978-3-030-63339-4_10

1 Introduction

Internet of Things (IoT) and fog computing attracted a great deal of interest in contact with the Unmanned Ariel Vehicle (UAV). UAV has remotely interacted with fog computing, web technologies, and service-oriented architecture (SOA) through newly developed IoT. Mainly, the concept of the Internet of Drones (IoD) is framed to access and control the moments of drones in airspace using layered control architecture with navigation services between locations [1]. The three major layered architectures are mobile network, air traffic control network, and the IoT, which are provided for different UAV applications that are present implementation of the architecture. The context of fog computing robotics has been coined that are an effort to incorporate robotics across the internet with fog computing. The drawbacks of low-cost UAVs are processing and storage capacities and battery-powered UAVs that are efficient in computing specific applications with real-time data and reliability constraints.

UAV is a flying detected object identification as an autonomous driving capabilities that can perform particular tasks for simple operations of complex objects [2–6]. IoD packages that are monitor to transmission delivery in coastal areas for the identification of wood, fire prevention.

IoD drones that are feasible related to remote-controlled operations that are automatically adjust the speed parameters and to detect directions like planes. UAV is also working as drones mainly are investigate such kind of applications used in for security purposes such as military, medical, live stream operations, and agriculture field [7, 8]. The optimization of IoT security mechanisms for the direct adoption of IoD to feasible and secured protocols. Protocols can operate in various network layers by providing services such as mobility, real-time, and clock synchronization communication. In the recent study, some of the authors investigated and found the vulnerabilities attacks on IoT applications in existing security mechanisms and standards were implemented such kinds of systems.

The critical security mechanisms and optimized algorithm for variable cryptography solutions for IoDT [9]. Some researchers found security attacks in the IoD, however, reported successful hacking and hijacking of an AR. Drone 2.0 (equipped with a 32-bit ARM processor) resulted by the crystallographic lack of secure communication channels is suggesting that the majority of commodity drone telemetry systems do not use cryptography to protect your correspondence [10]. Thus, they suggested a method of fingerprinting that could provide some protection for drones. A lot of advanced work is to identify the security attacks of the current drone configurations [11–13], they work on how to determine the frequency of the FHSS series of drone controllers used to describe radio frequencies [14], and to demonstrate the attacks due to the cause of drones to crash by radiofrequency of MEMS gyroscopes. MAV link protocols to get results that disable the drone attack mission.

UAVs can be considered a crucial solution in many areas of medical monitoring, agriculture, and transport [15, 16]. UAV-assisted can operate wireless network connectivity where to development of an expensive physical infrastructure. Drones are robotic flying network systems, which can be considered dynamics of drone flight

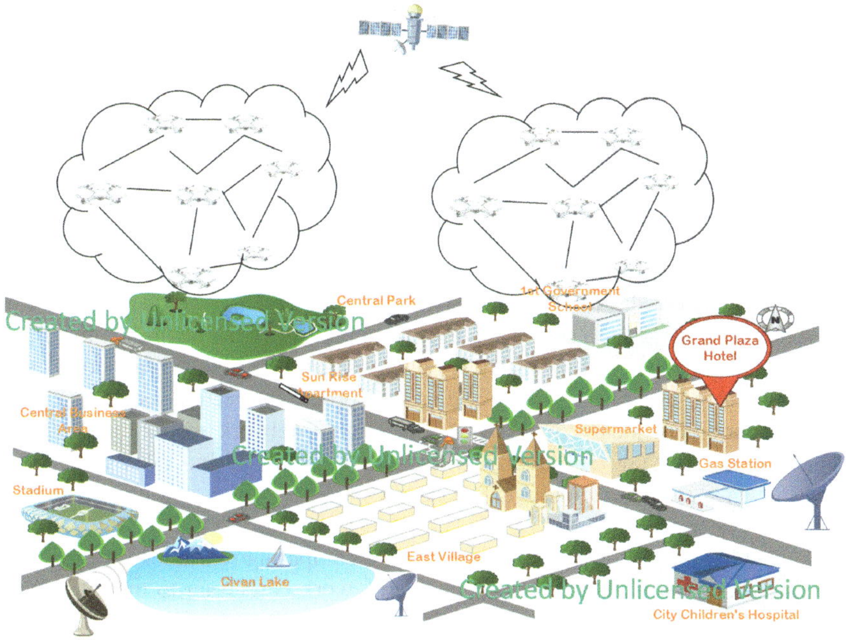

Fig. 1 Communication of the societal environment through the Internet of Drones (IoD) to Geo Satellite

that are mathematical representations by physical laws [17]. UAV modelling is strongly enhanced deep learning by solving connected robotic operations like recovering navigation, path planning, localization, and control [18]. Remote control systems [19] refer to a class of software that interfaces the on-board UAV between the operator and the pilot to ensure that each flight operation corresponds to what the pilot intended.

Figure 1 shows the communication of societal environment like central military, agriculture, stadium, central Park, supermarket through IoD to Geo Satellite. The application of IoTs to transmit and processes critical information. It ensures that the security and network privacy issues related to malware/ vulnerable attacks and distributed denial of service (DDoS) attacks are highly energy-constrained, which are the direct standard of cryptography protocols and secured IoD key algorithms. IoD is capable of enhanced state-of-proof of drone's communication existing through mobile network services. IoD is vulnerable to malicious attacks over radio waves frequency space due to the increasing number of attacks and threats that are alert to the attention of security measures for IoD networks.

Low cost of UAV, known as Drones, to enable various IoT applications [20–24]. UAVs are also used in many rescue operations, environmental disaster management, and surveillance applications. However, point-to-point communication is limited and may not be suitable for distributed applications in multi-UAV scenarios. UAV has limited processing and storage abilities with enormous quantitative prerequisites of specific applications. In this chapter, the Drone-map organizer is a service-oriented fog-based automation framework that controls to communicate with drones over the network. A Drone-map organizer is allowed to communicate with multi drones over the internet, which empowers them to control anywhere and anytime without any restrictions for long-distance.

The emerging IoT technology is continuous with fog edge computing environment by special software that generates their behavior and objectives in which the local concept of orchestration for large smart objects utilized and global users. IoT enables the objects and everyday activities of various communications tools in the context of smart-home vision [25]. IoT objects are more complex than the flexibility and transformation with a low-cost, fusion of heterogeneous networks highly distribution security problems in sensor networks, challenges the privacy protection specific network issues that are generated [26]. IoT needs to support network security and privacy challenges that are present features in handling of internal and external security threats/attacks, authentication, access control, data protection, malware detection, and high-level authorization [27].

Drone-Map planner gives admittance to fog computing resources for drones to heavy load computations. Air traffic control network is relevant to the Internet of drones it provides the security to maintain the free collisions navigation to utilize any types of drones. The main role of the air traffic controller to keep all drones in place as new systems motivate the the scalability of drones to be more autonomous. Automatic dependent surveillance-broadcasting (ADS-B), uses the navigation of drones and broadcast air crafts position [28]. The attacks are classified on the basis of a quantitative analysis of threats and vulnerabilities associated with the networking of drones and their incorporation in to existing mobile setups. This chapter summarizes the overview of the Internet of Drones (IoD) , UAV applications domains and IoT 5G challenges and analyses the IoT.

The privacy and security of the dronecommunication require specified sensors mechanisms that focus on the view of IoD. Recently established 3GPP service standardization with mission-critical push (MC-PTT) [29] features that allow D2D communication to create more robust and heterogeneous UAVs. D2D enables devices to communicate directly through D2D cell spectrum sharing to increase spectrum performance.

2 Overview of IoD and UAV

The IoD architecture for controlling and access over the internet was shown in Fig. 2. Drones are becoming increasingly common the various aspects of missions of drones multiple controlled connected to IoT Gateway, whereas the security terminology of UAV components includes processes, storage, sensors as well as increasing the efficiency of battery life and drone components and lower expectations. IoD supplies a vehicle of the Internet of things with robotics technology to allow remote control

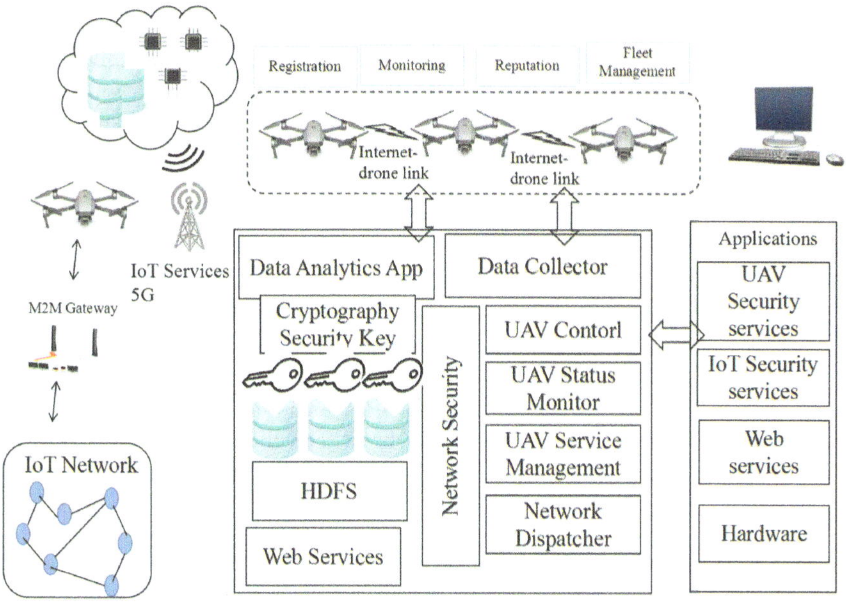

Fig. 2 Internet of Drones (IoD) security architecture

of drones as the seamless scalable of remote storage. UAV challenges associated with the point-to-point communication connected to the internet of drone wireless connectivity through the efficiency of resources (Fig. 2).

IoD security and quality of service (QoS) is criticle and big challenge in accessing drone resources with an authenticated and secured. IoD system various attacks like drone impersonate, sniffing, flooding, and technical details using a service-oriented approach, usually implemented the applications are UAV and IoT security services, and SOAP or REST web services [30]. Users may not be used as a technical programme or as missions to develop web services based on on-board access resources through various APIs. This architecture represents the Internet of Drone planning system to find out the system that addresses functional specifications.

The service layer provides services using a collection of resources represented by the UAV exposed at the end of the user. On top of the hardware, the robot operating system (ROS) and MAV link layer are used to develop the robotics application abstracts for network and processing systems like navigation, movements of object organizing, communication of information control at low-level devices. MAV link built over various transmission protocols such as TCP and UDP, which permits the exchange of predefined drones and ground station data to a high-level application interface for developers to automatically control and direct drones without any hardware intervention.

The server-side storage components for data streams captured by each IoT climate change, societal data, and transmission of information sources including sensor and time stored image data. Storage is a distributed system that assists with performs enormously manage stored data on a scale using software like the Hand-hoop chart. The real-time stream processing sample images are obtained from IoT and process data streams to detect possible attacks or threats requiring immediate action to perform a dynamic distribution environment. The real-time processing of data may be the necessary bases for the application, or new events are detected that required dynamic rescheduling of drones. Network interfaces to implement a server-side network sockets and web socket interfaces that manage sent UAV protocol can handle stream applications, context of drone map, and MAV communication information is collected via network sockets and sent to client device web sockets. The web service interface permits users to control drones and their applications. Web services are utilized to give the end-clients and users application services to control commands and observed the drones through parameters. The client layer can provide interfaces for both end-users and drones web applications that provide interfaces for both the IoT service layer and the UAV layer. The client layer provides several users who have registered multiple UAVs, transform parameters, and data analysis and decision making based on APIs for various programming languages to interact with drones.

2.1 Layers of UAV Drone

A specialized onboard data acquisition software is used to record the GNSS data automatically, which will not require manual intervention to on/off logging and can operate robustly in unattended mode. Figure 3 shows the layered architecture for UAV Drones IoT communication. The GPS recorded every tracing point with the coordinates of GNSS geo-tagged data and communicates wirelessly to the ground station as shown in Fig. 3. However, the drone is arranged with RTK GPS, and the exact coordinates are used for geotaging data. Drones are located in large-area surveys to collect the data and transform to ground stations via wireless communication. The drone moments are controlled by the operator from the ground station using GNSS and are able to monitor how the GNSS works to trace current data, record the data of start/stop, settings and mode changes of GPR, etc. GNSS software interfaces with the hardware of drone and ground station data log and wireless communication BUS, which can transform the data link to ground control in to task management and information monitoring.

UAV enables communication networks to suffer an allocations dedicated spectrum that is a high impact on energy efficiency. IoD enables the connectivity of drones via mobile networks in order to facilitate the data gathering in airspace and enhance storage capabilities and information management service. UAVs are remotely controlled and commanded by the operator from base stations using coordinated IoD through various channels deployed in UAVs. IoD acquires the related information from the sensors arranged on drones. IoD is being deployed for next-generation applications like smart car parking, smart health, smart military, and

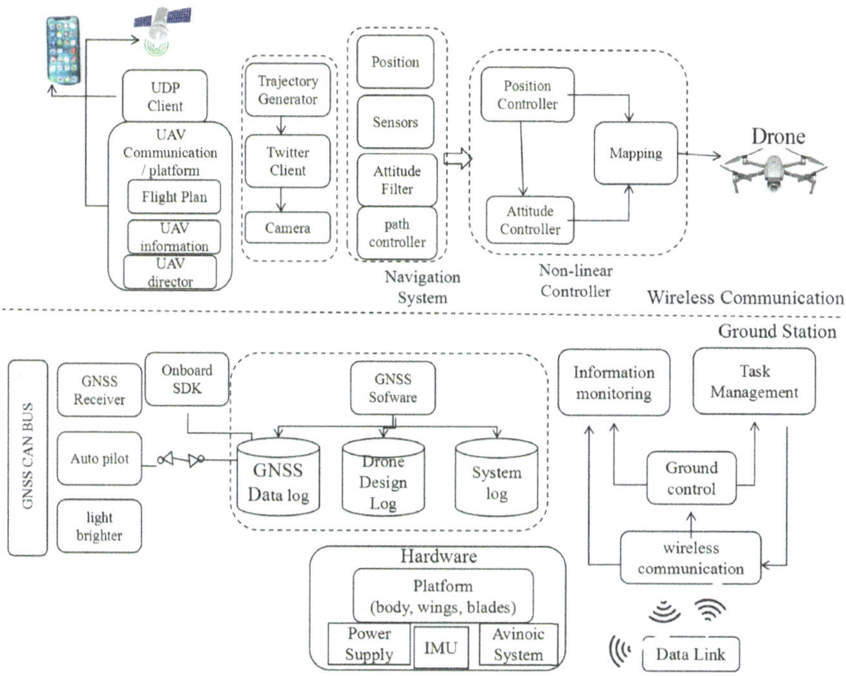

Fig. 3 Layered architecture for UAV drones IoT communication to a ground station

cellular networks. The next generation of IoD is integrated into third party object-oriented systems with the variety of tasks in collaboration modes and abilities for multitasking and operate the rural and urban environments. The attacks or threats are considered to be important factors for exploiting and vulnerabilities sensitive pieces of information are gathered to the IoD security and protect information. The drone controller of the central processing unit of the IoD and the communication to read process the data provided by the various sensors is useful it in the information. The drone controller implements the commutation interface between drones and the ground control station. The ground control station is located on the ground floor and provides human operations to monitor their activities in order to communicate with drones to send commands and receive data in real-time. The data link communicates to the wireless network that is controlled information between the drones and the ground control station via the communication network.

UAVs are used to provide individual Internet access to rural areas that are not protected by traditional communication areas. In Fig. 3, the four function layers are defined by various levels of quantitative complexity that can be described by the layer operator as a task related to the current flight status of the camera. The flight machine can be launched by installing a ground control station that is monitoring on smart-phones, PCs, and tablets. The simple guidance framework produces a route based on the different paths of the twitter client to the camera. UAV produces the route based on many ways that the first layer in this system can provide robots with the ability to automatically track the location of the multiple drones by mapping the direction.

2.2 UAV Sensor Technologies in 5G Networks

5G Technology is designed to improve the network traffic and high availability security control and mobile broadband applications that are required in a very low latency, supported by industrial applications, remote manufacturing, and tracking. UAV drones can be used for smart agriculture, smart buildings, virtual and augmented reality without limitations of scope including home enterprises, critical machine-to-machine communication. In the sense of 5G devices that can be integrated with deferential devices, communications may be implemented in appropriate gateways in the context of 5G architecture that is presented. Accelerometers are used to define flight control sensors to determine the location and orientation of the drone flight. Drones and UAV manage to flight.

Figure 4 demonstrates the sensor technologies that are assisted as part of drones that can be separated into three categories. The main categories are drone power, data collection, and communication sensors. The evaluation of drone internal control

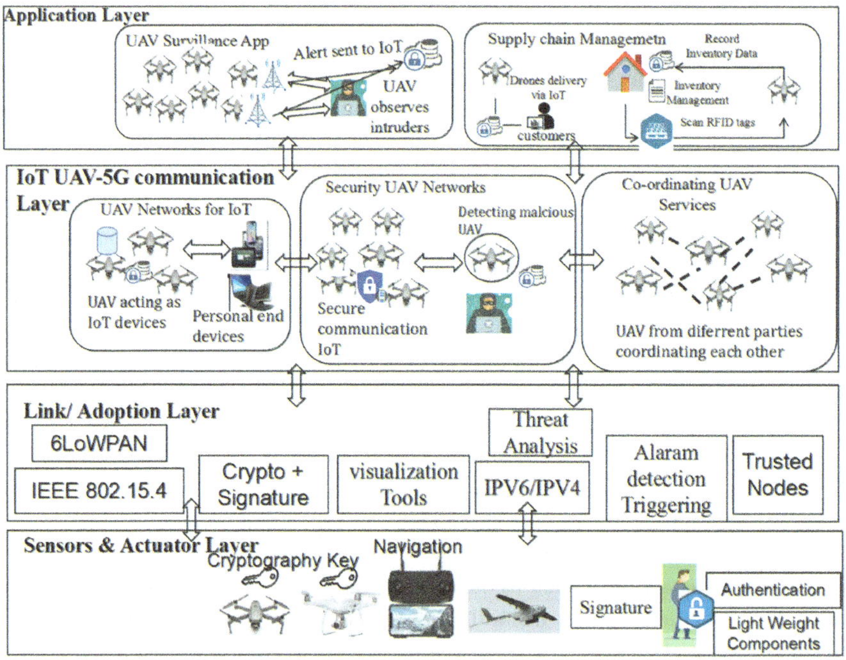

Fig. 4 Secured 5G mobile communication layers in UAV drones IoT

sensors that are accelerometers is used to measureof drone's position of the trajectory in the flight. The 5G technology is designed to enhance mobile broadband applications that requiring very low latency and to enhance the traffic protection and control provided by industrial applications, remote monitoring, security logistics, and fleet management. It also utilizes the smart agriculture, a virtual augmentation without the limitations of scope of enterprises that are vital device-to-device communications. Drones assisted by 5G network use cases linked to communication / linking involved automatic detection trajectory paths and robotics. Accelerometers are used to determine the position and oriented flight of the drones can manage to flight paths and directions using the flight control sensors with GNSS navigation. The flight control framework is designed to maintain the level of flight feedback from the tilting sensors coupled with accelerometers and gyroscopes. When the applications are required at a high level of scalability and effectively monitor and sensors to estimate the proper reduces estimations and consumption are reduced.

Data Acquisition Sensors: Drones are equipped with many sensors to collect the data needed for certain activities, depending on the function of the sensor to be used for data acquisition by drones. UAV may be equipped with high-end sensors and radar of air-bone systems that provide resolutions, surveillance and monitoring of applications sensors environmental and weather parameters sensors are used in the control of disaster.

UAV Communication systems: Manage and control tasks performed in communication systems and multiple network technologies of drones that are required to communicate with one another. A comprehensive list of network protocols and communication methods allowing multiple applications for IoT architecture complex network segments interconnected to multimedia streaming surveillance. It is based on the data rates and latency specifications available in 5G drone communications technology, which are allowed worldwide applications. In the conceptual model, UAVs can interconnect complex network segments of IoT networks for demanding services. Security services must be implemented in the Link/Approval layer of IPV6. The adoption layer that provides end-to-end encryption, continuous security key management mechanisms for authentication schemes, is designed to provide robust communication at the endpoint security channels protocols were proposed. Secure connections with firewalls link and intrusion protection system are allowed. Physical losses are the key security risks in the data link layer that include routing attacks such as selective forwarding, sinkhole, and hot-hole attacks [31]. Primary security threats in the 5 G network communication layer that cause physical damage to the hardware framework, like intrusion and detection of processes, should ensure that only users can access sensitive data generated by physical devices.

Privacy and protection for UAVs: Privacy of the IoT domain and protection for receiving, storing, and transmitting sensitive object information and building patterns and a variety of links that can be used as IoT device signatures. One of the core challenges in the design of cooperative applications involving several UAV networks that can simultaneously provide coordination between the different types of protection and the fleet task of developing and sustaining scalable aerial networks, which are fleet management techniques. Drone Map applications support various drone applications over the internet and we need to build some real-world applications with drone map planners to demonstrate the efficiency of IoT applications enabling. All services of the Drone map planner are designed to destroy applications with predefined IP address port number and primary entity, including IP address, port number, and MAV link device ID. When a drone communicates over the internet, it communicates the drones automatically and is shown on the web interface. The web interface contains all data about drones, including altitude, air/ground speeds, battery level, GPS coordinate location address. These real-time data are accessed via the web services interface using a remote device.

3 Security Threats and Attacks of IoDT (Internet of Drones Things)

Figure 5 shows the vulnerability of IoT drones to detect contact pathways and attack as a variety of vulnerabilities. These are the techniques used to hacking the UAV drones from channel jamming and to Spoof malware, such as the Middle-Man attack

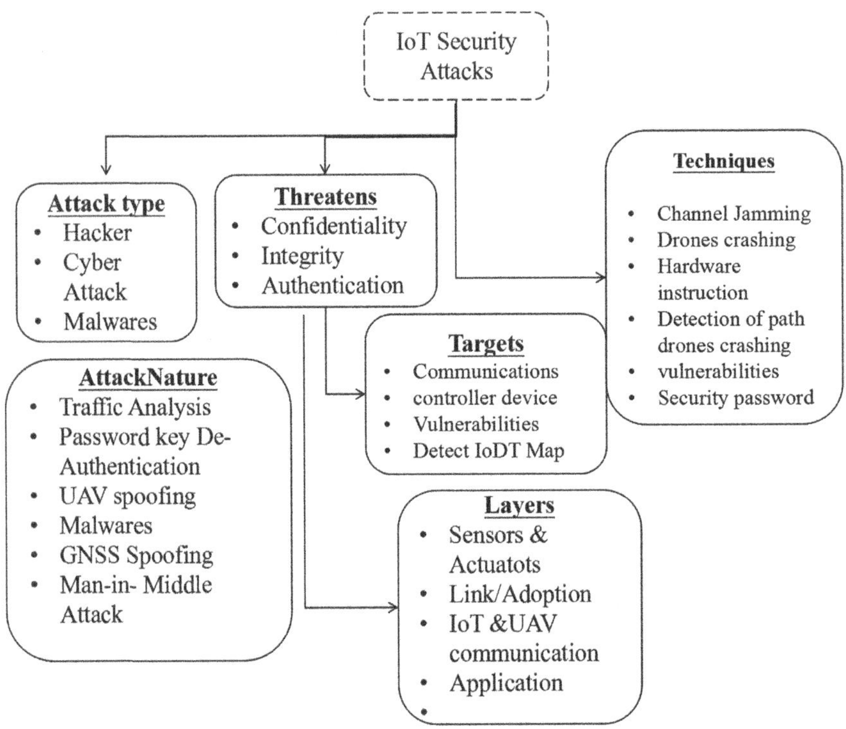

Fig. 5 IoT security attacks for communication with IoT

and the GNSS spoofing. The adoption of link layers non-lethal solutions to counter these threats to be various malware such as highly inefficient and the data presents some of the issues related to drone communication pathways that threats are in highly-ineffective and unreliable. UAV fault-tolerant control present in the device architecture, using a neural network adaptive framework for the identification and isolation of the network design. In this scheme, real-time detection in drones ensures real-time identification and isolation of faults in actuators to configure network issues that are tolerant in order to reconfigure the controller or have an impact on efficiency. In this Wi-Fi jamming, the approach is observed to be implemented as these drones use a 2.4 GHz frequency.

All these jams are wireless contact within a specific area of coverage. However, very small jamming capacity cannot be easily identified in the environment, and other nearby frequencies are jammed. This approach is based on a three-way handshake router and newly installed rogue computers. It allows the attacker to de-authenticate, or jam, the connection between the drones and the control unit. The Wi-Fi attack that was present enables the attacker to search for drones to communicate the DDoS attack, which interprets the transfer of the particular data, either delaying it, which allows the attacker to leads the de-authenticated attack. A DOS assault that intercepts network traffic and floods with a request to interrupt a drone/device link. Denial of service will be performed either by de-authenticated of the UAV drones that access can be sent periodically to the drone network security event commands . This leads to the estimation of the location of the drone unit for the GNSS signal simulator used by drones to launch a GPS spoofing attack, which transmits false signals to the control system of each drone, normally more powerful than the fake signals instead of the original ones.

GNSS allows drone navigation non-encryption of easily spoofed signals that are directly managed by anti-spoof algorithms operator that can help mitigate GPS spoof attacks. A drone that uses GPS could be targeted by jamming the GNSS signal that makes the drones unable to determine their location. Jamming the objective of disrupting all satellite communication during antenna selection and orientation can help to minimize jamming attacks. Eavesdropping successfully dealt with in Man-in-Middle attacks allows the attacker to track violation of drone confidentiality. Some of the confidential information that collects through IoT when it classifies them in terms of privacy and trust, with their respective tasks. If the data is interferes with the data access to adjust the cluster controller and the malicious actions of the controller to gain control of the drones. The main task is to safely store IoT data integrity, data protection, and encrypted data that is not available to anyone without any key for decryption (Table 1).

4 Security Issues Associated by IoD

The IoT security attacks and Jamming attack, which is interference with radio signals that causes communication problems for drones to function efficiently as well as the

Table 1 IoDT security detects malware, threats, and vulnerabilities

Attacker	Security impact parameters	Target components
An attacker signature key and capture the network communications	Privacy, integrity, and confidentiality	Network platforms
An attacker can produce transmission information	The user can misguide the falsified integrity data	Network platforms services
By obtaining access control an attacker may alter authentication or access permission	Non-legitimate users can be caught and permission alerted which leads to system crash	Network platforms services
DOS and physical attacks by excessive false request	The main target of the attacker is to disable the IoT devices used for some applications	Network platforms services
IoT coordination can be warned by hacking or GNSS spoofing	The attacker caught by the IoT devices or altered the coordination which could result in collision attacks	Network platforms services
The attacker can affect the performance of IoT by increasing the resource depletion rate.	The performance of IoT in the target missions which may lead to mission failure	Network platforms services
The attacker can capture the IoDT services	The IoT is assigned for dedicated tasks of component services lead to facilities	Network platforms services

effect of energy usage. Data exploitation is an intrusion into the sensitive data of the server, and access to confidential information. Some of the collisions when two or more drones operate at the same frequency in order to operate the unreliability of the network. Attackers can degrade the entire system by causing other drone nodes to do real-time based transmissions. Malicious node that causes all types of packet disruption/corruption to pass through them and other associated packet loss attacks that impact packets and through put delivery ratio methods are needed.

A flooding attack can cause huge packet flow resulting in a complete network congestion De-synchronization attack is a malicious node that moves messages by transmitting sequence numbers to more than one network operating node. Network Jamming attacks are disrupt some of the nodes on the Man-in-Middle attacker network, which deploy a malicious drone between the operating network, causing leakage of information, disruption, and other security attacks. Drones instability and other non-compatibility with IoT-based operations. Apart from other problems, limited capital, less processing power, and limited storage space, the overall IoT functions as a back door. Service Denial attacks, privacy threats, malicious access points, unauthorized threats, IoT application services intrusion attack. Real-time applications also need to be addressed, depending on the application scenario of the drone fleet management. Device control, considering into account the quality of the service parameter, will become a key issue and a further significant obstacle to 5G network access.

5 Evaluation of IoD Control Sensors

The accelerometer used to determine the direction and orientation of the drone in flight, and one of the technologies used to detect embedded devices, which did not have any moving drones, and the UAV systems managed to maintain the drone route measurement units together with the GNSS. As a results, this process helps to evaluate the paths of the drone.

UAVs monitor the airflow sensors effectively to estimate the correct rate at the required speed in order to reduce the overall optimizing power consumption and to detect system engine failures. IoT based UAV is a more complex concept architecture in which machines can interact with dynamic processes of smart objects and technological ecosystems. Security mechanisms such as restricted application protocol access control and user application protection that allow for protected messages and are minimal configurations such as filtering and perimeter protection [32]. Privacy security is a wide range of sensors that measure different types of information to ensure the group signature and data of all communication protocol devices with an authentication mechanism.

The integrated community of signature methods tends to be effectively securing and preserving the identity of data encrypted measuring devices to prevent interpretation of dummy data by the reporting service. Most of these devices have focused on the processing of video content through computer vision. In this case, privacy-protected solutions concerns are handheld devices that can be used from drone targets and displays from multiplepoints of view, introducing a new dimension. Jamming is used for the proposed scheme that goes beyond malicious transmission activities to all radio communications. The transmitter and receiver can be exchanged while cooperative jamming enables the content of the two parties to communicate without restoring the encryption of data. The GNSS jammer could be adopted to attack the signal correlation mechanism by transmitting the location, navigation, and timing (PNT) capabilities of a particular receiver [33]. Conflicting concepts in many safety scenarios to prevent unintended communication between the different methods used to determine the position of a radio receiver by manipulating the distribution of radio signals and inferring the distance to be a transmitter following well-known propagation models [34, 35]. In most cases, the implementation of the above techniques or drones is feasible ; such techniques rely on available radio-level knowledge. One of the most reliable Received Signal Strength (RSS) information, without loss of attack and scenario in general. Several strategies were conceived to counteract malicious jammers. The jamming attacks are intended to mitigate the harmful impact on frequency spread spectrum of satellite communication by exploiting the solution provided by a cooperative spatial algorithm that allows drones to cooperatively mitigate jammer action [36].

Figure 6 shows the drones that use the Jamming signal to estimate the range, the secret behind the solution that enables a drone to push forward its mission by the presence of a jammer that can be used by any other radio communication to calculate the distance to the transmitting source. Although typical localization techniques suffer from traditional propagation phenomena such as multi-path transmission, the jamming case is radically different, enabling drone communication capabilities to access the GPS location service.

The scenario mentioned above is the ideal situation for estimating the distance between drones and the estimated direction of transmission of the jammer signal, at which distance d can be calculated as [37, 38].

$$d = \frac{\lambda}{4\pi}\sqrt{G\frac{T}{R}} \quad (1)$$

where λ is the radio frequency wavelength, T and R are transmitted and receiver power by jammer and drone, G sums up the transmitter-receiver gains. In the presence of a more practical pathway, the transmitter power and transmitter antennas have again been evaluated for localization variation as the drone moves closer to the jamming signal. We note that the model from Eq. 1 could be strengthened by further statistical definition of the channel due to the multi-path fading of the approximate width. The jamming signal is used to estimate the line-of-sight portion and to follow a tracking mechanism by which the drone first estimates the obtained power

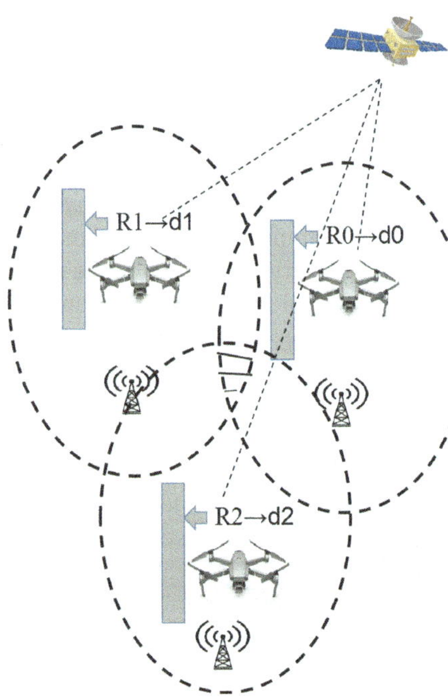

Fig. 6 Drones exploiting the Jamming signal to estimate the range

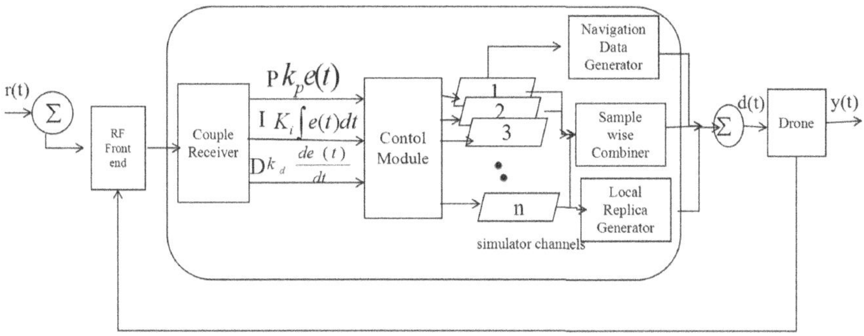

Fig. 7 UAV flight controller spoofing and navigation attack system

$R_1 = 2 \times R_0(+3\,\text{dB})$, the drone can be estimated jammer by a factor of $\frac{d_1}{d_2} = \frac{1}{\sqrt{2}}$, according to the firmware done that has been compromised by a malicious entry. We can implement a navigation device that acts as odd when drones receive radio commands from the remote controller while changing the mode when a jamming attack is detected. It can be presumed that the mode is triggered by the data interruption and communication links of the GNSS [31, 39]. The jammer will deploy various techniques to interrupt contact with the drone, while the same jamming pattern broadcasts the signal in front of the antenna. We believe that our drone is a typical commercial radio feature and that the antenna of the drone is perfectly omnidirectional or a command antenna for more information about the jammer and its location [40, 41].

Figure 7 shows the UAV flight navigation path of the target system controller block diagram, as the drone approximate RSS (Received Signal Strength) of the jammer controller c(t) depends on the output process variable y(t). Y(t) = r(t) is an error e(t) when the signal control variable r(t) is converted . As the error control increases, c(t) compensates for the return to the original value r(t) = y(t) of the output y(t). The three main elements are described by the PID controller: proportional, integrative, and derivative controller via [37, 38]:

$$C(t) = k_p e(t) + k_i \int_0^t e(T)dT + K_d \frac{\partial}{\partial t} e(t) \tag{2}$$

where K_p, K_i, and K_d respectively represent the additive, integral derivative gains. Lastly, we follow methods for tuning the above mentioned parameters. The architectural drone controller must be able to targeting a drone in the presence of a received signal resistance (RSS) jammer and a standard loop control system consisting of a proportional integral and derivative controller (PID) and a drone controller. The GPS systems together have a high degree of complexity of the different subsystems , while the operation of the satellites retains the monitoring and updating tasks of the stations. IoT cryptography energy consumption for various primitives such as efficient elliptical curve storage techniques while avoiding the overheads cryptography systems and optimising digital key signature and public-key encryption.

6 Conclusions

In this book chapter, we describe the spoofing techniques for the development of a portable system, and the recreation of unauthorized UAV signal form the unauthorised UAV communication system using the low cost SDR equipment, and GNSS receiver vulnerability with partially applicable. We have defined the UAV drones for fleet navigation and coordination for different layers of IoT applications that this framework needs to implement. We suggested an operating model that recognizes the role of private and public entities as an appropriate IoT framework was addressed. We used the information gained from three main networks, the cellular network, air traffic control, and the Internet. Lastly, addressed the gaps and potential studies that could benefit from the large current literature on the solutions.

References

1. Philip, K., Nagi, M.: A framework for sensing radio frequency spectrum attacks on medical delivery drones. IEEE Access (2020). https://www.researchgate.net/publication/341148312
2. Schmidt, E., Akopian, D., Pack, D.J.: Development of a real-time software-defined GPS receiver in a LabVIEW-based instrumentation environment. IEEE Trans. Instrum. Meas. **67**(9), 2082–2096 (2018). https://doi.org/10.1109/TIM.2018.2811446
3. Shvetsova, S.V., Alexey, V.: Safety analysis of goods transportation by unmanned aerial vehicles. World Transp. Transp. **17**(5), 286–297 (2020). https://doi.org/10.30932/1992-3252-2019-17-5-286-297
4. Sciancalepore, S., Ibrahim, O., Oligeri, G., Pietro, R.D.: Picking a needle in a Haystack: detecting drones via network traffic analysis. arXiv: 1901.03535v1 [cs.CR] (2019). https://www.researchgate.net/publication/33035767
5. Khan, N.A., Jhanjhi, N.Z., Brohi, S.N., Usmani, R.S.A., Nayyar, A.: Smart traffic monitoring system using unmanned aerial vehicles (UAVs). Comput. Commun. (2020)
6. Khan, N.A., Jhanjhi, N.Z., Brohi, S.N., Nayyar, A.: Emerging Use of UAV's: Secure Communication Protocol Issues and Challenges. Elsevier (2020)
7. Guntur, S.R., Gorrepati, R.R., Dirisala, V.R.: Internet of medical things remote healthcare and health monitoring perspective. Medical Big Data and Internet of Medical Things: Advances, Challenges, and Applications, chap. 11. CRC Press Taylor & Francis Group, Boca Raton (2018)
8. Guntur, S.R., Gorrepati, R.R., Dirisala, V.R.: Robotics in healthcare: an Internet of Medical Robotic Things (IoMRT) perspective. Machine Learning in Biosignal Analysis and Diagnosis Imaging, chap. 12. Elsevier, Amsterdam (2019)
9. Nayyar, A., Bao-Le, N., Nguyen, N.G.: The Internet of Drone Things (IoDT): future envision of smart drones. In: First International Conference on Sustainable Technologies for Computational Intelligence. Advances in Intelligent Systems. Springer (2020). https://doi.org/10.1007/978-981-15-0029-9_45
10. Caparra, G., Ceccato, S., Formaggio, F., Laurenti, N., Tomasin, S.: Low power selective denial of service attacks against GNSS. In: Proceedings of the 31st International Technical Meeting of

the Satellite Division of the Institute of Navigation (ION GNSS + 2018). Institute of Navigation (2018). https://doi.org/10.33012/2018.15909
11. Wang, Q., Nguyen, T., Khanh, P., Kwon, H.: Mitigating jamming attack: a game-theoretic perspective. IEEE Trans. Veh. Technol. **67**(7), 6063–6074 (2018)
12. Jameel, F., Wyne, S., Kaddoum, G., Duong, T.Q.: A comprehensive survey on cooperative relaying and jamming strategies for physical layer security. IEEE Commun. Surv. Tutor. **21**, 2734–2771 (2018)
13. Perez Marcos, E., Caizzone, S., Konovaltsev, A., Cuntz, M., Elmarissi, W., Yinusa, K., Meurer, M.: Interference awareness and characterization for GNSS maritime applications. In: 2018 IEEE/ION Position, Location and Navigation Symposium (PLANS), pp. 908–919 (2018)
14. Shi, X., Yang, C., Weige, X., Chen, J.: Anti-drone system with multiple surveillance technologies: architecture, implementation, and challenges. IEEE Commun. Mag. (2018). https://doi.org/10.1109/MCOM.2018.1700430
15. Son, Y., Noh, J., Choi, J., Kim, Y.: Gyrosfinger: fingerprinting drones for location tracking based on the outputs of MEMS gyroscopes. ACM Trans. Priv. Secur. **21**(2), 10:1–10:25 (2018)
16. Sanjab, A., Saad, W., Baskar, T.: Prospect theory for enhanced cyber-physical security of drone delivery systems: a network interdiction game. arXiv preprint arXiv:1702.04240 (2018)
17. Khan, M.A., Alvi, B.A., Safi, E.A., Khan, I.U.: Drones for good in smart cities: a review. In: International Conference on Electrical, Electronics, Computers, Communication, Mechanical and Computing (EECCMC) 28 & 29 Jan 2018. https://www.researchgate.net/publication/31684633
18. Mabodi, K., Mehadi, Y., Zandiyan, S.: Multi-level trust-based intelligence schema for securing of the internet of things (IoT) against security threats using cryptographic authentication. J. Supercomput. (2020). https://doi.org/10.1007/s11227-019-03137-5
19. Fotohi, R.: Securing of unmanned aerial systems (UAS) against security threats using the human immune system. Reliab. Eng. Syst. Saf. **193**, 106675 (2020)
20. Qin, T., Wang, B., Chen, R., Qin, Z.: Wang L IMLADS: intelligent maintenance and lightweight anomaly detection system for internet of things. Sensors **19**(4), 958 (2019)
21. Zhang, J., Rajendran, S., Sun, Z., Woods, R., Hanzo, L.: Physical layer security for the internet of things: authentication and key generation. IEEE Wirel. Commun. **26**(5), 92–98 (2019). https://doi.org/10.1109/mwc.2019.1800455
22. Carrio, A., Sampedro, C., Rodriguez-Ramos, A., Campoy, P.: A review of deep learning methods and applications for unmanned aerial vehicles. J. Sens. **2017** (2017)
23. Fotouhi, A., Ding, M., Hassan, M.: Understanding autonomous drone maneuverability for the internet of things applications. In: 2017 IEEE 18th International Symposium on A World of Wireless, Mobile and Multimedia Networks (WoWMoM), pp. 1–6 (2017)
24. Motlagh, N.H., Bagaa, M., Taleb, T.: UAV-based IoT platform: a crowd surveillance use case. IEEE Commun. Mag. **55**, 128–134 (2017)
25. Kersnovski, T., Gonzalez, F., Morton, K.: A UAV system for autonomous target detection and gas sensing. In: Proceedings of the Aerospace Conference, Big Sky, MT, USA, pp. 1–12 (2017)
26. Kumbhar, A., Guvenc, I., Singh, S., Tuncer, A.: Exploiting LTE-advanced HetNets and FeICIC for UAV-assisted public safety communications. IEEE Access **6**, 783–796 (2018)
27. Butun, I., Österberg, P., Song, H.: Security of the internet of things: vulnerabilities, attacks, and countermeasures. IEEE Commun. Surv. Tutor. (2019). https://doi.org/10.1109/COMST.2019.2953364
28. Eldosouky, A., Ferdowsi, A., Saad, W.: Drones in distress: a game-theoretic countermeasure for protecting UAVs against GPS spoofing. arXiv:1904.11568v1 [cs.SY] 16 (2019). https://www.researchgate.net/publication/332726565
29. Jansen, K., Schafer, M., Moser, D., Lenders, V., Popper, C., Schmitt, J.: Crowd-GPS-Sec: leveraging crowdsourcing to detect and localize GPS spoofing attacks. In: IEEE Symposium on Security and Privacy (SP), San Francisco, CA, pp. 1018–1031 (2018)
30. French, A., Mohammad, M., Eldosouky, A., Saad, W.: Environment-Aware Deployment of Wireless Drones Base Stations with Google Earth Simulator (2018). https://www.researchgate.net/publication/325414049

31. Mozaffari, M., Saad, W., Bennis, M., Nam, Y.-H., Debbah, M.: A Tutorial on UAVs for Wireless Networks: Applications, Challenges, and Open Problems (2018)
32. Mozaffari, M., Kasgari, A.T.Z., Saad, W., Bennis, M., Debbah, M.: Beyond 5G with UAVs: foundations of a 3D wireless cellular network. IEEE Trans. Wirel. Commun. **18**(1), 357–372 (2019)
33. Mozaffari, M., Saad, W., Bennis, M., Debbah, M.: Wireless communication using unmanned aerial vehicles (UAVs): optimal transport theory for hover time optimization. IEEE Trans. Wirel. Commun. **16**(12), 8052–8066 (2017)
34. Zhang, A., Liu, X., Gros, A., Tiecke, T.: Building detection from satellite images on a global scale (2017)
35. Granjal, J., Monteiro, E., Silva, J.S.: Security for the internet of things: a survey of existing protocols and open research issues. IEEE Commun. Surv. Tutor. **17**, 1294–1312 (2015)
36. Caparra, G., Ceccato, S., Formaggio, F., Laurenti, N., Tomasin, S.: Low power selective denial of service attacks against GNSS. In: Proceedings of the 31st International Technical Meeting of The Satellite Division of the Institute of Navigation (ION GNSS + 2018). Institute of Navigation (2018)
37. Pietro, R., Oligeri, G., Tedeschi, P.: JAM-ME: exploiting jamming to accomplish drone mission. In: IEEE Conference on Communications and Network Security (CNS) (2019)
38. Tedeschi, P., Oligeri, G., Pietro, R.: Leveraging jamming to help drones complete their mission. IEEE Access **4**, 1–16 (2016)
39. Zhang, Q., Mohammad, M., Saad, W.: Machine Learning for Predictive On-Demand Deployment of UAVs for Wireless Communications. arXiv:1805.00061v1 [eess.SP] (2018)
40. Mohammad, M.: Performance optimization for UAV-enabled wireless communications under flight time constraints. In: IEEE Global Communications Conference (GLOBECOM) (2018)
41. Zeng, Y., Zhang, R.: Energy-efficient UAV communication with trajectory optimization. IEEE Trans. Wirel. Commun. **16**(6), 3747–3760 (2017)

The manufacturer's authorised representative in the EU is Springer Nature Customer Service Centre GmbH, Europaplatz 3, 69115 Heidelberg, Germany. If you have any concerns regarding our products, please contact ProductSafety@springernature.com

Printed and bound by CPI Group (UK) Ltd, Croydon, CR0 4YY
28/09/2026
02224090-0014